Electrode Development for Lithium-ion Batteries

Xiu Li · Chao Lai · Jianmin Ma

Electrode Development for Lithium-ion Batteries

Springer

Xiu Li
School of Electronic and Information Engineering
Tiangong University
Tianjin, China

Chao Lai
School of Chemistry and Materials Chemistry
Jiangsu Normal University
Xuzhou, China

Jianmin Ma
School of Chemistry
Tiangong University
Tianjin, China

ISBN 978-981-96-3020-2 ISBN 978-981-96-3021-9 (eBook)
https://doi.org/10.1007/978-981-96-3021-9

The original submitted manuscript has been translated into English. The translation was done using artificial intelligence. A subsequent revision was performed by the author(s) to further refine the work and to ensure that the translation is appropriate concerning content and scientific correctness. It may, however, read stylistically different from a conventional translation.

This Springer imprint is published by the registered company Springer Nature Singapore Pte Ltd.
The registered company address is: 152 Beach Road, #21-01/04 Gateway East, Singapore 189721, Singapore

Preface

Lithium-ion batteries have become one of the main parts in our daily life. The components of lithium-ion batteries are not familiar to our common people, even our material scientists who are not in battery field. Under such background, we wrote this book on *Electrode Development for Lithium-ion Batteries* for letting more people know which materials are used in our batteries. This book will cover a series of commercial anode and cathode materials including lithium anode, carbon anode, silicon anode, lithium titanate anode, sulfur cathode, lithium manganese oxide cathode, ternary layered oxide cathode, lithium iron phosphate and lithium iron manganese phosphate cathode, lithium-rich manganese-based cathode, and spinel lithium nickel manganese oxide cathode. Moreover, we hope that this book will help the students from all the levels and researchers to learn the development of electrodes for lithium-ion batteries.

Tianjin, China — Xiu Li
Xuzhou, China — Chao Lai
Tianjin, China — Jianmin Ma

Contents

Lithium Anode

1

1.1 Physical Properties of Lithium Metal

As a silvery-white metal, lithium (Li) has a density of 0.534 g/cm^{-3}, a melting point of 180.54 °C, and a boiling point of 1317 °C. It is the lightest metal. It can react with a large number of inorganic and organic reagents. Its reaction with water is very violent. Around 500 °C, it easily reacts with hydrogen, being the only alkali metal that can form a hydride stable enough to melt without decomposing. Its ionization energy is 5.392 eV. It can combine with oxygen, nitrogen, sulfur, etc., and is the only alkali metal that reacts with nitrogen at room temperature to form lithium nitride (Li_3N). Because it is easily oxidized and turns dark, and its density is less than that of kerosene, it should be stored in liquid paraffin. Li has seven isotopes, two of which are stable, namely Li^6 and Li^7. Apart from the stable ones, the longest half-life is Li^8, with a half-life of 838 ms, followed by Li^9, with 187.3 ms, and all other isotopes have a half-life of less than 8.6 ms. Li^4 has the shortest half-life of all isotopes, only 7.58043×10^{-23} s. Li^6 has a strong ability to capture slow neutrons and can be used to control the speed of nuclear reactions in uranium reactors. It can also be used in radiation protection, extending the life of nuclear missiles, and future applications in nuclear-powered aircraft and spacecraft. After being irradiated with neutrons in a nuclear reactor, Li^6 can produce tritium, which can be used to achieve thermonuclear reactions. Li^6 can be used as a coolant in nuclear devices.

1.2 Chemical Properties of Lithium Metal

Because Li has a high charge density and a stable helium-type double electron layer, it easily polarizes other molecules or ions, but it is not easily polarized itself. This affects the stability of it and its compounds. Although the standard hydrogen potential of Li is the most negative, reaching −3.04 V (vs. SHE), but because the solubility of lithium hydroxide is not large and the heat released by the reaction of lithium with water cannot melt lithium, the reaction of Li with water is not as violent as sodium. After the reaction has been going on for a while, the nitride oxide film on the surface of Li is dissolved, making the reaction more violent. Li easily reacts with hydrogen around 500 °C to produce lithium hydride, which is the only alkali metal that can form a hydride stable enough to melt without decomposing. Its ionization energy is 5.392 eV, and it can combine with oxygen, nitrogen, sulfur, etc. It is the only alkali metal that reacts with nitrogen at room temperature to form lithium nitride (Li_3N). It darkens easily when oxidized. If lithium is thrown into concentrated sulfuric acid, it will quickly float on the sulfuric acid, burn and explode.

1.3 Extraction of Lithium Metal

Li is known as the most important energy metal of the twenty-first century. Li and its compounds are important strategic resources for national economy and national defense construction, and are known as the "element that drives the world forward" [1]. Li and its compounds are important strategic resources for the national economy and national defense construction, with important applications in fields such as Li batteries, ceramic industry, glass manufacturing, lubricants, and atomic energy thermonuclear fusion. In the early days, Li was mainly used as an additive in the ceramic and glass industries to improve product performance. With the global trend of electrification, Li, as an important raw material for new battery technology, is widely used in the energy industry. It is not only used in batteries, but also in controlled nuclear fusion. Controlled nuclear fusion, known as the "artificial sun", is considered a key technology to fundamentally solve energy problems. Therefore, Li is not only a key element in the development of the national economy and the promotion of modern high-tech industry progress, but also a national strategic element

X. Li et al., *Electrode Development for Lithium-ion Batteries*, https://doi.org/10.1007/978-981-96-3021-9_1

in the field of nuclear energy. The reserves of lithium resources and their separation and extraction technologies directly affect the strategic security of the country and have important strategic significance for the national economy and national defense construction.

Li resources in nature mainly exist in two forms: brine (existing in the ionic state in salt lakes), and pegmatite (lithium mica, lithium feldspar, etc.). Ore Li is mainly used in the preparation of high-quality battery-grade lithium carbonate, while seawater or brine lithium is mainly used in the preparation of industrial-grade lithium carbonate. According to the data from the U.S. Geological Survey (USGS) 2022 report, the proven lithium reserves worldwide have increased to 89 million tons [2]. Global lithium resources are mainly distributed in South America, North America, Australia, and China. Bolivia, Argentina, and Chile, with their salt flats, hold over 55% of the world's lithium resources, forming the so-called "Lithium Triangle". Australia's lithium resources are mainly lithium feldspar, possessing the world's largest and best lithium mines. China has abundant brine and lithium mica resources.

1.3.1 Extraction of Salt Lake Brine Lithium Resources

Over 60% of the world's lithium reserves are in brine, making brine one of the most important resources for lithium extraction. However, due to the complex composition and numerous impurities in brine, which contains various elements such as Mg, Na, K, Ca, and B, the process is complex and variable, especially the presence of Mg impurities severely affects lithium extraction. The Mg/Li ratio has always been an important indicator for evaluating the feasibility of lithium extraction from salt lakes. Most of China's brine has a relatively high Mg/Li ratio. Traditional methods have difficulty separating Mg and Li, which severely restricts the efficiency of lithium mining. Currently, the main methods for extracting lithium resources from brine include precipitation, calcination, extraction, adsorption, electrochemistry, electrodialysis, and membrane separation.

1.3.1.1 Precipitation and Calcination

Precipitation is the earliest method used to extract lithium resources from salt lake brine and is the simplest method. This method is suitable for brine with a lower Mg/Li ratio and magnesium-free brine. Precipitation enriches the lithium ions (Li^+) in the brine to a certain concentration. This process uses natural solar energy to evaporate and concentrate the brine, remove Mg^{2+}, Ca^{2+} and other impurities, and finally add a precipitant to extract lithium. Due to different precipitants, precipitation lithium extraction can be divided into carbonate precipitation and aluminate precipitation. Carbonate precipitation is currently the most mature industrial method. It involves evaporating industrial brine to obtain a concentrated solution with a high concentration of lithium ions. At this time, the impurity Mg^{2+} concentration also increases. By adding sodium carbonate to the concentrated solution, Mg^{2+} first reacts with CO_3^{2-} to form magnesium carbonate ($MgCO_3$) precipitate, and excess sodium carbonate will react with lithium ions to form lithium carbonate (Li_2CO_3) precipitate. This method is mainly suitable for brine with a low magnesium/lithium ratio, and when the magnesium/lithium ratio is high, it will cause a large waste of sodium carbonate. As shown in Fig. 1.1, the process of producing lithium carbonate by precipitation is to put industrial pure alkali into the concentrated brine of the salt lake, so that lithium is precipitated in the form of lithium carbonate.

The main principle of the aluminate precipitation method is to prepare aluminum hydroxide [$Al(OH)_3$] using aluminate, and add aluminum hydroxide to the solution to form $LiCl \cdot 2Al(OH)_3 \cdot nH_2O$ precipitate. The precipitate is roasted, leached, and impurities removed to obtain LiCl, and then sodium carbonate is added to prepare lithium carbonate. This method uses the high selectivity of aluminum hydroxide for lithium chloride to achieve efficient magnesium-lithium separation and can produce high-purity

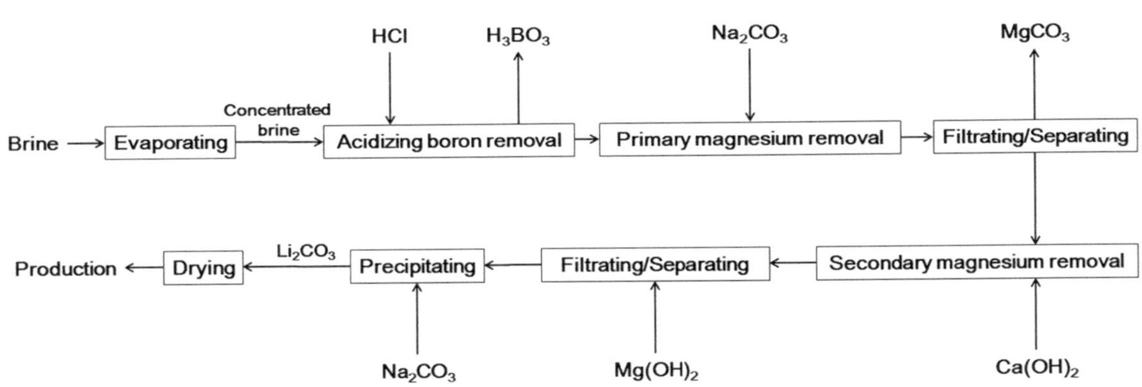

Fig. 1.1 Process flow diagram of lithium carbonate production by precipitation method

lithium carbonate. The main principle is to precipitate lithium ions in brine with alkali of carbonate or aluminate, which is a widely used and technically mature method in industry.

The precipitation process is feasible, but the process is long, the recovery rate is low, the material turnover is large, and it is only suitable for salt lake resources with a low magnesium/lithium ratio. Moreover, a large amount of waste residue will be produced during the process, which poses a great threat to the environment. Recently, new precipitants have been invented. Liu et al. [3] prepared Al/Na_2SO_4 composite material for sulfate-type brine, and lithium was precipitated in the form of $Li_2Al_4(OH)_{12}SO_4 \cdot xH_2O$. However, the presence of magnesium is not conducive to the precipitation of lithium. When the magnesium/lithium ratio reaches 20, the lithium precipitation rate drops from 89.2 to 54.7%. Liu et al. [4] proposed a method of activating Li_3PO_4 to induce the precipitation of carbonate rock brine, and successfully prepared active Li_3PO_4 with high surface energy. The results show that the precipitant greatly reduces the required temperature (from 90 to 30 °C), thereby reducing the energy consumption of the process.

The calcination method re-evaporates the brine, and the magnesium and lithium in it are precipitated in the form of salts, and the obtained substance is roasted. Specifically, at 600 The crude product of magnesium oxide is obtained by calcining at ~700 °C for 2 h, then the calcined product is leached to separate the lithium salt. Alkali is added to remove the excess calcium and magnesium ions in the solution. Then, by evaporation and concentration, a high-purity lithium-containing solution can be obtained, and then lithium carbonate products are produced by adding pure alkali through the precipitation method. However, a large amount of hydrogen chloride gas and heavy magnesium oxide are produced as by-products during the calcination process. These by-products have low economic value, and hydrogen chloride severely corrodes equipment, consumes a lot of energy, and the operation is complicated, so it is less used in practice.

1.3.1.2 Extraction Method

Solvent extraction has been widely used in the metallurgical and chemical industries, and research on extracting lithium from brine using solvent extraction has also developed rapidly. The extraction of lithium has developed since the 1960s and is a relatively mature traditional process, suitable for the separation of lithium and magnesium in high magnesium/lithium ratio brine. The main principle is that the complex formed by the selective complexation reaction of lithium ions and organic solvents is easily soluble in the organic phase, separating lithium ions from other impurity ions, and then lithium ions are extracted through counter-extraction. The research on lithium extractants is mainly based on tributyl phosphate (TBP), and more efficient extraction is achieved through the combination of dispersants, diluents, etc. The most commonly used system is the combination of TBP and $FeCl_3$ [5]. Under the synergistic effect of Fe^{3+}, TBP has a high selectivity for Li^+. The biggest problem with this process is that it usually requires high concentrations of 6 ~ 9 mol L^{-1} hydrochloric acid for stripping. Researchers have recently studied this issue, and Yu et al. [6] proposed a new technology for multi-stage centrifugal extraction using tributyl phosphate isomers. After 5 stages of centrifugal extraction, the extraction rate of lithium reached 90.1%, and the extraction rate of lithium under 1 mol L^{-1} hydrochloric acid conditions was close to 100%. Bai et al. [7] developed a new solvent extraction system: trialkylmethyl di(2-ethylhexyl) ammonium normal phosphoric acid + TBP + $FeCl_3$, the extraction rate of lithium slightly dropped to 70%, and the concentration of hydrochloric acid required for lithium extraction also dropped to 1 ~ 1.5 mol L^{-1}. Cai et al. [8] prepared a functional extractant, named 3-methyl-1-octylimidazole trifluoroacetone, this extractant targets high-concentration sodium salt water, the separation coefficient of lithium and sodium reached 227, which is the maximum value that a single extractant can achieve. The following is the lithium extraction mechanism based on TBP mixed extractant [9].

$$Fe^{3+} + 4Cl^- \rightarrow FeCl_4 \quad (1.1)$$

$$Li^+ + FeCl_4^- + 2TBP \rightarrow Li[FeCl_4 \cdot 2TBP] \quad (1.2)$$

$$Li[FeCl_4 \cdot 2TBP] + H^+ \rightarrow H[FeCl_4 \cdot 2TBP] + Li^+ \quad (1.3)$$

$$H[FeCl_4 \cdot 2TBP] + Na^+ + OH^- \rightarrow Na[FeCl_4 \cdot 2TBP] + H_2O \quad (1.4)$$

$$H[FeCl_4 \cdot 2TBP] + Li^+ \rightarrow Li[FeCl_4 \cdot 2TBP] + Na^+ \quad (1.5)$$

1.3.1.3 Adsorption Method

This adsorption method uses highly selective adsorbents to adsorb lithium ions in brine. For the extraction of lithium from high magnesium/lithium ratio salt lake brine, the essence is the separation of lithium ions from other impurity elements, where the adsorbent with selective adsorption capacity for lithium ions is key. By using the selective separation of the adsorbent for lithium ions, the lithium ions in the salt lake brine are selectively adsorbed and separated, and then the lithium ions are eluted with the eluent. For example, using manganese-based adsorbent Li_2MnO_3 [10]. However, the loss of manganese is inevitable during the acid washing process. Titanium-based adsorbents are gradually receiving attention from researchers due to their strong chemical stability. Such as Li_3TiO_3 [11] and $Li_4Ti_5O_{12}$

[12] have excellent selective adsorption performance for lithium. However, titanium-based adsorbents are prone to agglomeration, leading to a decrease in adsorption capacity. Therefore, the selection of binders and pore-forming agents is crucial. Ryu et al. [13] combined the advantages of manganese-based and titanium-based adsorbents to prepare $Li_{1.33}(Ti_{0.1}Mn_{0.9})_{1.67}O_4$ composite adsorbent. Compared with the single manganese-based adsorbent, its structure is more stable, effectively reducing the loss of manganese. In recent years, lithium aluminum layered double hydroxides (LDHs) adsorbents have been discovered. Although their adsorption capacity is not as good as traditional adsorbents, they are still receiving attention due to their easy elution characteristics. Paranthaman et al. [14] synthesized LDHs with different Li/Al molar ratios. Preliminary experiments confirmed that the adsorbent synthesized with a Li/Al molar ratio of 1:25 has the highest selectivity for lithium, and the extraction rate of lithium reaches about 91%.

The currently developed adsorbents are mostly in powder form, with poor fluidity and permeability, which is not conducive to industrial applications, and heavy metal ions can easily cause environmental pollution.

1.3.1.4 Electrochemical Method

The electrochemical method is an emerging lithium extraction technology that uses brine as an electrolyte to construct a battery system. The main principle, simply stated, is to reverse the principle of a lithium battery. By controlling the potential and simulating the charging process of the battery, lithium in the brine enters the lithium-free cathode, achieving selective extraction of lithium. It has the advantages of being efficient, green, and low in energy consumption.

Commonly used electrode materials for electrochemical lithium extraction include $LiFePO_4$ [15], $LiMn_2O_4$ [16] and so on. Zhao et al. [17] showed that lithium ions are easily embedded in the $FePO_4$ lattice and have excellent reversibility. However, only a small amount of Mg^{2+} is embedded at higher voltages. Therefore, by controlling the working voltage of the system, selective extraction of lithium from brine can be achieved. Liu et al. [18] proposed a chemical precipitation method to convert lithium ions in the lithium-containing cathode liquid into lithium phosphate precipitate. High concentration Fe^{3+} solution is used to convert lithium phosphate, resulting in a high concentration lithium ion solution and iron phosphate precipitate. The final product, Li_2CO_3, is obtained by carbonizing the precipitate. Xu et al. [19] demonstrated the possibility of separating magnesium and lithium using a $LiMn_2O_4/\lambda\text{-}MnO_2$ electrode. When treating low-lithium brine, the magnesium/lithium ratio dropped from 147.8 to 0.37, and when treating high-lithium brine, the magnesium/lithium ratio dropped from 58.8 to 1.7, similar to the separation effect with a $LiFePO_4$ / $FePO_4$ electrode.

The electrochemical method has good application prospects, but the chemical reactions in the system have not been fully explored, and the suppression of side reactions are important issues. The main influencing factors of the electrochemical method are the electrode itself, lithium capacity, cycle stability, etc.

1.3.1.5 Electrodialysis and Membrane Separation Methods

Electrodialysis and membrane separation are two rapidly developing new separation technologies used for lithium extraction from brine. High magnesium/lithium ratio brine can be effectively separated by selective electrodialysis. By using special ion exchange membranes with monovalent and divalent selective separation capabilities, combined with single or multi-stage electrodialysis coupling processes, and the use of bipolar membranes, battery-grade purity lithium products can be obtained. The whole process is low in energy consumption, environmentally friendly, has a high magnesium-lithium separation coefficient, high lithium recovery rate, and has great application potential. Zhao et al. [20] and Liu et al. [21] An improved solution has been proposed: a sandwich liquid membrane electrodialysis system composed of two cation exchange membranes and an organic liquid membrane loaded with lithium ions. This system achieves the recognition and rapid electromigration of lithium ions with the assistance of an electric field. Under optimal conditions, the magnesium/lithium mass ratio in brine can be reduced from 100 to less than 2. The system is adaptable and can separate impurities such as K, Mg, Ca from lithium, with significantly lower energy consumption than traditional electrodialysis methods. Membrane separation technology has significant advantages such as high separation efficiency, low energy consumption, simple operation process, high equipment integration, and environmental protection and energy saving. It has great technical advantages compared to traditional separation technologies. The membrane processes used in the lithium extraction process from salt lake brine mainly include nanofiltration, a selective electrodialysis process based on the selective passage of multivalent ions through the membrane, and a membrane capacitive deionization process (electrosorption) etc. The hydrated ion radii of Li^+, Na^+, K^+, Mg^{2+}, and Ca^{2+} follow the rule: $Mg^{2+} > Ca^{2+} > Li^+ > Na^+ > K^+$, so separation can be achieved based on the different ion radii and charge numbers of each ion using nanofiltration or selective electrodialysis. Hou et al. [22] summarized the current technology of separating lithium ions using metal–organic framework (MOF) based membranes. In recent years, with the continuous expansion of the scale of the new energy vehicle industry, the lithium battery industry has developed rapidly. Lithium, as a key raw material for batteries, securing lithium resources is a daunting task. The cost of salt lake lithium extraction technology is relatively low, and the deep development of this technology has far-reaching strategic significance for ensuring the safety of domestic lithium supply. The main research direction of brine lithium extraction is still to achieve lithium-magnesium separation under high magnesium-lithium ratio conditions. Among the existing lithium extraction technologies, the adsorption method has a simple process and good selectivity, and is more suitable for high magnesium-lithium ratio brine; the precipitation method is suitable for low magnesium-lithium ratio brine, but the lithium recovery rate is low and it is time-consuming; the extraction method can be applied to different salt lake brines depending on the extractant; the calcination method has gradually lost competitiveness due to its high energy consumption; the membrane method represented by nanofiltration and electrodialysis is suitable for high magnesium-lithium ratio brine; the electrochemical method still needs further in-depth exploration. Enterprises, universities, and research institutes are also vigorously innovating technology, through innovative technical methods, coupling and matching separation technology outside the membrane process system, complementing each other's strengths, playing the advantages of different technologies, looking for the best technical solutions, and starting from process intensification, new process route design, system internal integration coupling, etc., to achieve effective coupling and matching between different technologies, and solve the prominent engineering problems faced by salt lake lithium extraction. The development of new materials can effectively improve the efficiency of lithium-magnesium separation, but the economic, effective, and stability issues of the salt lake lithium extraction process need to be considered. At the same time, special attention should be paid to the research and development of new technologies, and efforts should be made to break through the key "bottleneck" technology of salt lake lithium extraction.

1.3.2 Extraction of Lithium from Minerals

Currently, the main lithium-bearing resources are pegmatite deposits, gas-thermal deposits, and brine deposits. Among them, pegmatite lithium deposits are widely distributed, mainly mining lithium mica, lithium feldspar, phosphorus aluminum lithium stone, transparent lithium feldspar, etc. In recent years, the recovery of lithium and its valuable metals from lithium-bearing minerals (lithium feldspar, lithium mica, and zinc ore) has attracted widespread attention in the industrial and academic worlds.

1.3.2.1 Extraction of Lithium from Lithium Feldspar

Lithium feldspar ($LiAlSi_2O_6$) typically appears grayish-white, slightly yellow or green, often associated with quartz, feldspar, and mica. Lithium feldspar has α, β, γ three-phase structures. Natural lithium feldspar mainly exists in the form of α-lithium feldspar, which is a monoclinic crystal structure with extremely low chemical activity. Apart from hydrofluoric acid and some fluorides, α-lithium feldspar hardly reacts with other compounds. Theoretically, the chemical composition (mass fraction) of lithium feldspar is 8.07% Li_2O, 27.44% Al_2O_3, 64.49% SiO_2. Lithium feldspar is currently the most important resource in lithium extraction technology. The main methods of lithium feldspar extraction include limestone roasting, sulfuric acid roasting, chlorination roasting, alkali pressure cooking, and salt roasting.

(1) **Limestone Roasting Method**

The limestone roasting method is the earliest method of extracting lithium from lithium feldspar. In industrial applications, potassium feldspar and lime (or limestone) are roasted at high temperatures, leached with water, filtered, concentrated, and purified. Then, a certain amount of Na_2CO_3 is added to the solution to react with lithium ions to precipitate Li_2CO_3. A sufficient amount of CaO destroys the mineral structure at temperatures above 1100 °C. As shown in reaction (1.6), the roasted material is leached to obtain a LiOH solution. However, due to the extremely low solubility of $Li_2O \cdot Al_2O_3$, it is difficult to extract with water leaching. During the leaching process, an excess of CaO is needed to convert aluminum into insoluble $3CaO \cdot Al_2O_3 \cdot 6H_2O$. Lithium reacts to form soluble LiOH, separating lithium and aluminum. The leaching reaction is shown in reaction (1.7), and the leachate is evaporated, concentrated, and crystallized to obtain $LiOH \cdot H_2O$ product. The advantage of the limestone method is its strong universality and the easy availability of reaction raw materials. The downside is high equipment energy consumption, large amount of slag, and a large amount of solid waste. At the same time, compared with other methods, the extraction rate of lithium is relatively low, and the limestone roasting method has been basically eliminated.

$$2LiAlSi_2O_6 + 8CaO \rightarrow Li_2O \cdot Al_2O_3 + 4[2CaO \cdot SiO_2] \tag{1.6}$$

$$Li_2O \cdot Al_2O_3 + 3CaO + 6H_2O \rightarrow LiOH + 3CaO \cdot Al_2O_3 \cdot 6H_2O \tag{1.7}$$

(2) **Sulfuric acid roasting method**

At present, most major production companies use the sulfuric acid method to extract lithium. The phase change-sulfuric acid dissolution method is currently the most mainstream method for treating lithium spodumene. First, the lithium spodumene is calcined at a high temperature above 1000 °C, which transforms it from the α-type monoclinic system to the β-type tetragonal system, as in reaction (1.8). The activity of β-spodumene is higher, making it easier for H^+ to enter the crystal and react with Li^+. Then, β-spodumene is mixed with concentrated H_2SO_4 and roasted at around 300 °C to obtain a lithium sulfate solution by water leaching. Lime powder is then added to neutralize the excess H_2SO_4 and adjust the pH to neutral, removing impurities such as Fe, Al, Mg, Ca from the leachate, as in reaction (1.9). The lithium sulfate solution is then evaporated and concentrated to obtain a purified lithium sulfate solution, and then a sodium carbonate solution is added to undergo a double decomposition reaction. The precipitate from the reaction is lithium carbonate, as in reaction (1.10).

$$\alpha - \mathrm{LiAlSi_2O_6} \rightarrow \beta - \mathrm{LiAlSi_2O_6} \quad (1.8)$$

$$2\beta - \mathrm{LiAlSi_2O_6} + \mathrm{H_2SO_4} \rightarrow \mathrm{Li_2SO_4} + 2\mathrm{HAlSi_2O_6} \quad (1.9)$$

$$\mathrm{Li_2SO_4} + \mathrm{Na_2CO_3} \rightarrow \mathrm{Na_2SO_4} + \mathrm{Li_2CO_3} \downarrow \quad (1.10)$$

The biggest advantage of using this process to extract lithium is that the process is simple, the process controllability is strong, and the lithium recovery rate is high (up to 80% or more), but the problem is that a large amount of sulfuric acid will be consumed in production, and the acid corrosion resistance of the equipment is required to be high. Although researchers have been trying to solve the high consumption of energy and sulfuric acid, high temperatures above 1000 ℃ and excessive concentrated sulfuric acid are still necessary. The research direction of sulfuric acid roasting method mainly focuses on the development of low-temperature conversion technology of lithium spodumene and the recovery and cascading use of residual acid.

(3) **Chlorination roasting**

Research has found that β-spodumene is roasted with chlorine gas at high temperature to extract lithium chloride. At a high temperature of 1100 ℃, spodumene undergoes a chlorination reaction to produce LiCl [23], as shown in reaction Eq. (1.11).

$$12\mathrm{LiAlSi_2O_6} + 6\mathrm{Cl_2} \rightarrow 12\mathrm{LiCl} + 2\mathrm{Al_6Si_2O_{13}} + 3\mathrm{O_2} + 20\mathrm{SiO_2} \quad (1.11)$$

β-spodumene is roasted with a chlorinating agent (Cl_2 or $CaCl_2$ etc.) at high temperature to produce lithium chloride. The main steps are: (1) $CaCl_2$ is roasted with β-spodumene at high temperature (about 900 ℃) to produce lithium chloride; (2) Water leaching separates to obtain lithium chloride solution. The chlorination roasting process is simple and the slag amount is small. However, this process causes severe corrosion to the equipment, the product LiCl is a gaseous product that is difficult to collect, and Cl is difficult to remove from the system. The free Cl not only corrodes the equipment, but also may contaminate the salt, affecting the product quality.

(4) **Alkali pressure cooking method**

The pure alkali pressure cooking method is also a typical alkali method for lithium extraction, and has become a new research hotspot in recent years. The principle of the alkaline process is generally to replace the Li^+ in spodumene with Na^+ under high pressure conditions. First, α-spodumene is crushed and then roasted at high temperature (about 950 ~ 1100 ℃) to transform into β-spodumene. Then β-spodumene reacts with sodium carbonate solution to form Li_2CO_3 (temperature about 225 ℃), then carbon dioxide is introduced to prepare $LiHCO_3$, and finally Li_2CO_3 crystals are precipitated by heating, as shown in reaction Eq. (1.12).

$$\beta - \mathrm{Li_2O} \cdot \mathrm{Al_2O_3} \cdot 4\mathrm{SiO_2} + n\mathrm{H_2O} + \mathrm{Na_2CO_3} \rightarrow \mathrm{Na_2O} \cdot \mathrm{Al_2O_3} \cdot 4\mathrm{SiO_2} \cdot n\mathrm{H_2O} + \mathrm{Li_2CO_3} \quad (1.12)$$

The lithium slag produced by this process is sodium aluminosilicate, a high-value-added product that can be used in the food industry and has certain economic benefits. In addition, compared with the acid method, the alkali pressure cooking process has less corrosiveness to the equipment. However, the high-pressure alkali leaching process is still unavoidable for β-lithium feldspar during the high-temperature transition process.

(5) **Salt Roasting Method**

The principle of the salt roasting method is similar to that of the alkali pressure cooking method, mainly using alkali metal ions such as Na^+, K^+ to replace the Li^+ site. For β-lithium feldspar that has undergone high-temperature transformation, the temperature required for salt roasting is lower, about 600 ℃. At high temperatures, α-lithium feldspar is transformed into β-lithium feldspar, and β-lithium feldspar and K_2SO_4 undergo ion exchange to form soluble Li_2SO_4. After water leaching, a lithium-containing solution is obtained, and lithium is finally extracted. This method is simple and can avoid some drawbacks of the sulfuric acid method, but due to the high price of potassium sulfate materials, the economic feasibility is low. Although researchers have tried to replace expensive potassium sulfate with cheap sodium sulfate, if all are replaced with sodium sulfate, the leaching of lithium will decrease. In addition, a large amount of Na^+ or K^+ enters the leaching liquid system, greatly affecting the quality of the lithium product.

1.3.2.2 Extraction of Lithium from Lithium Mica

The chemical formula of lithium mica is $K(Li, Al)_3(Si, Al)_4O_{10}(OH, F)_2$, which is one of the most common minerals, a monoclinic mineral, and flaky. The chemical composition of lithium mica varies greatly. Generally, the content of Li_2O is 1.2 wt.%~5.9 wt.%, K_2O is 4.8 wt.%~13.8 wt.%, Al_2O_3 is 11.3 wt.%~28.8 wt.%, and SiO_2 is 46.9 wt.%~60 wt.%. In addition, lepidolite also contains iron, calcium, magnesium, rubidium, and cesium. With the rapid growth of new energy electric vehicles, the demand for lithium is increasing, and the extraction of lithium from spodumene can no longer meet market demand. Therefore, lepidolite is attracting more and more attention. The main methods of extracting lithium from lepidolite are limestone roasting, sulfuric acid method, alkali method, chlorination roasting, and sulfate roasting.

(1) **Limestone Roasting Method**

The limestone method is the oldest method of extracting lithium from ore. Lepidolite concentrate and limestone are mixed in a certain ratio and ball milled. After mixing, they are roasted at a high temperature of 800~900 °C to mature, and then water quenched, finely ground, leached, filtered or centrifuged to obtain leachate and residue. The leachate is evaporated, crystallized, and centrifuged to obtain lithium hydroxide. The leach residue mainly consists of calcium silicate and calcium fluoride. Its reaction mechanism [24] is shown in Eq. (1.13):

$$\begin{aligned} &2MeAl_2Si_3O_9(FOH) + 14CaCO_3 \rightarrow 2(Me_2O \cdot Al_2O_3) \\ &+ CaF + H_2O + 4CaO \cdot 2SiO_2 \cdot CaCO_3 \\ &+4(2CaO \cdot SiO_2) + 13CO_2 \uparrow \end{aligned} \tag{1.13}$$

The limestone method has the advantages of mature technology and simple process, but due to the low lithium content in the leachate, the large energy consumption, and the low lithium recovery rate, it has gradually been phased out in industrial production with the emergence and improvement of other processes.

(2) **Sulfuric Acid Method**

Researchers have imitated the traditional β-spodumene method and proposed a method of dissolving spodumene with sulfuric acid. Lepidolite and concentrated sulfuric acid are acid-roasted at an appropriate temperature to obtain acidified clinker, and then water leaching is used to obtain a lithium sulfate solution. After acid treatment, the accompanying valuable metals rubidium and cesium become soluble salts. The reaction mechanism is as follows in reaction Eq. (1.14) [25]. Vieceli et al. [26] reported a method for extracting lithium from lepidolite through mechanical activation, acid digestion, and water leaching. Under the condition of 165 °C, with a 65 w.% concentrated sulfuric acid dosage for digestion for 4 h, the lithium extraction rate is about 85%. Zhang et al. [27] focused on the influence of sulfuric acid concentration based on the above research. As the concentration of sulfuric acid increases, the extraction rates of lithium, rubidium, and cesium show a trend of first increasing and then decreasing. Among them, mica hardly reacts with concentrated sulfuric acid, which is different from the research results of Vieceli et al. This may be due to H^+ It does not dissociate in concentrated sulfuric acid. The dissolution reaction is shown in reaction (1.15). The action of sulfuric acid on β-lithium feldspar only replaces Li^+ with H^+, without destroying the structure of aluminum silicate. During the leaching process, the extraction rate of aluminum is very low. However, during the sulfuric acid dissolution process, the mineral structure of lithium dolomite is destroyed, and a large amount of aluminum and iron are leached into the solution.

$$\begin{aligned} &3SiO_2 \cdot Al_2O_3 \cdot 2[(Li, K)(F, OH)] + H_2SO_4 \rightarrow Li_2SO_4 \\ &+ Al_2(SO_4)_3 + K_2SO_4 + H_2SiO_3 + SiF_4 + H_2O \end{aligned} \tag{1.14}$$

$$\begin{aligned} &K(Li, Al)_3(Si, Al)_4O_{10}(OH, F)_2 + H_2SO_4 + H_2O \rightarrow Li_2SO_4 \\ &+ KAl(SO_4)_2 + Al_2(SO_4)_3 + SiO_2 + HF + SiF_4 \end{aligned} \tag{1.15}$$

The extraction of lithium by sulfuric acid method needs to be carried out under high temperature conditions. The fluorine-containing tail gas produced by high temperature treatment will cause environmental pollution and needs a special adsorption process for recovery treatment. In the water leaching process, the leaching rates of lithium, potassium, rubidium, and cesium are relatively high, which can better realize the comprehensive utilization of resources and improve the total recovery rate of lithium. However, the composition of the leachate is complex, the residual acid content is large, which results in a large amount of gelatinous precipitate in the process of extracting valuable metals, making the subsequent purification complex. In addition, sulfuric acid is corrosive to equipment, and the equipment requires high corrosion resistance. Moreover, due to the difficulty of purification, the waste residue produced is difficult to be utilized, and there are currently no industrial application examples.

(3) **Alkali method**

The limestone roasting method is the earliest developed and typical alkali method, and the process has gradually been replaced in industrial production. Therefore, researchers are constantly developing new methods to extract lithium. Jandova et al. [28] used iron-lithium mica concentrate (containing 5.19% Li_2O) and $CaCO_3$ are mixed and roasted, and lithium carbonate is prepared by CO_2 carbonization or solvent extraction after water leaching. The leaching rate of Li is 90%, and the lithium mother liquor can be further recovered with potassium and rubidium. In addition, Wang Ding and others [29] used high-pressure steam method to extract lithium from lithium mica. The water in high-pressure steam can quickly dissolve strong alkali, allowing the reaction to occur under local high-concentration strong alkali conditions. Lithium mica can quickly react with strong alkali, making lithium mica quickly alkali soluble, and the conversion rate of lithium in lithium mica reaches 96.9%. This process reduces the amount of alkali used and shortens the time, effectively reducing production costs and reducing the output of polluting waste slag. Alkali pressure cooking method is a widely studied alkali method process. Yan and others [30] under the condition of 100 wt.% CaO concentration, high-pressure leaching at 150 °C for 60 min, the extraction rate of lithium reached 98.9%.

Overall, the metal leaching rate of the alkali pressure cooking method is high, the impurities in the mother liquor are less beneficial to purification, the process flow is short and the equipment corrosion is small, and the solution is alkaline, reducing the alkali adjustment process of the traditional extraction method. This is the most prominent advantage of this process. If high-pressure equipment can be avoided, it will be beneficial to the industrial implementation of this process.

(4) **Chlorination roasting**

The chlorination roasting method is to convert the lithium and other valuable metals in lithium mica into soluble chlorides, and after leaching, a lithium-containing solution is obtained to realize the extraction of lithium and alkali metals. The chlorination process of calcium chloride or sodium chloride with lithium mica is shown in Formulas (1.16)–(1.18).

$$2NaCl + 6SiO_2 + M_2O + Al_2O_3 \rightarrow 2NaAlSi_3O_8 + 2MCl \tag{1.16}$$

$$CaCl_2 + SiO_2 + M_2O \rightarrow CaSiO_3 + 2MCl \tag{1.17}$$

$$CaCl_2 + 2SiO_2 + Al_2O_3 + M_2O \rightarrow CaAlSi_2O_8 + 2MCl \tag{1.18}$$

In addition, researchers have studied the chlorination roasting method to achieve the purpose of synergistic extraction of lithium, rubidium, cesium and other high-value elements. Zhang and others [31] studied the intensified chlorination roasting process to effectively extract lithium, rubidium, cesium and potassium from lithium mica. Considering the use of $CaCl_2$ When NaCl is used as the only chlorinating agent, the extraction effects of Li, Rb, Cs, and K are poor. Therefore, a mixture of $CaCl_2$ and NaCl is used. Under optimal conditions, the extraction efficiencies of Li, Rb, Cs, and K reach 92.49%, 98.04%, 98.33%, and 92.90% respectively. Yan et al. [32] studied the extraction of lithium under different auxiliary material ratios. When the total amount of chlorinating agent added is 100 wt.%, as the amount of NaCl increases, the extraction rate of lithium shows a trend of first increasing and then decreasing. The maximum values are reached when NaCl and $CaCl_2$ are 60 wt.% and 40 wt.% respectively. Under the best roasting conditions (880 °C, 30 min), the extraction rates of lithium, rubidium, and cesium are 92.9%, 93.6%, and 93.0% respectively. The reaction process of the chlorination roasting method is summarized as shown in reactions (1.19).

$$K(Li, Al)_3(Si, Al)_4O_{10}(OH, F)_2 + NaCl + CaCl_2 \rightarrow LiCl + KCl + NaAlSi_3O_8 + CaAl_2Si_2O_8 + CaSiO_3 + CaF_2 + H_2O \tag{1.19}$$

This process can achieve the synergistic extraction of lithium, rubidium, and cesium, while effectively shortening the time, producing less silicon-aluminum slag that is easy to use. However, the large amount of chlorine salt added will severely corrode the equipment, and inevitably, a lot of Cl^- will escape in the form of HCl and Cl_2. The volatilization of chlorides will cause environmental pollution, the use of chlorine gas has a higher operational risk, the requirements for equipment corrosion resistance are higher, resulting in increased production costs.

(5) **Sulfate Roasting**

Sulfate roasting is currently the most commonly used method for extracting lithium from lepidolite. Lithium and lepidolite are added to the sulfate roasting, converting lithium into soluble lithium sulfate. The principle is to exchange the cations in the sulfate with lithium ions, the lithium ions are displaced, and $Li_2\ SO_4$ solution is obtained by leaching. As shown in reactions (1.20)–(1.21)

$$Li_2O \cdot Al_2O_3 \cdot SiO_2 + K_2SO_4 \rightarrow K_2O \cdot Al_2O_3 \cdot 3SiO_2 + Li_2SO_4 \tag{1.20}$$

$$Li_2O \cdot Al_2O \cdot SiO_2 + CaSO_4 \rightarrow CaO \cdot Al_2O_3 \cdot 3SiO_2 + Li_2SO_4 \tag{1.21}$$

Luong et al. [32] roasted lithium mica with Na_2SO_4 at 1000 °C for 0.5 h, then immersed it in water, achieving a lithium extraction rate of 90.4%. The main lithium-containing products during the roasting process were $LiKSO_4$ and $Li_2NaK(SO_4)_2$. Setoudeh et al. [33] mechanically activated lithium mica and Na_2SO_4 in a planetary ball mill using zirconia media for 5 h. Through the study of the amount of Na_2SO_4 used and the roasting temperature, it was found that the lithium extraction rate could reach over 99%. In addition, Yan et al. [34] combined sulfate roasting and chloride roasting to obtain a more suitable process. Using Na_2SO_4 and $CaCl_2$ to co-roast with lepidolite, the reaction process is as shown in reaction (1.22). Under the same roasting conditions of 880 °C and 30 min, the extraction rates of lithium, rubidium, and cesium were 94.8%, 93.5%, and 90.1% respectively. Compared with the chloride roasting method, the lithium extraction rate slightly increased, while the extraction rates of rubidium and cesium slightly decreased. The reduction of chloride salts greatly reduced the production of corrosive gases such as HCl and Cl_2. The process operating conditions were optimized, providing more possibilities for equipment selection without significantly affecting extraction.

$$\begin{aligned} &K(Li, Al)_3(Si, Al)_4O_{10}(OH, F)_2 + Na_2SO_4 + CaCl_2 \\ &\rightarrow Li_2SO_4 + K_2SO_4 + LiCl + KCl + NaAlSi_3O_8 \\ &+ CaAl_2Si_2O_8 + CaF_2 + H_2O \end{aligned} \quad (1.22)$$

The sulfate method has less impurity content, a higher leaching rate, better comprehensive recovery effect of metals, and a simple process, which has been widely used in the industrial production of lithium extraction from lithium mica. However, the sulfate method also needs to be completed after high-temperature calcination. The high-temperature calcination process will volatilize fluorine and sulfur-containing waste gas, causing serious environmental pollution. Minerals are prone to stick together during the calcination process, reducing the calcination efficiency, and lithium products are easily contaminated by potassium.

The development of lithium extraction technology can not only adapt to the growing demand for lithium consumption, but also alleviate the problem of China's high dependence on lithium resource imports. In order to achieve a clean, efficient, and comprehensive utilization of the lithium extraction process, it is urgent to develop efficient and sustainable lithium extraction technology. Due to the diversity and complexity of solid waste, various technologies can be used in the future to extract lithium from minerals. Research in technology should focus on industrial production practices, while also strengthening theoretical research, so that production costs can be effectively reduced, resources can be saved, and high-efficiency environmental protection requirements can be met. It is also necessary to improve the existing process from the following aspects. (1) Reduce environmental pollution and simplify process steps; (2) Improve lithium extraction efficiency; (3) Realize the recycling and utilization of waste slag; (4) Realize the comprehensive utilization of associated resources.

1.4 Lithium Metal Anode

1.4.1 Overview of Lithium Batteries

Since its introduction to the market in 1991, lithium-ion batteries have been widely used in electronic devices, new energy vehicles, and other fields. Lithium-ion batteries are favored for their high energy density and long cycle life [35]. However, as the energy density of traditional graphite anode lithium-ion batteries is constantly approaching its theoretical value, and people's requirements for the endurance of electronic products and electric vehicles are getting higher and higher, it is necessary to further improve the battery energy density to meet market demand. Improving the energy density of the battery lies in changing the battery positive and negative electrode materials. Using high energy density electrode materials can directly increase the energy density of ion batteries. Metal batteries are expected to break through the current energy density bottleneck. Metal secondary batteries with metal anodes have higher energy density than ion batteries, such as lithium-sulfur (Li–S), lithium-air (Li–Air), zinc-air (Zn–Air) batteries, etc. [36]. Metal lithium anodes and metal zinc anodes with high theoretical specific capacity have become the preferred anode materials for current metal batteries. Among them, metal lithium has a high theoretical specific capacity (3860 mAh g^{-1}) and low oxidation reduction potential (−3.04 V vs. SHE) and is considered the "holy grail" of secondary lithium battery anode materials. It can greatly improve the energy density of lithium batteries and has become a research hotspot for secondary lithium battery anode materials in recent years.

1.4.2 Development of Lithium Batteries

American physical chemist G. N. Lewis first proposed the concept of a lithium metal anode in 1912. He measured the standard electrode potential of lithium metal as −3.04 V (vs. SHE) in 1913, laying the foundation for the future development of lithium metal batteries [37]. As one of the most active metals, lithium metal cannot be compatible with aqueous electrolytes, so the research on lithium metal anodes stagnated for a while. Not until the 1960s, the application of non-aqueous solvent electrolytes allowed for extensive research on lithium metal primary batteries, such as Li‖$(CF)_n$, Li‖$CuCl_2$, Li‖MnO_2, Li‖$Ag_2V_4O_{11}$ and other primary battery systems [38]. The subsequent discovery of intercalation reaction cathode materials promoted the further development of rechargeable lithium metal batteries. In the 1970s, M. Stanley Whittingham, the father of rechargeable lithium batteries, first discovered the intercalation cathode material TiS_2, which matched with the lithium metal anode, and developed a lithium metal secondary battery, laying the theoretical foundation for the rechargeability of lithium metal batteries. Based on Whittingham's invention, Exxon developed a secondary lithium metal battery with a lithium-aluminum alloy anode and a TiS_2 cathode [39]. In the 1980s, Moli developed a high-energy-density Li||MoS_2 battery, which could cycle hundreds of weeks and achieved commercialization, quickly occupying a large market share. However, due to the serious problem of dendrite growth of the lithium metal anode, the safety issue of battery fire and explosion forced a large-scale recall. After that, the research on lithium metal secondary batteries once again fell into a trough. Because the fatal safety hazard of the lithium metal anode could not be completely solved, research turned to other anode materials. In 1991, Sony successfully commercialized lithium-ion batteries with a lithium cobaltate cathode, a carbon-based anode, and a carbonate electrolyte, marking the beginning of the lithium-ion battery era [40]. Until the twenty-first century, as the energy density of common anodes of lithium-ion batteries has approached their theoretical values, people's focus has returned to the research on high-energy-density lithium metal anodes. During this period, lithium-air batteries and lithium-sulfur batteries showed extremely high theoretical energy densities and were widely studied. Therefore, in pursuit of higher energy density, future lithium metal batteries will develop in the direction of lithium-sulfur batteries, lithium-oxygen batteries, etc. Such high-energy-density batteries have re-injected new vitality into the research on lithium metal anodes.

1.4.3 Lithium Battery Structure and Principle

There is a fundamental difference between lithium metal batteries and lithium-ion batteries. Lithium-ion batteries use cathode and anode materials that can intercalate and deintercalate lithium ions, while lithium metal batteries use lithium metal as the anode material. The working principle of the anode of a lithium metal battery is that during the discharge process, lithium metal loses electrons to undergo electrochemical stripping, and during the charging process, lithium ions gain electrons to form lithium metal. The electrochemical reaction equations of lithium-ion batteries and lithium metal batteries are as follows.

The expression for the $LiCoO_2$/C battery is [41]:

$$\text{Anode: } LiCoO_2 \rightleftharpoons Li_{(1-x)}CoO_2 + xLi^+ + xe^- \quad (1.23)$$

$$\text{Cathode reaction: } 6C + xLi^+ + xe^- \rightleftharpoons C_6Li_x \quad (1.24)$$

$$\text{Overall reaction: } LiCoO_2 + 6C \rightleftharpoons Li_{(1-x)}CoO_2 + C_6Li_x \quad (1.25)$$

The expression for the Li/S battery is [42]:

$$\text{Anode: } S_8 + 16Li^+ + 16e^- \rightleftharpoons 8Li_2S \quad (1.26)$$

$$\text{Cathode: } 16Li \rightleftharpoons 16Li^+ + 16e^- \quad (1.27)$$

$$\text{Overall reaction: } S_8 + 16Li_2 \rightleftharpoons 8Li_2S \quad (1.28)$$

Batteries such as Li/S and Li/O_2 demonstrate extremely high theoretical energy densities. These lithium metal batteries can well meet the urgent needs of the new industry for high energy density storage devices, and can fully utilize the advantages of lithium metal, thus showing great potential.

1.4.4 Challenges of Metallic Lithium Anodes

Before the practical application of metallic lithium anodes, it is necessary to overcome the problems they present, among which safety and recyclability are the biggest challenges faced by metallic lithium anodes. Due to issues such as interface side reactions, dendrite growth, and large volume changes during the charging and discharging process of the lithium anode, the cycle life of the battery is significantly shortened, the Coulombic efficiency of the battery is reduced, and even safety hazards such as explosions may occur, limiting the practical application of metallic lithium batteries.

1.4.4.1 Interface Side Reactions

Due to the instability of the metallic lithium anode interface, interface side reactions occur when it comes into contact with the electrolyte during the battery cycle. In commonly used carbonate or ether electrolytes for lithium batteries, metallic lithium, due to its high reactivity, forms an SEI film on its surface. The formation of SEI on metallic lithium consists of two main steps (Fig. 1.2) [43]. The first step is the spontaneous chemical reaction between metallic lithium and the electrolyte when the electrolyte is added during the battery assembly process. Therefore, insoluble products form an interface layer on the surface of the metallic lithium, effectively inhibiting further chemical reactions between the metallic lithium and the electrolyte. The SEI formed by spontaneous chemical reactions is called the "original SEI". The second step is that after the battery cycle, the original SEI cannot effectively prevent the electrochemical decomposition of the electrolyte. In addition, due to significant morphological changes in the lithium electrode during the cycle, the original SEI formed by chemical reactions may break, exposing fresh lithium to the electrolyte. The exposed fresh lithium can further react with the electrolyte to produce SEI. Both of these processes promote the growth of the original SEI during the cycle.

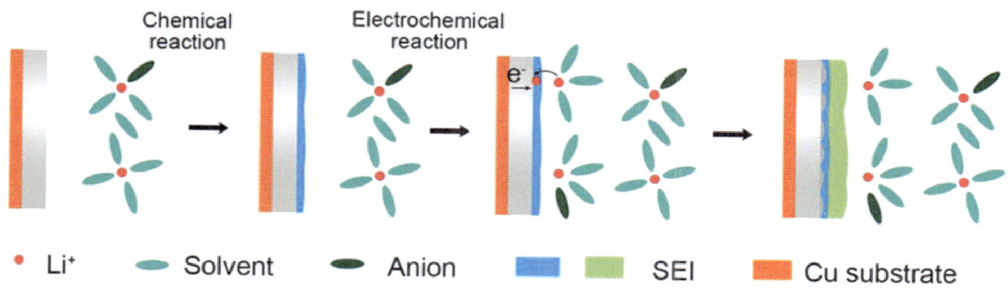

Fig. 1.2 The formation process of SEI on the surface of metallic lithium [43]

During the charging and discharging process of the battery, the repeated formation and breaking of the SEI film cause the consumption of metallic lithium and the electrolyte. In the lithium battery cycle, this continuous electrochemical decomposition of the electrolyte is usually manifested as low Coulombic efficiency.

1.4.4.2 Dendrite Growth

Metallic lithium shows various deposition methods during the deposition process, such as dendritic, mossy, etc., as shown in Fig. 1.3a. The physical and chemical characteristics of the metal anode interface, the mass transfer control of the solid–liquid interface, and the driving of the metal deposition crystal type all affect the shape of the metal deposition. Doll **é** and others [44]. The phenomenon of lithium evolving from mossy to dendritic with the increase of current density has been observed. During the charging and discharging process of the battery, an SEI film is formed on the surface of the lithium anode. Due to the poor elasticity of the common SEI, it is prone to cracking under stress, as shown in Fig. 1.3b. Once a crack is formed in the SEI film, the exposed fresh lithium surface has higher conductivity and surface area, thus becoming a preferred deposition site. Then, lithium ions diffuse to these deposition sites and concentrate nearby to form tips. In the absence of mechanical constraints, this process will accelerate, leading to uneven electric fields at the electrode and electrolyte interface, resulting in uneven lithium deposition and the formation of lithium dendrites. The formation of dendrites leads to repeated destruction of the SEI film and the increase in the specific surface area of the electrode caused by the dendrites further causes more side reactions between metallic lithium and the electrolyte. When the dendrites grow large enough to pierce the separator and cause a short circuit, a large current passes through the dendritic dendrites, rapidly generating heat, leading to the risk of fire and explosion.

1.4.4.3 Volume Change

When a metallic lithium battery deposits on the metallic anode, due to the irregular growth of dendrites, the

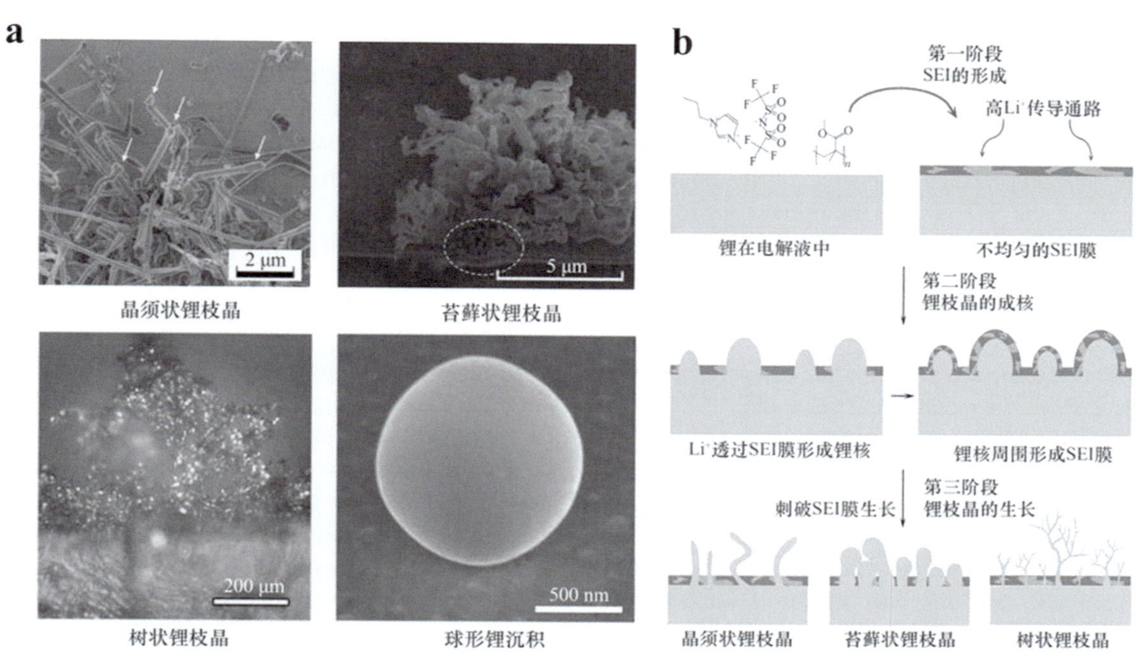

Fig. 1.3 **a** Different lithium deposition shapes and **b** the mechanism of lithium dendrite growth in lithium batteries [45]

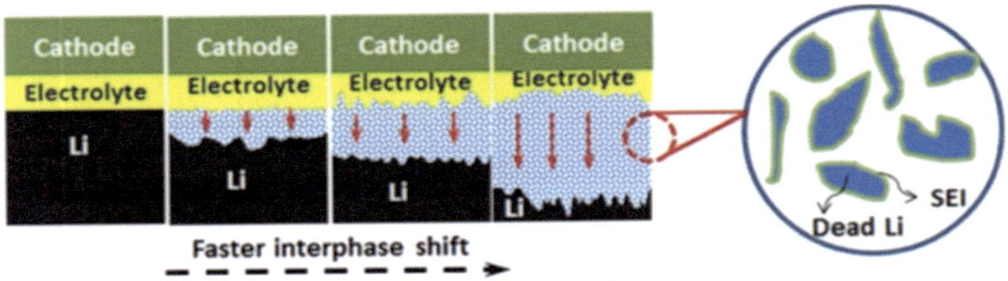

Fig. 1.4 Anode volume change and the production of "dead lithium" [46]

volume continues to increase as the charging and discharging process continues, which destroys the fragile SEI film and causes interface stress changes (Fig. 1.4). The stress changes further promote the formation of lithium dendrites. Among them, the twisting or peeling of dendrites from their roots causes a loss of anode capacity. After continuous cycling, porous lithium electrodes and excessive "dead lithium" are produced. In addition, the non-electron-conductive SEI film wraps around the metallic lithium to form an isolation effect, causing the pulverization of lithium and accelerating the formation of "dead lithium", hindering ion transport, attenuating battery capacity, and greatly affecting the electrochemical performance of the battery.

1.4.5 Modification of Metallic Lithium Anode

Research on the interface of metallic lithium provides new insights into the formation and evolution of the SEI film and the construction of a stable and efficient SEI to suppress the growth of lithium dendrites. To stabilize the SEI film, the design of the interface between the metallic lithium anode and the electrolyte is also crucial. So far, researchers have proposed many strategies and innovative ideas to improve the stability of the metallic lithium anode, such as electrolyte design. The composition of the electrolyte is an important factor affecting the composition and structure of the SEI film on the lithium anode surface. Electrolyte design mainly includes electrolyte additives [47], improvements to the new type of electrolyte [48]. In addition, artificial SEI films [49] can also alleviate interface side reactions and suppress the growth of lithium dendrites through electrode surface pretreatment, thereby improving the performance of the metallic lithium anode.

1.4.5.1 Electrolyte Additives

Currently, common commercial electrolytes such as $LiPF_6$-EC/DEC or $LiPF_6$-EC/DMC, these types of carbonate electrolytes, although they have good compatibility with high-pressure cathodes, their stability towards metallic lithium is poor, easily forming a fragile SEI film, leading to a decrease in battery cycle performance. Since electrolyte additives also react, participating in the formation of the SEI film, they can change the composition and structure of the surface SEI film, promote the uniform deposition of lithium, and inhibit the formation of lithium dendrites. Therefore, the introduction of additives can improve the stability of the electrolyte with metallic lithium, thereby improving the cycling performance of the metallic lithium anode. In addition, inorganic additives and organic additives are commonly used electrolyte additives. Inorganic additives include HF, LiF, CO_2, SO_2, AlI_3, $AlCl_3$, nitrates, borates, metal cations, etc. Stassen et al. [50] generated a protective film containing Li_2SO_4 on the lithium surface by adding SO_2, this film further improved the stability of the SEI film and promoted the performance of the lithium anode. $LiNO_3$ additive is commonly used in lithium-sulfur battery ether electrolytes, it can form a stable SEI film on the surface of metallic lithium to inhibit the dissolution of polysulfides. In 2018, $LiNO_3$ was first used in carbonate electrolytes [51], as shown in Fig. 1.5a, after adding $LiNO_3$, the morphology of lithium deposition changed from dendritic to uniform blocky; NO_3^- can decompose to form a SEI film rich in Li_3N, changing the composition of the SEI film, making the SEI film have high ionic conductivity, changing the deposition of lithium, and inhibiting the growth of lithium dendrites. Ding et al. [52] proposed a new theory (SHES theory), as shown in Fig. 1.5b, by adding some metal cations, such as Cs^+ and Rb^+, to the electrolyte, these cations have a lower reduction potential than lithium ions, accumulate at the edge of lithium during the lithium deposition process, forming a self-healing electrostatic shielding layer that repels positively charged lithium ions, thereby inhibiting the formation of lithium dendrites.

Because the SEI film formed by inorganic additives has poor mechanical properties, it cannot adapt well to the volume changes during the deposition and dissolution process of lithium. In contrast, the elasticity of the SEI film formed by organic additives on the surface of metallic lithium is better than that of inorganic additives, and can better adapt to the volume changes of the metallic lithium electrode during the battery cycling process. Organic additives generally include FEC [53], VC [54], tetrahydrofuran [55], organic aromatic compounds [56] etc. Gong et al. [57] found that

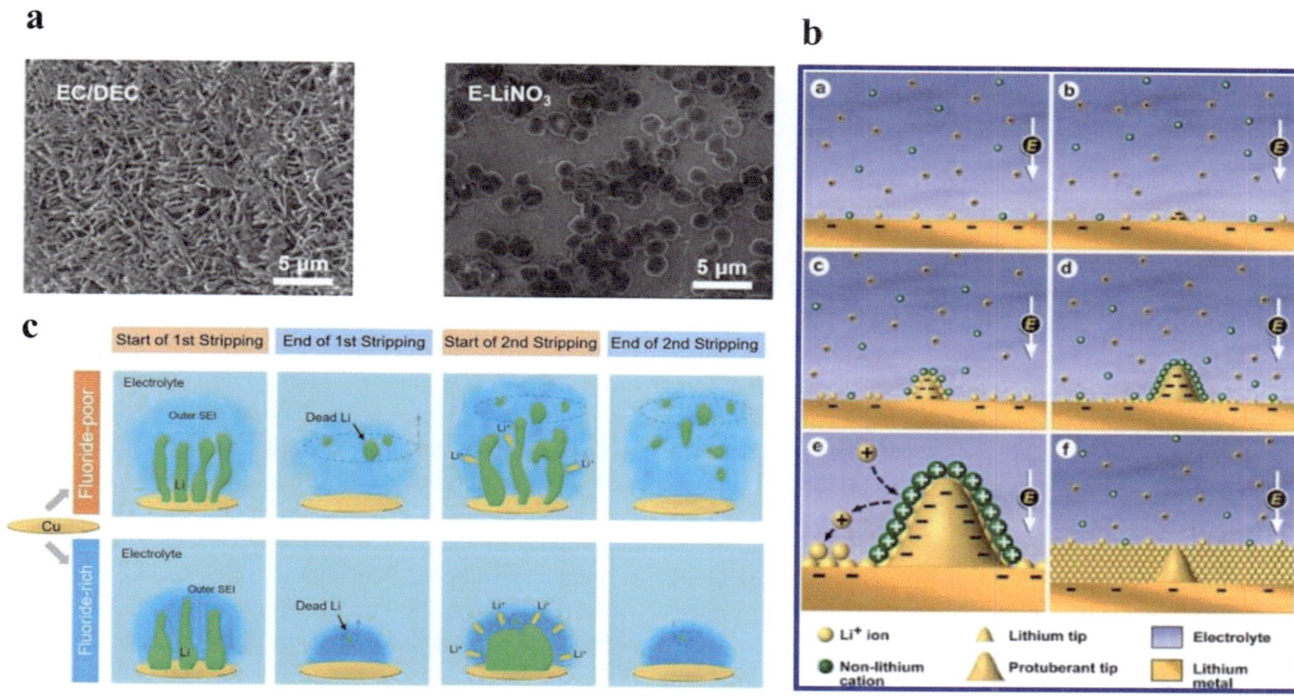

Fig. 1.5 **a** Lithium deposition morphology with and without the addition of $LiNO_3$ in carbonate electrolyte [58]; **b** lithium deposition under the self-healing electrostatic shielding mechanism of SHES theory [52]; **c** the effect of FEC additive on the lithium metal anode [57]

a fluorine-rich SEI can effectively inhibit dendrite growth, suppress electrolyte decomposition, and thus improve battery efficiency. As shown in Fig. 1.5c, in the presence of FEC, all lithium is deposited within the SEI film generated in the initial cycle. The lithium metal deposition is dense and uniform, and only a very small amount of "dead lithium" is produced in the end. Without the addition of FEC, large isolated dendrites will appear during the deposition process. As the charging and discharging proceed, some lithium will eventually lose its connection with the electrode, resulting in a large amount of "dead lithium".

1.4.5.2 New Type of Lithium Salt

The most commonly used lithium salt in lithium-ion batteries is $LiPF_6$. Although $LiPF_6$ has a relatively high ionic conductivity, it easily reacts with water and has poor thermal stability. $LiPF_6$ easily decomposes in lithium metal batteries to produce harmful by-products such as HF and PF_5^-, which cannot form a stable SEI film on the surface of lithium metal, leading to poor cycle stability. Therefore, it is necessary to develop other lithium salts to stabilize the interface of the lithium metal anode. Many lithium salts have been developed to replace $LiPF_6$, such as borates, lithium bis(oxalato)borate (LiBOB) [59], lithium difluoro(oxalato)borate (LiDFOB) [60]. Schedlbauer et al. [60] compared the cycling performance of the lithium metal anode in $LiPF_6$ and LiDFOB electrolytes. The battery containing LiDFOB can maintain a Coulombic efficiency in the range of 95–97% after 500 cycles, while the battery containing $LiPF_6$ only has a Coulombic efficiency of 75% after 500 cycles. Another type is sulfonimide salts, such as lithium bis(fluorosulfonyl)imide (LiFSI), lithium bis(trifluoromethanesulfonyl)imide (LiTFSI), etc. Wang's group [61] replaced $LiPF_6$ with LiFSI to improve the cycling stability of the lithium metal anode in carbonate electrolytes. It was found that LiFSI forms a dense LiF-rich SEI film on the surface of the deposited lithium, effectively suppressing side reactions and the formation of "dead lithium". Compared with $LiPF_6$, more anions, FSI^- It decomposes into LiF and other components from FSI^- during the cycle process, forming a dense and stable SEI film. Although LiTFSI lithium salt has good chemical stability and higher lithium ion conductivity, it has strong corrosiveness to aluminum foil at high potentials, and FSI^- anions can replace $TFSI^-$ to inhibit the corrosion of aluminum foil [62]. There are also other new types of lithium salts such as lithium perchlorate ($LiClO_4$) [63], lithium difluorophosphate (LiDFOP) [64] and many other lithium salts. Despite the development of various new lithium salts, there are still some problems with the single salt system, such as LiFSI can corrode Al substrate. Therefore, combining different lithium salts together to form a dual salt system can take advantage of each to improve the cycling performance of lithium metal batteries, such as LiTFSI–LiBOB [65], LiFSI–LiTFSI [66], LiDFOB–LiBF [67], LiDFP–LiBOB [68] and so on.

1.4.5.3 High Concentration Lithium Salt

Due to the concentration polarization of the electrolyte, the lithium ion content on the surface of the negative electrode will decrease and the distribution will be uneven, which ultimately leads to the appearance of lithium dendrites during the lithium deposition process. Using high concentration lithium salt can increase the concentration of lithium ions in the electrolyte, reduce the concentration polarization on the electrode surface, and provide more lithium ions during the reaction process. In addition, ultra-high salt concentration can quickly form a dense SEI film. On the other hand, high concentration lithium salt will reduce the concentration of solvated lithium ions, so the increased lithium ion concentration will lead to efficient lithium deposition and stripping, and the HOMO and LUMO energy levels of the electrolyte will drop to the level of the salt, promoting the pre-decomposition of the salt anions on the negative electrode. Fan et al. [69] used high concentration 10 M LiFSI EC/DMC electrolyte, and found that FSI^- preferentially decomposes on the surface of the lithium metal negative electrode and $LiNi_{0.6}Mn_{0.2}Co_{0.2}O_2$ (NMC622) positive electrode to form a fluorine-rich SEI. This battery can maintain 99.2% Coulombic efficiency after 200 cycles. At the same time, under the condition of a high charging cut-off voltage of 4.6 V with a capacity of 2.5 mAh cm^{-2}, the capacity retention rate is still 86% after 100 cycles. Zhang et al. [70] developed a new type of "local high concentration electrolyte" by diluting LiFSI DMC high concentration electrolyte with bis(2,2,2-trifluoroethyl) ether BTFE, where 2.5 M LiFSI/DMC-BTFE (molar ratio 1:1) electrolyte was used for Li||$LiNi_{1/3}Mn_{1/3}Co_{1/3}O_2$ The battery has a Coulombic efficiency of up to 99.5%, and the capacity retention rate is still 80% after 700 cycles. Although increasing the concentration of lithium salts can improve the performance of the lithium metal anode, the increase in lithium salt concentration means that the cost of the electrolyte will also increase correspondingly. Moreover, a higher concentration of lithium salts will increase the viscosity of the electrolyte, which is not conducive to the cycle of the battery.

1.4.5.4 New Organic Solvents

Organic solvents are an important component of the electrolyte and also affect the composition of the SEI film. The main solvents for lithium batteries are currently carbonates and ethers, such as EC, DEC, DOL, DME, etc. The fluorination of carbonate solvents is another emerging method for developing solvent structures based on carbonates. By adjusting the solvation structure of lithium ions, the fluorinated components weaken the interaction with lithium ions, promoting the desolvation of lithium ions on both electrodes. The performance of this electrolyte can far exceed that of traditional EC electrolytes. Among various fluorinated carbonate solvents, FEC is the most commonly used and will preferentially decompose at the anode to form a fluorine-rich SEI. Yang et al. [71] replaced EC with FEC in 1 M $LiPF_6$ FEC/DMC electrolyte. Under the conditions of a current density of 2 mA cm^{-2} and a capacity of 3.3 mAh cm^{-2}, the Li||Li battery cycled stably for 3500 h, proving that FEC can improve battery performance. In addition, difluoroethylene carbonate (DFEC) is another emerging film-forming agent synthesized by further fluorination. Amine et al. [72] studied the effect of different substituents of cyclic carbonate solvents on the performance of lithium metal anodes, including EC, FEC, DFEC, etc. The results found that electrolytes with fluorine-substituted FEC and DFEC bases have excellent stability, and EC cannot form a stable SEI on the surface of lithium metal. In addition, the performance of Li||NMC622 batteries using DFEC-based electrolytes in Li||$LiNi_{0.6}Mn_{0.2}Co_{0.2}O_2$ (NMC622) batteries is significantly better than that of FEC-based electrolytes. After 400 cycles, the average Coulombic efficiency is as high as 99.95%, and the capacity retention rate is greater than 82%, showing better high-voltage compatibility. This makes the cyclic carbonate DFEC considered for the first time as an important solvent for the stable SEI of lithium metal batteries.

1.4.5.5 Artificial SEI Film

The reaction between lithium metal, organic solvents, lithium salts, and electrolyte additives allows the in-situ formed SEI to protect the lithium metal anode. However, the composition and structure of this in-situ formed SEI film are complex and unstable, so it is necessary to prepare a dense protective layer on the surface of the lithium metal anode as an artificial SEI film to stabilize the interface between the lithium metal anode and the electrolyte to reduce the occurrence of side reactions, thereby improving the stability of the lithium metal anode. Cui et al. [73] Innovatively created an artificial SEI film with interlinked hollow carbon spheres to protect the lithium metal anode. As shown in Fig. 1.6, the SEI film is generated on the surface of the skeleton structure, ideally moving together with it during the battery cycling process without separation. Meanwhile, lithium ions can easily pass through the skeleton structure, and lithium deposition will occur underneath, without dendrite formation. The hollow carbon sphere coating layer is weakly bonded to the metal current collector, and the interlinked hollow carbon spheres used for copper foil modification will form columnar lithium metal instead of lithium dendrites. The electrode modified by this method has good Coulombic efficiency and cycle stability. Cao et al. [74] first constructed a LiF–Li_2C_2 artificial SEI film. LiF has high interfacial energy and mechanical strength, and Li_2C_2 has a lower lithium ion nucleation barrier, inducing rapid and uniform lithium deposition. In addition, LiF and Li_2C_2

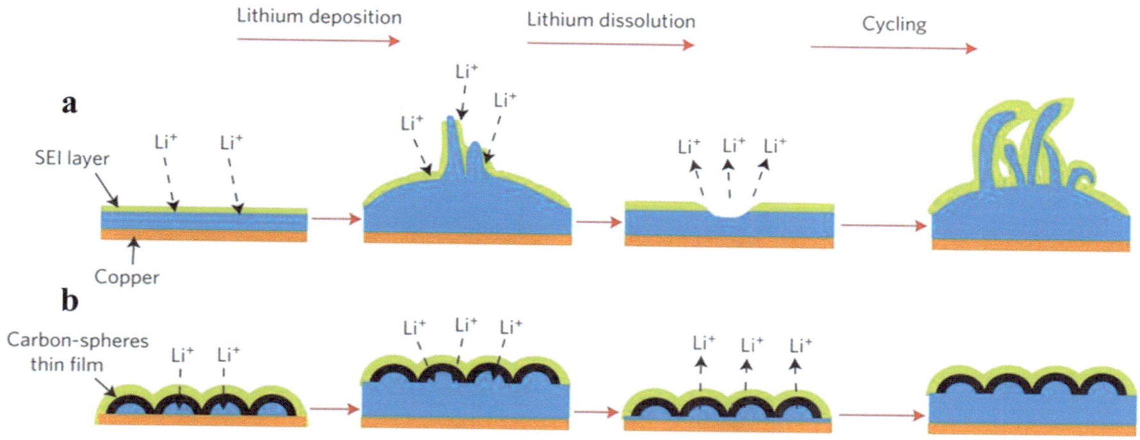

Fig. 1.6 Flexible coating of hollow carbon nanospheres adapts to the volume change in the lithium deposition process [73]

are key components in the SEI film, with ideal interface characteristics and high Young's modulus, which can effectively induce uniform lithium deposition and planar growth, and inhibit the growth of lithium dendrites. Jia et al. [75] used an iodate solution soaking method to non-in situ coat the lithium metal to form an artificial SEI film, the main components of which are $LiIO_3$ and LiI. The iodate-treated lithium metal anode exhibits higher ionic conductivity, smaller charge transfer impedance, not only can inhibit the side reactions between the electrolyte and lithium metal, but also make the lithium deposition more uniform and inhibit the growth of lithium dendrites. Li et al. [76] constructed an artificial solid electrolyte interface phase (ASEI), by introducing a boron nitride fluoride (F-BN) coating, an in-situ constructed SEI containing LiF and Li_3N was built, which is beneficial for stabilizing the electrode and electrolyte interface.

Although surface modification of lithium metal can further enhance the electrochemical performance of the lithium anode, the growth of lithium dendrites is uncontrolled, and there is no three-dimensional structure to accommodate the deposition of lithium metal, which cannot solve the volume change problem of the lithium metal anode, and it is still difficult to achieve dendrite-free lithium deposition. Therefore, introducing a three-dimensional structure to alleviate the volume change of the lithium anode is an effective method to limit the growth of lithium dendrites.

1.4.5.6 Three-Dimensional Skeleton

Introducing a three-dimensional porous skeleton into the anode structure of the battery, during the lithium deposition/dissolution process, the current collector has a three-dimensional porous structure and a larger specific surface area, making the spatial distribution of charge on the surface of the lithium metal electrode uniform, reducing the production of lithium dendrites. The three-dimensional skeleton can limit the deposition and dissolution process of lithium metal within the three-dimensional porous skeleton, effectively alleviating the volume change of the electrode. As shown in Fig. 1.7a, Guo et al. [77] A three-dimensional porous copper foil was prepared, and ammonia was reacted with the copper foil to form $Cu(OH)_2$ submicron fibers on the surface of the copper foil, which were then reduced to form a three-dimensional porous copper foil. The results showed that on the three-dimensional copper foil, during the deposition of metallic lithium, the charge was evenly distributed along the copper fiber skeleton, which could effectively confine the metallic lithium within the skeleton and effectively inhibit the growth of lithium dendrites. Dendritic growth was formed on the flat copper foil. Lu et al. [78] used a self-supporting skeleton composed of copper nanowires (Cu-NWs) as a three-dimensional current collector for the deposition of metallic lithium, and inhibited the growth of lithium dendrites by confining the metallic lithium in the three-dimensional nanostructure (Fig. 1.7b). The metallic lithium anode based on Cu-NWs exhibited high Coulombic efficiency and excellent rate performance.

Non-conductive skeletons, due to their insulating properties, are usually decorated with polar functional groups on the surface, which can homogenize the lithium ion flux in the pores, thereby inhibiting the formation of lithium dendrites during the cycle. As shown in Fig. 1.7c, Cheng et al. [79] covered the negative electrode current collector surface with a glass fiber cloth. The polar functional groups (Si–O, O–H, O–B) on the glass fiber cloth can adsorb a large amount of lithium ions to compensate for the electrostatic interaction and concentration diffusion between the lithium ions and the protrusions of the traditional copper foil negative electrode, avoiding the accumulation of lithium ions around the protrusions. Liang et al. [80] covered the current collector with an oxidized polyacrylonitrile (PAN)

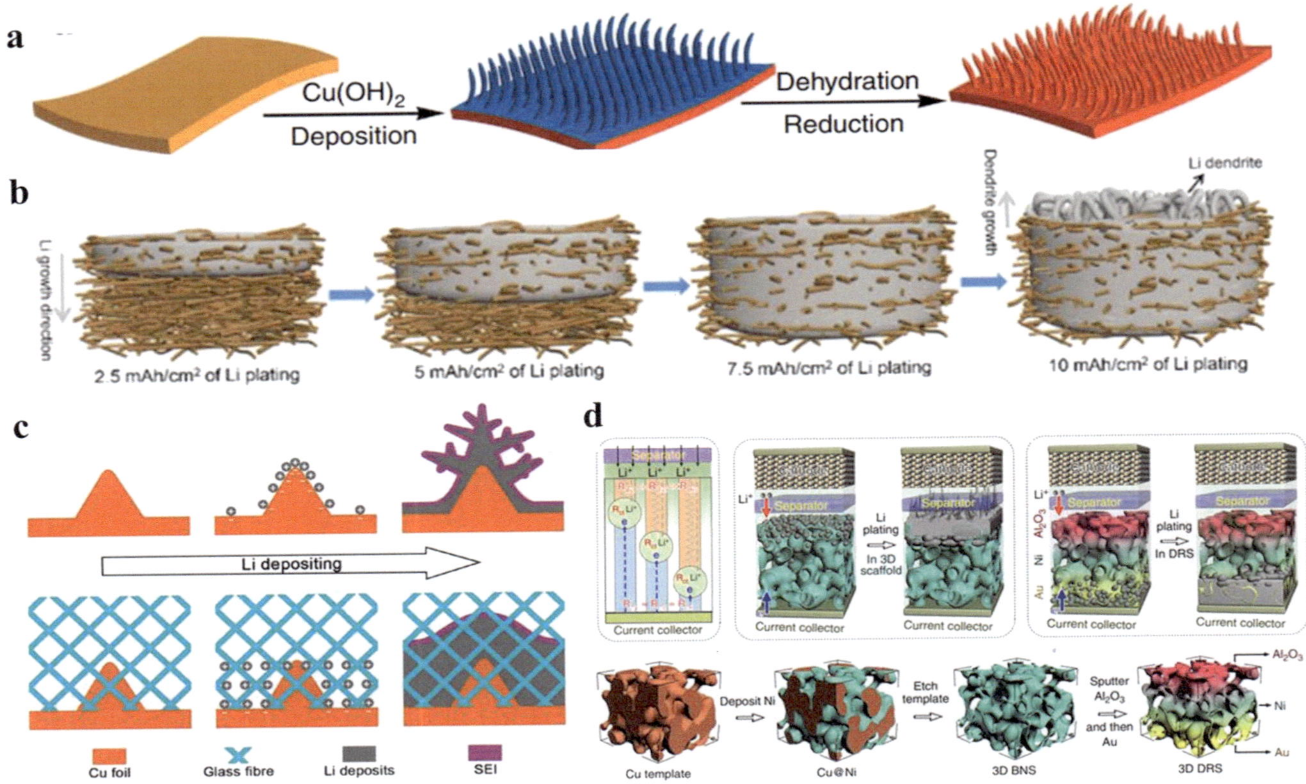

Fig. 1.7 **a** Preparation of three-dimensional copper foil [77]; **b** schematic diagram of lithium metal deposition on copper foil and Cu-NWs respectively [78]; **c** schematic diagram of lithium deposition on glass fiber cloth [79]; **d** schematic diagram of lithium metal deposition from bottom to top [85]

nanofiber layer. The polar functional groups (C=N, C–N, C=O, O–H) on the surface of the polymer nanofibers act as lithium ion binding sites. However, because the polymer fibers cannot support electron transport, the electrochemical process can hardly occur in most areas, so the pore chambers of the skeleton cannot be well utilized. Moreover, due to the insulating nature of the current collector, the actual current density on the electrode surface has not changed.

According to Sand's time theory [81], the conductivity and large specific surface area of the conductive skeleton can reduce the local current density inside the electrode. However, due to the concentrated electric field [82], metallic lithium tends to deposit at the top of the skeleton. Therefore, after cycling, lithium still evolves into a dendritic shape and accumulates in the upper region, thereby blocking the porous structure and leading to poor cycle stability. Non-conductive materials cannot support electron transport, and the porous structure cannot be well utilized. Lithium tends to grow on nuclei with low overpotentials or barriers (Au, Sn, Pt, Ag, Mg, ZnO, etc.), and the selective application of low-barrier materials to three-dimensional structured anodes can be considered an effective lithium-friendly modification strategy, to guide lithium deposition away from the separator, induce metallic lithium to preferentially deposit at the bottom of the skeleton, thereby achieving a bottom-up growth mode of metallic lithium in the three-dimensional skeleton, and further enhancing the modification effect of the three-dimensional skeleton on the metallic lithium anode. Xiang et al. [83] successfully achieved selective deposition of lithium ions on the back of the current collector by modifying one side of the carbon fiber felt as a three-dimensional skeleton with Au. Li et al. [84] A three-dimensional carbon nanofiber skeleton with gradient-distributed ZnO particles was proposed, which reduces the lithium nucleation barrier by introducing gradient-distributed ZnO. In addition, lithium reacts with ZnO before deposition, generating Li_2O and Zn, realizing uniform lithium deposition. Zhang's research group [85] prepared a gradient lithium-friendly porous nickel skeleton, as shown in Fig. 1.7d, which is passivated by coating with aluminum oxide to prevent top deposition, and the bottom Au acts as a lithium-friendly coating, forming a gradient lithium-friendly. The strategy of gradient lithium deposition can better control the deposition of lithium and greatly suppress the growth of dendrites, which is enlightening for the development of high-performance lithium metal anodes.

In summary, lithium metal has outstanding advantages as a lithium battery anode material due to its high specific capacity and low oxidation-reduction potential, which can greatly increase the energy density of the battery. A large amount of research has modified the lithium metal anode from different aspects, which can be roughly divided into surface modification and bulk phase modification. Surface modification mainly plays a role in stabilizing the interface, reducing side reactions, and suppressing lithium dendrites; while bulk phase modification mainly plays a role in stabilizing the electrode structure, alleviating volume changes, and suppressing lithium dendrites. However, the current problems faced by the lithium metal anode cannot be well solved, such as the problem of lithium dendrites, interface side reactions, electrode volume changes, etc., and a single modification method is difficult to solve multiple problems of the lithium metal anode at the same time, further research is still needed. It is necessary to develop new modification materials (such as new coating materials, new electrolyte additives, new three-dimensional current collectors, new lithium-rich alloys, etc.) and new modification strategies (such as the combination of multiple modification methods, new composite lithium metal preparation methods, etc.) to improve the electrochemical performance and safety performance of the lithium metal anode.

1.5 Lithium Alloy Anode

1.5.1 Overview of Alloy Anode

In recent years, high-capacity lithium-containing anode materials Li_xM_y (M is an element that can undergo alloying reaction with lithium) based on the alloying reaction mechanism are a new type of anode material, which has several times the lithium storage specific capacity of graphite, and can provide active lithium source for the battery. These characteristics enable it to be used with high-capacity lithium-free cathode materials (such as S, O_2, FeF_3, V_2O_5 to match, and construct a new system for the next generation of high energy density lithium-ion batteries. Alloy materials are formed by lithium metal and one or more allotrope elements (M=Zn, Si, Mg, Sn, Bi, Au, Ag, Sb, C, etc.), which have metallic properties. The alloy anode, with its high theoretical specific capacity and low electrode potential close to that of the lithium metal anode, has become a highly potential next-generation lithium-ion battery anode material. Unlike the deposition and dissolution behavior of metallic lithium, the charging and discharging process of alloy materials is an alloying/de-alloying process, accompanied by phase changes. The reaction of metallic lithium with alloy substrate materials can be represented by the formula:

$$\mathrm{Li} + x\mathrm{M} \rightarrow \mathrm{LiMx}\ (\mathrm{M} = \mathrm{Al, Si, Sn, Zn, Mg, Ag, etc.}) \quad (1.29)$$

Many metals and non-metals can form alloys with lithium, such as Li–Mg, Li–B, Li–Ag, Li–Sn, Li–In. Lithium ions react electrochemically with alloy electrode materials, and different metals can reduce the nucleation barrier of lithium. During battery charging, it forms intermetallic compounds with metallic lithium, and during battery discharging, it de-alloys to form a three-dimensional porous structure. Lithium alloys have good lithium affinity, high ionic conductivity, and high mechanical strength, which are beneficial for their application in lithium metal anode modification. For example, by introducing lithium-friendly alloys or alloying elements that are solid-soluble with lithium to optimize the wetting behavior of the anode material to lithium, or by introducing high ionic conductivity lithium alloys to effectively guide the lithium ion diffusion behavior at the lithium metal anode surface interface, or in all-solid-state batteries, use its mechanical properties to improve the interface stability of lithium metal and solid electrolyte. Using alloys as lithium-friendly modification materials for lithium metal anodes has the following advantages: (1) improving the wettability to lithium, reducing the difficulty of preparing lithium composite electrodes; (2) most alloy materials have strong interactions with lithium ions, which can slow down the movement of lithium ions to the protruding parts of the structure, which is beneficial to ensure a uniform lithium ion flow at the interface and achieve uniform deposition; (3) alloy materials have a low nucleation barrier, which can reduce the overpotential of lithium deposition, serve as a preferred nucleation site for lithium, and induce directional deposition of lithium. Based on the above lithium-friendly characteristics, researchers use alloys as lithium-friendly materials to modify electrodes to suppress the growth of dendrites.

1.5.2 Li–Mg Alloy Anode

The synthesis method of Li–Mg alloy anode material: high-temperature melting method, lithium and magnesium metals are melted according to the designed ratio to form lithium-magnesium alloy under an inert atmosphere, and then it is rolled into a suitable thickness (1 mm) sheet. Cut the lithium-magnesium alloy sheet as the anode. Since Li–Mg anodes are easily oxidized in air, they are stored in a glove box filled with argon. Li–Mg alloy anode material, by using part of magnesium to replace lithium to form an alloy, can not only retain the advantages of pure lithium metal anode but also weaken the surface reactivity of the anode. By appropriately replacing Li with Mg, a lithium-rich state can be achieved, which helps to provide enough exposed surface for the anode reaction. Therefore, the lithium-magnesium alloy electrode shows good structural stability during the electrochemical process. In the Li-Mg alloy anode system, when Li is deposited on the anode surface, due to the good lithium affinity and large surface energy of the Li–Mg alloy, the lithium deposition morphology is film-like, which can suppress the growth of lithium dendrites. The theoretical specific capacity of Li–Mg alloy (3070 mAh g^{-1}), the alloying of Li-embedded occurs at a potential of 0.015–0.05 V, and the extraction potential of Li is at 0.2–0.4 V.

Kong and others [86] reported on the use of Li–Mg alloy as the anode for Li–S batteries. Compared to the metallic lithium anode, the Li–Mg alloy showed significantly improved stability on the surface and in the body during the cycling process. They also found that after stripping Li, a conductive Li-poor Li–Mg alloy matrix was formed, promoting subsequent plating and diffusion of Li ions. This alloy, which becomes a Li-poor Li–Mg alloy after lithium dissolution, can serve as a porous skeleton. Its internal skeletal structure can stabilize the electrode and maintain the integrity of the SEI on the surface during the lithium deintercalation process, effectively inhibiting the growth of lithium dendrites and improving the battery's cycling performance.

Yang and others [87] used a lithium-rich Li–Mg alloy as the anode material in a solid-state battery system, as shown in Fig. 1.8. The Li–Mg skeletal structure, which serves as a mixed ion–electron conductor, can maintain the interface contact between the anode and the solid electrolyte and prevent the collapse of the anode structure, thereby enhancing the cycling performance of the solid-state battery.

The Li-Mg alloy as an anode has problems of high interface impedance, easy dendrite formation, and low interface stability. In most electrolyte systems, a surface film spontaneously forms on the Mg electrode surface, leading to a surface impedance several orders of magnitude higher. At the same time, the oxide film and SEI film formed on the Mg electrode surface limit the kinetic characteristics, resulting in poor rate performance.

Lithium alloy materials have high specific capacity and low potential characteristics, and their alloying lithium storage mechanism can fundamentally solve the problem of lithium dendrites, promising to provide a stable anode for the next generation of high specific energy lithium-air batteries, lithium-sulfur batteries, and other new battery systems. Among them, magnesium-lithium alloy has high energy density, low lithium deintercalation potential, and good lithium affinity. In addition, its solid solution reaction mechanism can avoid frequent phase changes during the alloying process, maintaining structural stability. Moreover, magnesium is abundant in the earth's crust, so among many alloy materials, lithium-magnesium alloy is the most promising alloy anode. However, it currently faces some kinetic problems. The extremely low lithium deintercalation potential makes the lithium-magnesium alloy an equipotential body for lithium deposition more easily. The lithium diffusion coefficient of the material is not high enough, causing lithium to deposit on the electrode surface instead of dissolving into the body phase to form a solid solution. Shortening the lithium transport path and increasing the lithium diffusion rate, such as refining grains, constructing non-equilibrium defects, and element doping, can enhance the electrode reaction kinetics and provide new ideas for solving the above problems.

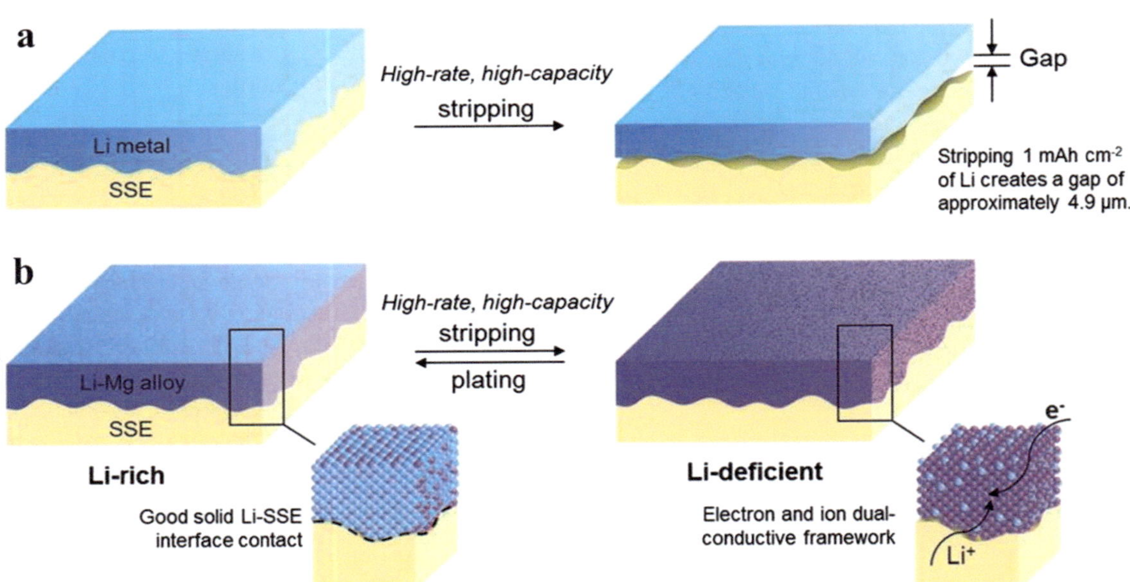

Fig. 1.8 Schematic of lithium deintercalation from **a** metallic lithium and **b** lithium-rich Li–Mg alloy in solid-state batteries [87]

1.5.3 Li–B Alloy Anode

The Li–B alloy material is prepared by heating and melting metallic lithium with elemental boron. The lithium ingot and boron powder are put into an iron crucible in a designed ratio, heated and stirred continuously under the protection of an inert gas until the temperature reaches 700 °C, forming a Li_xB_y with a stable structural skeleton.

The excess Li will embed into the skeleton, the solution solidifies and shrinks, and finally cools to obtain the Li–B alloy. The Li–B alloy skeleton with a certain hardness can suppress volume changes during the cycle. The Li–B alloy has a dual-phase structure composed of lithium boride compounds and free lithium, with Li^0 and Li^+ two states of Li, and discharge has two working voltage platforms. The lithium-rich dual-phase Li–B alloy as an anode material contains two phases, one is the intermetallic compound Li_7B_6 that can serve as a skeleton structure, and the other is metallic lithium filled in the gaps of the skeleton structure. Because the relative atomic mass of boron is small and the lithium-rich alloy contains metallic lithium, the lithium-rich dual-phase Li–B alloy has a high specific capacity. When the Li–B alloy is delithiated, the metallic lithium in it will preferentially undergo electrochemical reactions. After all the metallic lithium is dissolved, the intermetallic compound Li_7B_6 can continue to be delithiated. The Li–B alloy has physical properties similar to Li, the density of the LiB compound is only 1.5 g cm^{-3}, the theoretical specific capacity of Li_7B_6 is 3495 mAh g^{-1}, it has good conductivity (1.43×103 S cm^{-1}) and a high lithium ion diffusion rate. The Li–B alloy has a negative potential similar to lithium metal (0.4 V higher than Li/Li^+), which is beneficial for assembling high-voltage batteries. In 2013, Yang's team [88] first studied the Li–B alloy as the anode of a lithium/sulfur battery. Due to the above advantages, the Li–B alloy has good performance in suppressing the formation of lithium dendrites, reducing electrode interface impedance, and improving battery cycle performance. The Li–B alloy can effectively suppress the growth of Li dendrites, alleviate volume changes, and enhance battery safety performance. Due to the reduction of actual current density and interface impedance, the Li–B alloy can effectively suppress the growth of lithium dendrites and improve battery cycle performance. In 2014, Zhang's team [89] first used a three-dimensional lithium alloy (Li_7B_6) fiber matrix for ultra-stable lithium-sulfur batteries. The three-dimensional nanostructure Li_7B_6 skeleton has a high surface area and sufficient volume space, which can not only reduce the surface current density, but also fully accommodate the electrolyte and redeposited lithium ions to stabilize the concentration of lithium ions. By using the Li@Li_7B_6 anode, the Li–S battery can stably cycle up to 2000 times.

Chen et al. [90] The in-situ formed three-dimensional lithium-boron (Li–B) alloy is considered an excellent substitute for metallic lithium, which with an appropriate B content (15LiB), exhibits optimal electrochemical performance. As shown in Fig. 1.9, the three-dimensional LiB framework can mitigate the growth and volume change of lithium dendrites. Compared with pure lithium, the rich lithium alloy anode not only has a certain three-dimensional framework structure, but also compensates for the shortcomings of insufficient lithium source in the three-dimensional current collector. The rich lithium alloy has better structural stability and electrochemical performance, but the research on rich lithium alloys is still in the initial stage, and new alloy systems need to be developed for further research.

The Li–B/Li composite system is selected, in which the proportion of metallic Li accounts for 72% of the entire electrode. Even under high temperature conditions, the Li–B/Li composite electrode can still maintain its solid structure and exhibit good electrochemical stability. DFT calculations show that the high porosity of the three-dimensional framework of Li_5B_4 in the Li–B/Li alloy is conducive to enhancing the capillary force between Li and Li_5B_4, thereby improving the overall melting point of the electrode. In addition, the Li_5B_4 framework can also improve the conductivity of its electrode. However, after long-term electrochemical cycling, the battery assembled with Li–B alloy exhibits irreversible capacity decay, and the root cause is still elusive. Therefore, further research aimed at clarifying the degradation mechanism is highly desirable.

1.5.4 Ternary Alloys

Binary lithium-containing alloy systems have larger capacities than commercial graphite, but due to the catastrophic structural changes and large volume expansion of these binary lithium alloy electrodes, they usually exhibit quite poor cycling performance. To overcome the poor reversibility of these intermetallic lithium alloy electrodes, ternary lithium-containing alloy compounds have been studied. Choi et al. [91] adjusted the surface characteristics of the electrode material by doping a small amount of Mg into metallic lithium through the melting method. Doping Mg can increase the surface energy of the metallic lithium electrode, thereby making metallic lithium preferentially deposit on the electrode surface instead of on lithium dendrites, inducing more uniform deposition of metallic lithium. Rich lithium multiphase alloys can also serve as anode materials, such as rich lithium $Li_{2.6}BMg_{0.05}$ alloy, Li/Li_xC/Li_ySn alloy, etc. [92, 93]. Compared with pure lithium, rich lithium alloys have better structural stability and

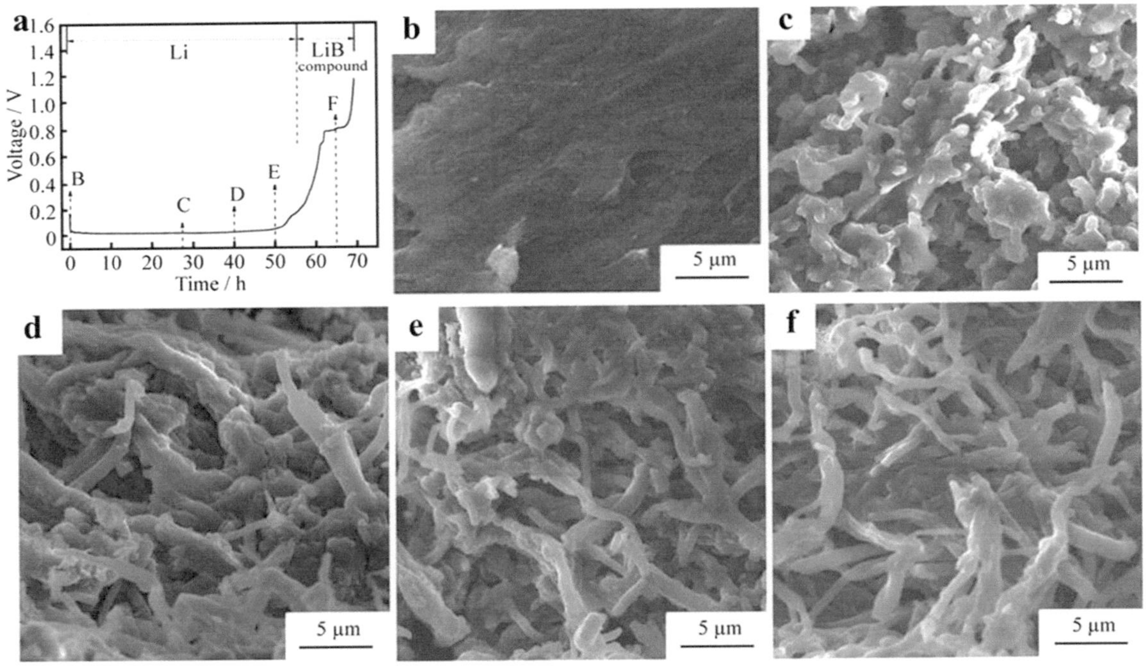

Fig. 1.9 **a** Electrochemical curve of Li–B alloy cycled at a current density of 0.5 $mAcm^{-2}$; **b** SEM images of Li–B alloy anode before electrochemical cycling and **c**, **d**, **e**, **f** after Li extraction [90]

electrochemical performance. For example, Yang et al. [92] reported a ternary $Li_{2.6}BMg_{0.05}$ alloy as a substitute for the original anode of metallic lithium $Li_{2.6}BMg_{0.05}$ is composed of rhombohedral Li_5B_4, cubic lithium, and Li_3Mg_7. Yan [94] reported a Mg-doped LiB framework for carrying metallic lithium and inhibiting lithium dendrite growth, as shown in Fig. 1.10a, b. The LiB framework can significantly reduce the volume change during the electrochemical dissolution/deposition process of lithium. Its superior lithium affinity and conductivity also help to reduce the local current density and homogenize the incoming Li^+ flux. More importantly, doping Mg elements in the LiB framework can enhance the adsorption energy of Li. The residual lithium-deficient lithium-magnesium alloy formed after lithium stripping helps to connect lithium fibers, stabilize the entire framework, reduce interface resistance, and effectively inhibit lithium dendrite growth, as shown in Fig. 1.10c.

The study of lithium-rich alloys as composite metallic lithium anodes is still in its initial stage, and most of the lithium-rich alloy systems studied are non-inert alloys, whose framework can reversibly deintercalate lithium, causing a certain volume change in the framework, which is not conducive to the stability of the electrode structure. Therefore, it is necessary to develop new alloy systems such as inert alloys and conduct more in-depth research on lithium-rich alloys.

With the development of technology, the application field of lithium is becoming more and more extensive, not only limited to the battery field. Lithium-ion batteries have become the mainstream batteries in the fields of consumer electronics, electric vehicles, etc. Due to the unique properties of lithium, it has a wide range of application prospects in the fields of medicine, chemical industry, nuclear reactors, etc. In the field of electric vehicles, lithium-ion batteries have become the most important type of battery. With the continuous expansion of the electric vehicle market, the demand for lithium is also increasing. In the future, with the continuous growth and popularization of the electric vehicle market, the demand for lithium will continue to increase. In addition, lithium has a wide range of application prospects in the medical field. Lithium compounds can be used to treat diseases such as malaria, depression, and multiple sclerosis. With the continuous development of medical technology, the application prospects of lithium will be even broader. In general, lithium is a very important metal element with a wide range of application prospects. In the future, with the continuous development of technology and the emphasis on environmental protection, the demand for lithium will continue to increase, and problems will be continuously discovered and solved.

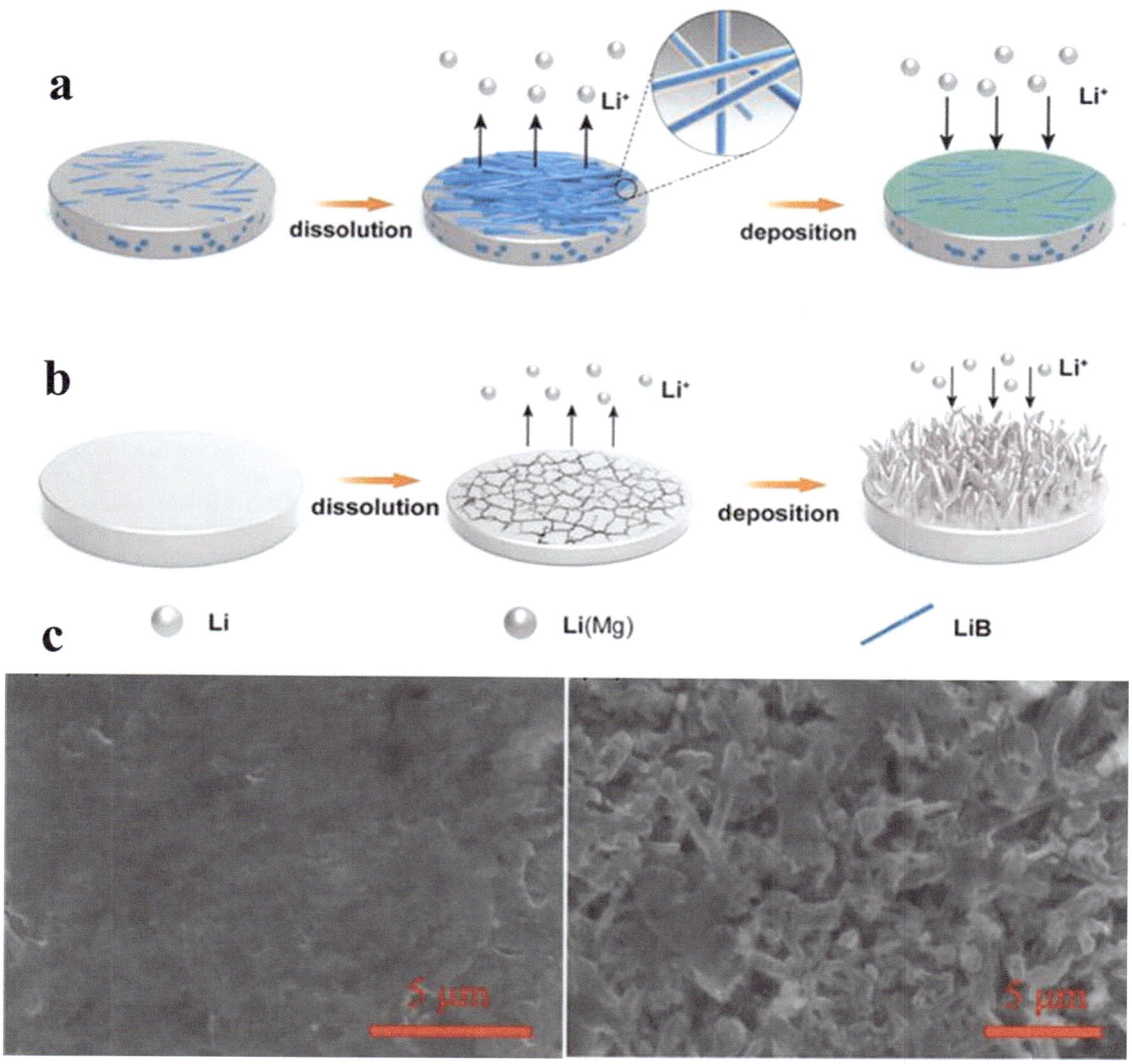

Fig. 1.10 Differences in lithium deposition and stripping between **a** Li–B–Mg composite material and **b** pure Li foil [94]

References

1. Wu J, Ren X, Wei Q (2020) Progress in the separation and extraction of lithium from salt lake brine. Inorgan Salt Ind 52(12):1–6
2. Callaghan RM (2022) Mineral commodity summaries. U.S Geological Survey 703:648–7709
3. Liu X, Zhong M, Chen X et al (2020) Enriching and separating lithium from magnesium in sulfate type salt lake brine. Hydrometallurgy 192:105247
4. Liu D, Li Z, He L et al (2021) Facet engineered $Li_3 PO_4$ for lithium recovery from brines. Desalination 514:115186
5. Sun Q, Chen H, Yu J (2022) Investigation on the lithium extraction process with the TBP–$FeCl_3$ solvent system using experimental and DFT methods. Ind Eng Chem Res 61(13):4672–4682
6. Yu X, Fan X, Guo Y et al (2019) Recovery of lithium from underground brine by multistage centrifugal extraction using tri-isobutyl phosphate. Sep Purif Technol 211:790–798
7. Goto M, Okumura K, Nakagawa S et al (2018) Nuclear and thermal feasibility of lithium-loaded high temperature gas-cooled reactor for tritium production for fusion reactors. Fusion Eng Des 136:357–361
8. Cai C, Hanada T, Fajar AT et al (2021) An ionic liquid extractant dissolved in an ionic liquid diluent for selective extraction of Li(I) from salt lakes. Desalination 509:115073
9. Song J, Huang T, Qiu H et al (2017) Recovery of lithium from salt lake brine of high Mg/Li ratio using Na [$FeCl_4$ * 2TBP] as extractant: thermodynamics, kinetics and processes. Hydrometallurgy 173:63–70
10. Pulido R, Naveas N, Martín-Palma RJ et al (2022) Experimental and density functional theory study of the Li^+ desorption in spinel/layered lithium manganese oxide nanocomposites using HCl. Chem Eng J 441:136019
11. Liu M, Wu D, Qin D et al (2022) Spray-drying assisted layer-structured $H_2 TiO_3$ ion sieve synthesis and lithium adsorption performance. Chin J Chem Eng 45:258–267

12. Li X, Chen L, Chao Y et al (2022) Highly selective separation of lithium with hierarchical porous lithium-ion sieve microsphere derived from MXene. Desalination 537:115847
13. Ryu T, Shin J, Ghoreishian SM et al (2019) Recovery of lithium in seawater using a titanium intercalated lithium manganese oxide composite. Hydrometallurgy 184:22–28
14. Paranthaman MP, Li L, Luo J et al (2017) Recovery of lithium from geothermal brine with lithium–aluminum layered double hydroxide chloride sorbents. Environ Sci Technol 51(22):13481–13486
15. Xiong J, Zhao Z, Liu D et al (2022) Direct lithium extraction from raw brine by chemical redox method with $LiFePO_4/FePO_4$ materials. Sep Purif Technol 290:120789
16. Luo G, Zhu L, Li X et al (2022) Electrochemical lithium ions pump for lithium recovery from brine by using a surface stability Al_2O_3–ZrO_2 coated $LiMn_2O_4$ electrode. J Energy Chem 69:244–252
17. Zhao Z, Si X, Liu X et al (2013) Li extraction from high Mg/Li ratio brine with $LiFePO_4/FePO_4$ as electrode materials. Hydrometallurgy 133:75–83
18. Liu D, Zhao Z, Xu W et al (2021) A closed-loop process for selective lithium recovery from brines via electrochemical and precipitation. Desalination 519:115302
19. Xu W, He L, Zhao Z (2021) Lithium extraction from high Mg/Li brine via electrochemical intercalation/de-intercalation system using $LiMn_2O_4$ materials. Desalination 503:114935
20. Zhao Z, Liu G, Jia H et al (2020) Sandwiched liquid-membrane electrodialysis: lithium selective recovery from salt lake brines with high Mg/Li ratio. J Membr Sci 596:117685
21. Liu G, Zhao Z, He L (2020) Highly selective lithium recovery from high Mg/Li ratio brines. Desalination 474:114185
22. Hou J, Zhang H, Thornton AW et al (2021) Lithium extraction by emerging metal-organic framework-based membranes. Adv Funct Mater 31(46):2105991
23. Barbosa LI, Valente G, Orosco RP et al (2014) Lithium extraction from β-spodumene through chlorination with chlorine gas. Miner Eng 56:29–34
24. Wu X (2012) Study on the process and mechanism of lithium extraction from Yichun mica. Central South University, pp 1–60
25. Li L, Hu N, Huang X (2008) Production method of refined lithium sulfate solution in the process of extracting lithium from lithium mica by sulfuric acid method 2006101453625
26. Vieceli N, Nogueira CA, Pereira MF et al (2017) Effects of mechanical activation on lithium extraction from a lepidolite ore concentrate. Miner Eng 102:1–14
27. Zhang X, Tan X, Li C et al (2019) Energy-efficient and simultaneous extraction of lithium, rubidium and cesium from lepidolite concentrate via sulfuric acid baking and water leaching. Hydrometallurgy 185:244–249
28. Jandová J, Dvořák P, Vu HN (2010) Processing of zinnwaldite waste to obtain Li_2CO_3 103(1–4):12–18
29. Wang D, Chen S (2020) Study on the process of extracting lithium from lithium mica by high pressure steam method. Inorgan Salt Ind 52(2):47–49
30. Yan Q, Li X, Yin Z et al (2012) A novel process for extracting lithium from lepidolite. Hydrometallurgy 121:54–59
31. Zhang X, Aldahri T, Tan X et al (2020) Efficient co-extraction of lithium, rubidium, cesium and potassium from lepidolite by process intensification of chlorination roasting. Chem Eng Process 147:107777
32. Luong VT, Kang DJ, An JW et al (2013) Factors affecting the extraction of lithium from lepidolite. Hydrometallurgy 134:54–61
33. Setoudeh N, Nosrati A, Welham NJ (2021) Lithium recovery from mechanically activated mixtures of lepidolite and sodium sulfate. Miner Process Extr Metall 130(4):354–361
34. Yan Q, Li X, Wang Z et al (2012) Extraction of valuable metals from lepidolite. Hydrometallurgy 117:116–118
35. Du C, Zhao Z, Liu H et al (2023) The status of representative anode materials for lithium-ion batteries. Chem Rec e202300004
36. Bruce PG, Freunberger SA, Hardwick LJ et al (2012) Li–O_2 and Li–S batteries with high energy storage. Nat Mater 11(1):19–29
37. Lewis GN, Keyes FG (1913) The potential of the lithium electrode. J Am Chem Soc 35(4):340–344
38. Liu Y, Su MY, Gu ZY et al (2022) Advanced lithium primary batteries: key materials, research progresses and challenges. Chem Rec 22(10):e202200081
39. Whittingham MS (1976) Electrical energy storage and intercalation chemistry. Science 192(4244):1126–1127
40. Zhang H, Li C, Eshetu GG et al (2020) From solid-solution electrodes and the rocking-chair concept to today's batteries. Angew Chem Int Ed 132(2):542–546
41. Guo Y, Li H, Zhai T (2017) Reviving lithium-metal anodes for next-generation high-energy batteries. Adv Mater 29(29):1700007
42. Liu B, Zhang J, Xu W (2018) Advancing lithium metal batteries. Joule 2(5):833–845
43. Wu H, Jia H, Wang C et al (2021) Recent progress in understanding solid electrolyte interphase on lithium metal anodes. Adv Energy Mater 11(5):2003092
44. Dollé M, Sannier L, Beaudoin B et al (2002) Live scanning electron microscope observations of dendritic growth in lithium/polymer cells. Electrochem Solid-State Lett 5(12):286–289
45. Feng S, Liu X, Guo S et al (2022) Nucleation, growth and suppression of lithium dendrites. J Chem Ind Eng 73(1):97–109
46. Lu D, Shao Y, Lozano T et al (2015) Failure mechanism for fast-charged lithium metal batteries with liquid electrolytes. Adv Energy Mater 5(3):1400993
47. Li P, Zhang H, Lu J et al (2023) Low concentration Sulfolane-based electrolyte for high voltage lithium metal batteries. Angew Chem Int Ed 354–361
48. Mo J, Yao Y, Li C et al (2022) A low-concentration all-fluorinated electrolyte for stable lithium metal batteries. ChemComm 58(89):12463–12466
49. Kim T-H, Lim J-H, Yim J-H et al (2023) A porous lithium metal anode with an artificial solid electrolyte interface fabricated by a facile process. J Alloys Compd 170424
50. Stassen I, Hambitzer G (2002) Metallic lithium batteries for high power applications. J Power Sour 105(2):145–150
51. Liu Y, Lin D, Li Y et al (2018) Solubility-mediated sustained release enabling nitrate additive in carbonate electrolytes for stable lithium metal anode. Nat Commun 9(1):3656–3665
52. Ding F, Xu W, Graff GL et al (2013) Dendrite-free lithium deposition via self-healing electrostatic shield mechanism. J Am Chem Soc 135(11):4450–4456
53. Piao N, Liu S, Zhang B et al (2021) Lithium metal batteries enabled by synergetic additives in commercial carbonate electrolytes. ACS Energy Lett 6(5):1839–1848
54. Mosallanejad B, Malek SS, Ershadi M et al (2022) Insights into the efficient roles of solid electrolyte interphase derived from vinylene carbonate additive in rechargeable batteries. J Electroanal Chem 116126–116141
55. Abraham K, Foos J, Goldman J (1984) Long cycle life secondary lithium cells utilizing tetrahydrofuran. EIC Labs Inc. Norwood, MA
56. Morita M, Aoki S, Matsuda Y (1992) AC impedance behaviour of lithium electrode in organic electrolyte solutions containing additives. Electrochim Acta 37(1):119–123

57. Gong C, Pu SD, Gao X et al (2021) Revealing the role of fluoride-rich battery electrode interphases by operando transmission electron microscopy. Adv Energy Mater 11(10):2003118
58. Yan C, Yao Y, Chen X et al (2018) Lithium nitrate solvation chemistry in carbonate electrolyte sustains high-voltage lithium metal batteries. Angew Chem Int Ed 57(43):14055–14059
59. Zheng J, Engelhard MH, Mei D et al (2017) Electrolyte additive enabled fast charging and stable cycling lithium metal batteries. Nat Energy 2(3):1–8
60. Schedlbauer T, Krüger S, Schmitz R et al (2013) Lithium difluoro(oxalato) borate: a promising salt for lithium metal based secondary batteries? Electrochim Acta 92:102–107
61. Yang G, Zhang S, Weng S et al (2021) Anionic effect on enhancing the stability of a solid electrolyte interphase film for lithium deposition on graphite. Nano Lett 21(12):5316–5323
62. Cho E, Mun J, Chae OB et al (2012) Corrosion/passivation of aluminum current collector in bis(fluorosulfonyl) imide-based ionic liquid for lithium-ion batteries. Electrochem Commun 22:1–3
63. Sasaki Y, Hosoya M, Handa M (1997) Lithium cycling efficiency of ternary solvent electrolytes with ethylene carbonate—Dimethyl carbonate mixture. J Power Sour 68(2):492–496
64. Guo L, Huang F, Cai M et al (2021) Organic-inorganic hybrid SEI induced by a new lithium salt for high-performance metallic lithium anodes. ACS Appl Mater Interfaces 13(28):32886–32893
65. Li X, Zheng J, Engelhard MH et al (2018) Effects of imide-orthoborate dual-salt mixtures in organic carbonate electrolytes on the stability of lithium metal batteries. ACS Appl Mater Interfaces 10(3):2469–2479
66. Miao R, Yang J, Feng X et al (2014) Novel dual-salts electrolyte solution for dendrite-free lithium-metal based rechargeable batteries with high cycle reversibility. J Power Sour 271:291–297
67. Weber R, Genovese M, Louli A et al (2019) Long cycle life and dendrite-free lithium morphology in anode-free lithium pouch cells enabled by a dual-salt liquid electrolyte. Nat Energy 4(8):683–689
68. Zheng H, Xiang H, Jiang F et al (2020) Lithium difluorophosphate-based dual-salt low concentration electrolytes for lithium metal batteries. Adv Energy Mater 10(30):2001440
69. Fan X, Chen L, Ji X et al (2018) Highly fluorinated interphases enable high-voltage Li-metal batteries. Chem 4(1):174–185
70. Chen S, Zheng J, Mei D et al (2018) High-voltage lithium-metal batteries enabled by localized high-concentration electrolytes. Adv Mater 30(21):1706102
71. Markevich E, Salitra G, Chesneau F et al (2017) Very stable lithium metal stripping-plating at a high rate and high areal capacity in fluoroethylene carbonate-based organic electrolyte solution. ACS Energy Lett 2(6):1321–1326
72. Su C, He M, Amine R et al (2019) Cyclic carbonate for highly stable cycling of high voltage lithium metal batteries. Energy Stor. Mater. 17:284–292
73. Zheng G, Lee SW, Liang Z et al (2014) Interconnected hollow carbon nanospheres for stable lithium metal anodes. Nat Nanotechnol 9(8):618–623
74. Cao S, He X, Nie L et al (2022) CF4 plasma-generated LiF–Li_2C_2 artificial layers for dendrite-free lithium-metal anodes. Adv Sci 9(21):2201147
75. Jia W, Wang Q, Yang J et al (2017) Pretreatment of lithium surface by using iodic acid (HIO3) to improve its anode performance in lithium batteries. ACS Appl Mater Interfaces 9(8):7068–7074
76. Li Y, Hu A, Gan X et al (2022) Synergy of in-situ heterogeneous interphases tailored lithium deposition. Nano Res 1–9
77. Yang C, Yin Y, Zhang S et al (2015) Accommodating lithium into 3D current collectors with a submicron skeleton towards long-life lithium metal anodes. Nat Commun 6(1):8058–8086
78. Lu L, Ge J, Yang J et al (2016) Free-standing copper nanowire network current collector for improving lithium anode performance. Nano Lett 16(7):4431–4437
79. Cheng X, Hou T, Zhang R et al (2016) Dendrite-free lithium deposition induced by uniformly distributed lithium ions for efficient lithium metal batteries. Adv Mater 28(15):2888–2895
80. Liang Z, Zheng G, Liu C et al (2015) Polymer nanofiber-guided uniform lithium deposition for battery electrodes. Nano Lett 15(5):2910–2916
81. Brissot C, Rosso M, Chazalviel JN et al (1998) In situ study of dendritic growth in lithium/PEO-salt/lithium cells. Electrochim Acta 43(10–11):1569–1574
82. Zou P, Chiang S, Li J et al (2019) Ni@Li_2O co-axial nanowire based reticular anode: tuning electric field distribution for homogeneous lithium deposition. Energy Stor. Mater. 18:155–164
83. Xiang J, Yuan L, Shen Y et al (2018) Improved rechargeability of lithium metal anode via controlling lithium-ion flux. Adv Energy Mater 8(36):1802352
84. Nan Y, Li S, Shi Y et al (2019) Gradient-distributed nucleation seeds on conductive host for a dendrite-free and high-rate lithium metal anode. Small 15(45):1903520
85. Pu J, Li J, Zhang K et al (2019) Conductivity and lithiophilicity gradients guide lithium deposition to mitigate short circuits. Nat Commun 10(1):1896–1905
86. Kong LO, Wang L, Ni Z et al (2019) Lithium-magnesium alloy as a stable anode for lithium-sulfur battery. Adv Funct Mater 29(13):1808756
87. Yang C, Xie H, Ping W et al (2019) An electron/ion dual-conductive alloy framework for high-rate and high-capacity solid-state lithium-metal batteries. Adv Mater 31(3):1804815
88. Duan B, Wang W, Zhao H et al (2013) Li–B alloy as anode material for lithium/sulfur battery 2(6):A47
89. Cheng XB, Peng HJ, Huang JQ et al (2014) Dendrite-free nanostructured anode: entrapment of lithium in a 3D fibrous matrix for ultra-stable lithium-sulfur batteries. Small 10(21):4257–4263
90. Huang H, Gui Y, Sun F et al (2021) In situ formed three-dimensional (3D) lithium-boron (Li-B) alloy as a potential anode for next-generation lithium batteries. Rare Met 40:3494–3500
91. Choi SH, Lee SJ, Yoo DJ et al (2019) Marginal magnesium doping for high-performance lithium metal batteries. Adv Energy Mater 9(41):1902278
92. Liu S, Yang J, Yin L et al (2011) Lithium-rich $Li_{2.6}BMg_{0.05}$ alloy as an alternative anode to metallic lithium for rechargeable lithium batteries. Electrochim Acta 56(24):8900–8905
93. Hafez AM, Jiao Y, Shi J et al (2018) Stable metal anode enabled by porous lithium foam with superior ion accessibility. Adv Mater 30(30):1802156
94. Wu C, Huang H, Lu W et al (2020) Mg doped Li–LiB alloy with in situ formed lithiophilic LiB skeleton for lithium metal batteries. Adv Sci 7(6):1902643

2 Carbon Anode

Since its first commercialization in the 1990s, lithium-ion batteries have made significant progress. Currently, lithium-ion batteries have become the main power source for mobile phones, laptops, and digital cameras, and are widely used in the field of electric vehicles. The widespread use of lithium-ion batteries is mainly due to their high energy density, long cycle life, and negligible memory effect. Carbon materials, due to their unique properties, including low cost, non-toxicity, high electronic conductivity, high structural stability, and high specific surface area, have many advantages in energy storage and conversion. Carbon materials exhibit excellent electrochemical performance when used as anode materials for lithium-ion batteries [1, 2]. In 1991, Sony technically realized the first lithium-ion battery prototype, with graphite used as the lithium-ion anode. Since then, researchers have conducted in-depth research and optimization on the principle and performance of graphite storing lithium ions. Currently, graphite is widely used as the commercial anode of lithium-ion batteries. Over the past 30 years, a large number of carbon nanomaterials have been discovered, including carbon nanotubes (CNTs), graphene, carbon fibers, biomass-derived carbon, etc., which have been studied for use in lithium-ion battery anodes. The differences in morphology and size of different types of carbon materials will lead to significant differences in their lithium-ion storage performance. This chapter will introduce the research progress of carbon materials anodes for rechargeable lithium-ion batteries. It details the synthesis, physical and chemical properties, and innovative significance of carbon materials in lithium-ion batteries.

2.1 Graphite

Graphite is the oldest and most common anode material for lithium-ion batteries, which has been widely commercialized, with a theoretical capacity of 372 mA h/g when used as a lithium-ion battery anode. As a naturally occurring mineral, graphite can be mined from metamorphic and igneous rocks. In addition, artificial graphite can also be obtained by graphitizing other carbon-based precursors (such as petroleum and coal) at high temperatures in an inert environment. When graphite is directly used as a lithium-ion battery anode, its electrode interface dynamics are relatively slow, and severe dendrites will appear under fast charging conditions, leading to a sharp decline in lithium-ion battery performance, and in severe cases, it may cause battery short circuits, posing risks of combustion or even explosion. Therefore, it is necessary to modify graphite. Currently, there are two main ways to modify graphite: (1) enhancing the migration and diffusion of lithium ions in graphite; (2) enhancing the interface reaction between graphite and the electrolyte to form a thinner and more stable SEI layer. Common graphite modification techniques include acid-base treatment, high-temperature expansion, etc.

The Zhang team synthesized a multi-channel graphite anode material using an air oxidation etching method (Fig. 2.1a) [3]. In this multi-channel graphite anode material, the lithium ion transmission distance is significantly shortened, which also promotes the penetration of the electrolyte, allowing lithium ions to quickly contact graphite particles. As shown in Fig. 2.1b, the multi-channel graphite has more lithium ion transmission paths and shorter distances, therefore it has faster electrode reaction kinetics. When multi-channel graphite is used as the anode of a lithium-ion battery, its performance is far superior to that of original graphite material (Fig. 2.1c). After 3000 cycles at a current density of 6 C, its capacity retention rate is still as high as 85%, demonstrating excellent lithium-ion storage performance. In addition, this unique multi-channel structure also increases the lithium-ion storage sites, effectively enhancing its lithium-ion storage capacity. This simple air oxidation etching process is very practical in actual production. Similarly, the Cheng team used a KOH activation etching process to synthesize porous graphite (Fig. 2.1d)

X. Li et al., *Electrode Development for Lithium-ion Batteries*, https://doi.org/10.1007/978-981-96-3021-9_2

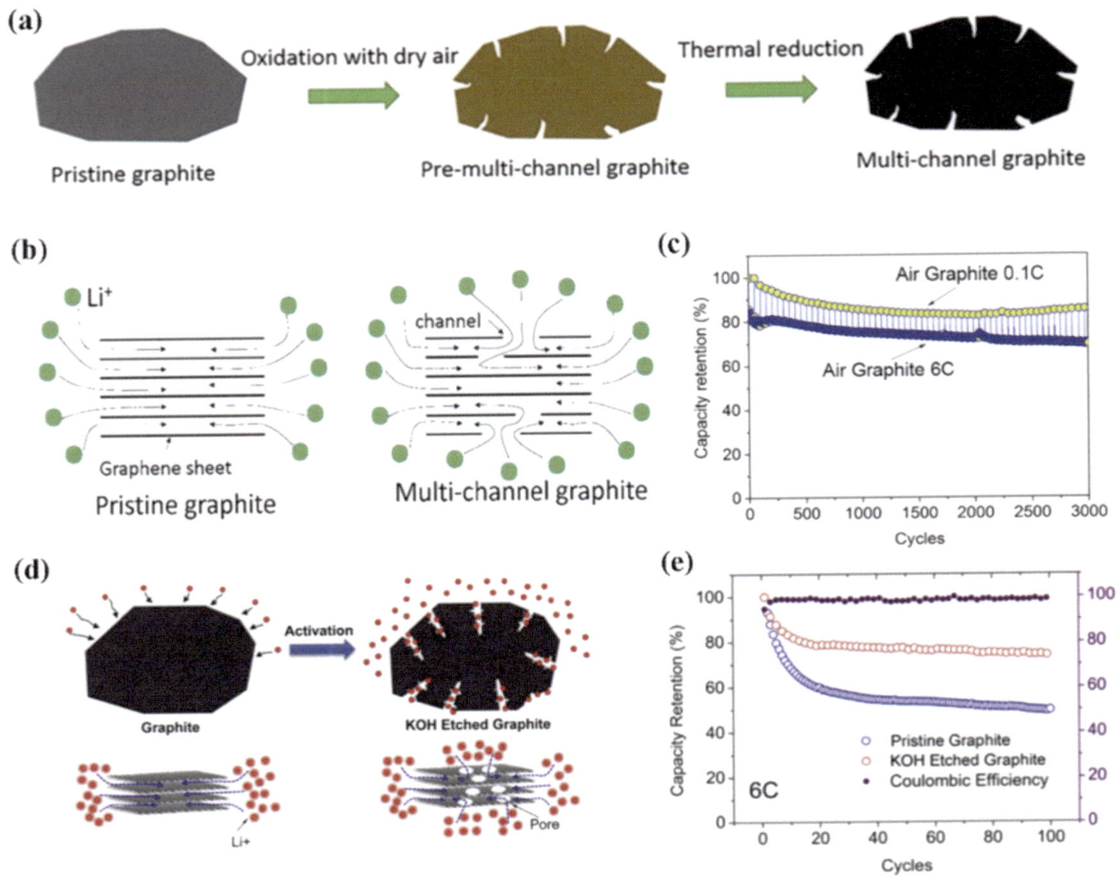

Fig. 2.1 **a** Synthesis pathway of multi-channel graphite; **b** schematic diagram of lithium ion transfer in graphite and multi-channel graphite; **c** cycling performance of multi-channel graphite produced by air oxidation as an anode for lithium-ion batteries [3]. **d** Schematic diagram of original graphite and KOH etched graphite and their cycling performance at 6 C as anodes for lithium-ion batteries (**d**) [4]

[4]. The holes etched on the graphite surface can also effectively increase the sites for lithium ion insertion and extraction, effectively shortening the lithium ion diffusion distance, thereby improving the charge and discharge rate of the graphite anode. The graphite etched by KOH, when used as an anode for a lithium-ion battery, has excellent rate performance and cycle stability. As shown in Fig. 2.1e, even at a higher current density (6 C), the capacity retention rate of the lithium-ion battery with a KOH etched graphite anode reaches 74%, which is far superior to the original graphite anode battery, with a Coulombic efficiency close to 100%, demonstrating excellent stability. These two examples show that constructing increased graphite porosity and specific surface area can effectively improve lithium ion transmission efficiency, and enhance its lithium ion storage capacity and cycle stability.

2.2 Graphene

Graphene is a single-layer carbon nanosheet. This unique structure gives graphene unique physical and chemical properties. Graphene, with its high electrical conductivity, large surface area, and high mechanical strength, is widely used in various secondary batteries and supercapacitors. Graphene is considered a potential alternative material for graphite anodes in lithium-ion batteries. Existing research shows that the reversible capacity of lithium-ion storage in graphene materials can reach 500–1000 mA h/g, which is far superior to the current graphite anodes, showing broad application prospects. In actual applications, graphene is usually modified to further enhance its lithium-ion storage performance. Common modification methods include synthesizing controllable nanostructures, increasing the porosity of graphene, and doping with heteroatoms. The Wook Ahn team synthesized graphene nanosheets through ultrasonic post-treatment, and controlled the morphology of graphene nanosheets by optimizing the reaction process. The modified graphene nanosheets have excellent lithium-ion storage performance, with an initial discharge capacity of 1079 mA at a current density of 40 mA/g h/g, with a capacity retention rate of 93% after 50 cycles [5]. Porous graphene is a new type of graphene nanostructure with unique electronic properties, which has a wide range of applications in many fields such as supercapacitors and catalysts. The Wei team used porous MgO sheets as templates and prepared porous graphene using chemical vapor deposition (CVD) technology [6]. As shown in Fig. 2.2a, this unique porous graphene has a pore size of about 3–8 nm. Porous graphene effectively solves the problems of large diffusion resistance and long diffusion distance between graphene and lithium ions. It enhances the reaction kinetics of lithium-ion storage and provides more active sites for lithium-ion storage. Based on these advantages, when porous graphene is used as the anode of a lithium-ion battery, the reversible capacity at a current density of 0.1 C is 1723 mA h/g, and the capacity is still 203 mA h/g at 20 C, showing excellent rate performance (Fig. 2.2b). Doping is another effective way to improve the lithium-ion storage performance of graphene. Heteroatom-doped graphene has multiple advantages: (1) improving its electronic conductivity; (2) providing additional active sites for lithium-ion storage; (3) reducing interface impedance; (4) enhancing the wettability of the material surface. Among various element dopings, nitrogen doping is most common in carbon-based anode materials for lithium-ion batteries. This is mainly because nitrogen atoms have properties similar to carbon atoms, but have higher electronegativity, and the interaction between lithium ions and nitrogen-doped carbon materials is stronger. The type and content of nitrogen doping have an important impact on the lithium-ion storage performance of graphene. Studies have shown that the higher the pyridine nitrogen content, the more reactive active sites will be exposed, so the electrochemical performance is more excellent. There are also various methods of nitrogen doping, the most common of which include annealing in an NH_3 environment, and mixing and annealing with substances such as melamine and dicyandiamide. The Liu team systematically studied the impact of different doping methods on the performance of lithium-ion batteries [7].

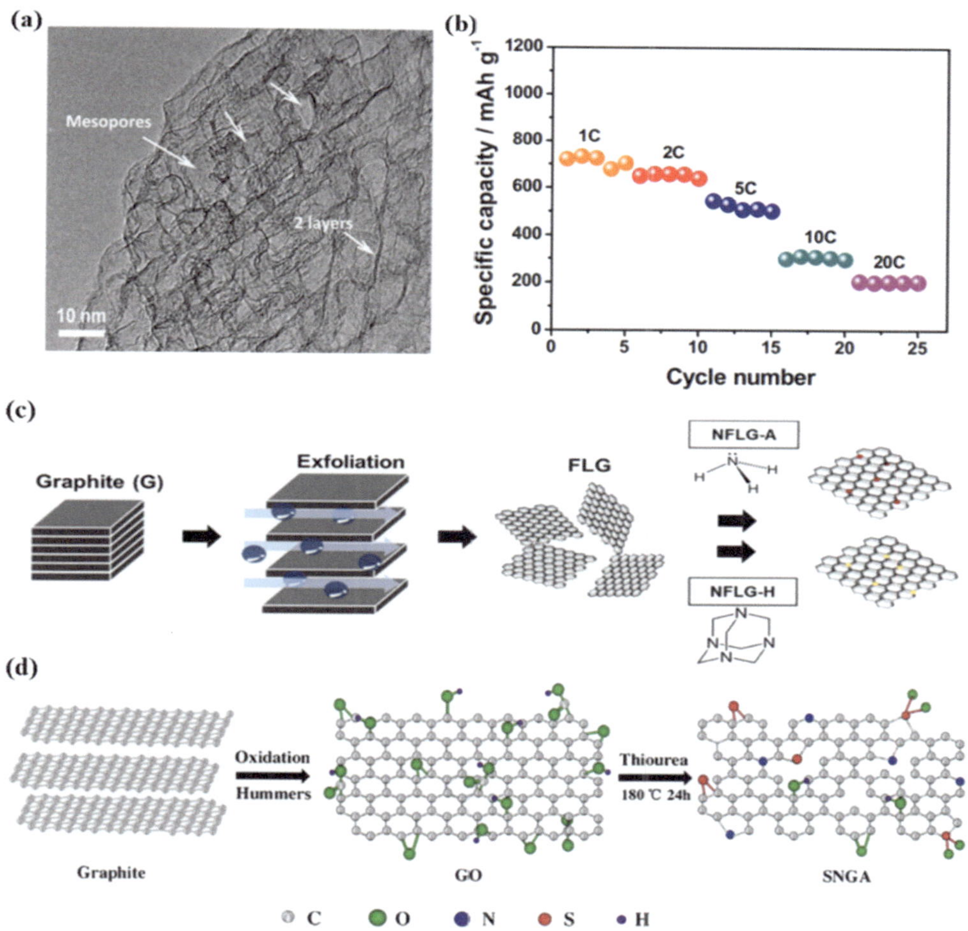

Fig. 2.2 **a** Transmission electron microscope image of porous graphene; **b** Rate graph when porous graphene is used as the anode of a lithium-ion battery [6]. **c** Schematic diagram of the synthesis of nitrogen-doped multilayer graphene [7]. **d** Synthesis pathway diagram of S/N dual-doped graphene [8]

As shown in Fig. 2.2c, researchers used graphite as the precursor material and prepared multilayer graphene through a simple and environmentally friendly method. Subsequently, nitrogen doping was carried out using hydrothermal liquid phase reaction and high-temperature solid phase reaction. The study showed that graphene doped by hydrothermal liquid phase method has a higher pyridine nitrogen content, while graphene doped by high-temperature solid phase method has higher pyrrole nitrogen and graphite nitrogen content. Graphene with a high pyrrole nitrogen content has more defects, which can provide more active sites for lithium ion storage, so the performance of graphene doped by hydrothermal liquid phase method is superior. The ability of graphene materials to accommodate single heterogeneous atoms is very limited, and single atom doping will inevitably encounter bottlenecks. In addition to single atom doping, multi-atom doping is also very common in improving the performance of graphene. The Sun team used a simple one-step hydrothermal method to synthesize three-dimensional S/N dual-doped graphene gel with thiourea as the raw material, and achieved precise control of the doping level (Fig. 2.2d) [8]. The rich porous structure and large specific surface area of the three-dimensional graphene gel accelerate the transmission of lithium ions and the penetration of the electrolyte. The synergistic effect of S/N dual-atom doping gives the graphene gel more defects and a favorable electronic structure. Based on these advantages, the S/N dual-doped graphene gel has excellent lithium ion storage performance, and after 400 cycles at a current density of 800 mA/g, the capacity is still 1109.8 mA h/g. In addition to S/N doping, N/B, N/P dual doping is also a good choice to improve the lithium ion storage performance of graphene.

2.3 Carbon Nanotubes/Carbon Fibers

Carbon nanotubes (CNTs) have become candidates for lithium-ion battery anodes due to their unique electrochemical and mechanical properties. Carbon nanotubes are divided into single-walled carbon nanotubes and multi-walled carbon nanotubes. There are significant van der Waals interactions between the sidewalls of multi-walled carbon nanotubes. Although carbon nanotubes have been proven to be a reliable lithium-ion battery anode material that can provide high specific capacity, there are certain limitations to applying carbon nanotubes to commercial lithium-ion batteries. For example, the inherent low volumetric capacity of carbon nanotubes.

To solve this problem, the Choi team designed a new three-dimensional (3D) structure, which increased its areal density to 34.9 mg/cm by stacking independent carbon nanotube layers, which is about three times higher than the current most advanced graphite anode (Fig. 2.3a) [9]. In addition, this three-dimensional stacked carbon nanotube has an interconnected structure, a large specific surface area, and a rich porosity. High porosity is beneficial to lithium ions diffuse through the electrolyte and transfer electron charges, improving the cyclic stability and rate performance of the electrode. The volumetric capacity of stacked carbon nanotubes is as high as 646 mA h/cm at 0.1 C, far exceeding the current commercial graphite lithium-ion storage performance. The flexibility and lightweight of lithium-ion batteries are a major trend for future development. Carbon nanotubes' excellent flexibility and mechanical stability have broad application prospects in the field of flexible batteries.

The Jeong team synthesized carbon nanotube films using chemical vapor deposition and direct spinning methods, followed by high-temperature treatment. The lithium-ion storage performance of the electrode made from the heat-treated carbon nanotube film is excellent [13]. The Choi team used three-dimensional (3D) standalone carbon nanotubes (CNTs) as the anode without a current collector to achieve high area and volumetric capacity flexible lithium-ion batteries [10]. Researchers grew carbon nanotubes on commercial-grade copper mesh substrates using catalytic chemical vapor deposition (CVD) method, then etched away Cu in $FeCl_3$ solution to obtain standalone carbon nanotube films. To prepare flexible lithium-ion batteries, three-dimensional carbon nanotubes on graphene/PET film were used as the working electrode, and lithium metal on PET film as the counter electrode (Fig. 2.3b). This carbon nanotube film-based flexible lithium-ion battery has excellent lithium-ion storage performance (0.1 C surface capacity is 0.25 mA h/cm, far exceeding other carbon-based material lithium-ion flexible batteries) and flexibility, with broad application prospects in wearable and new display fields in the future. Wearable lithium-ion batteries require excellent flexibility and safety during stretching and twisting. Carbon nanotube-based wearable lithium-ion batteries have attracted the attention of many researchers. The Peng team used two neatly arranged multi-walled carbon nanotube/lithium oxide composite yarns as the anode and cathode to prepare a stretchable linear lithium-ion battery without additional current collectors and binders. These two composite yarns can pair well to obtain a safe battery with superior electrochemical performance (Fig. 2.3c) [11]. These linear batteries have good flexibility (maintaining 97% of the capacity after 1000 bending cycles), lightweight, and an energy density an order of magnitude higher than lithium thin-film batteries. They also have excellent elasticity (maintaining 84% of the capacity after being stretched 200 times under 100% strain). This new type of fibrous battery will play an important role in smart clothing and medical testing fields in the future. Although carbon

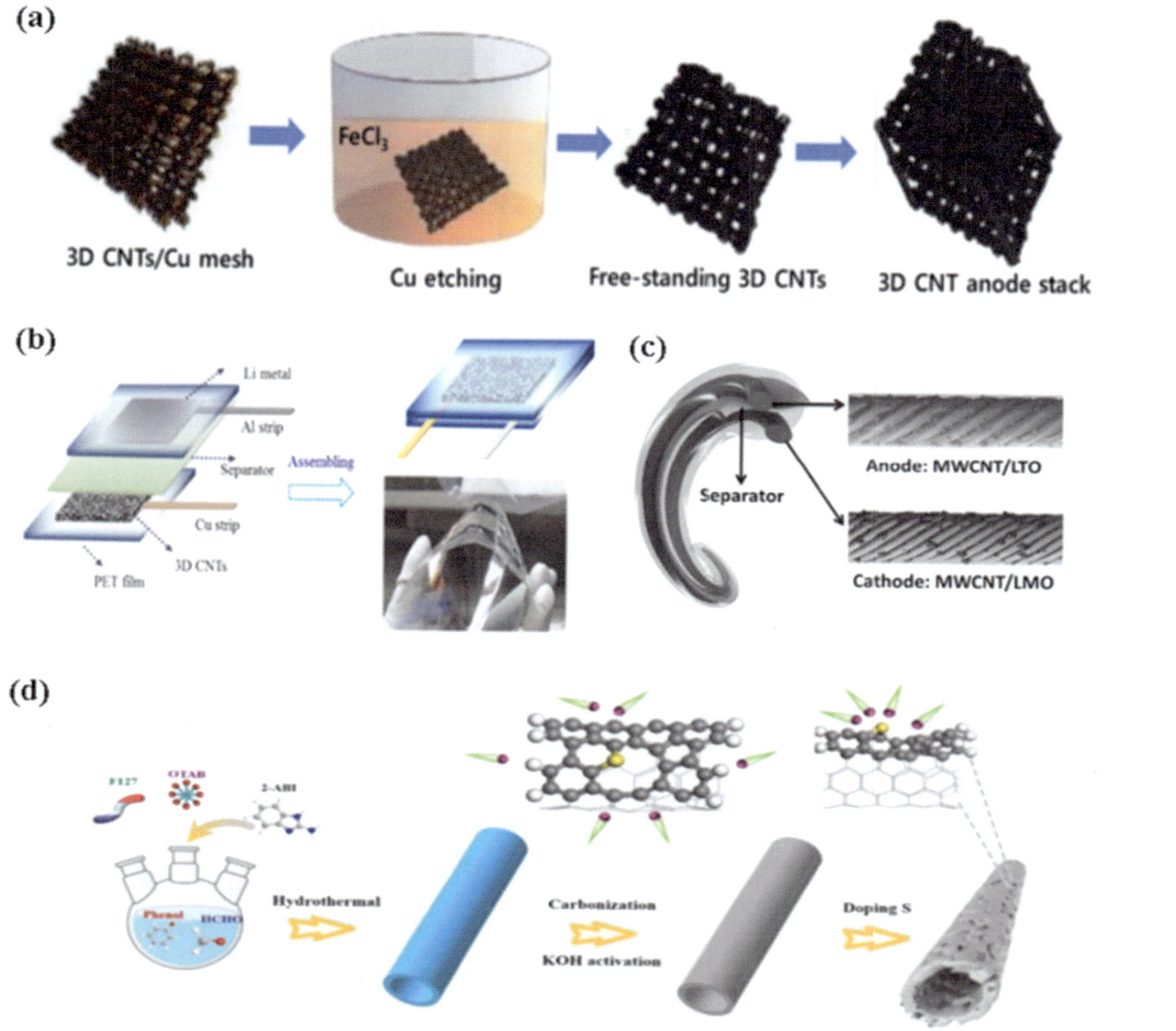

Fig. 2.3 **a** Preparation process of three-dimensional stacked carbon nanotube anode [9]. **b** Schematic diagram and actual photo of a flexible lithium-ion battery with a three-dimensional carbon nanotube as the working electrode [10]. **c** Structural diagram of a flexible linear lithium-ion battery [11]. **d** Schematic diagram of suitable S-doped porous carbon nanotubes [12]

nanotubes have made a series of important progress in lithium-ion battery anodes, the slow electron–ion transport dynamics of pure carbon nanotubes pose a severe challenge for further practical applications. To enhance the performance of carbon nanotubes, changing the morphology of carbon nanotubes, adjusting the crystallinity, and defect structure are some common strategies. The Zhang team synthesized a sulfur-doped carbon nanotube with a layered porous structure rich in defects and studied its lithium-ion storage performance (Fig. 2.3d) [12]. In this work, researchers used a simple hydrothermal strategy to dope sulfur into carbon nanotubes activated by KOH using phenolic resin as the raw material. The sulfur-doped carbon nanotubes exhibited excellent electrochemical performance, with a reversible capacity of up to 743.4 mA at a current density of 500 mA/g h/g. The doping of sulfur in carbon nanotubes has generated many defect structures, which not only endow the carbon nanotubes with additional active adsorption sites and good electrolyte/electrode interface compatibility, but also improve the diffusion kinetics of lithium ions.

Carbon fibers are a unique one-dimensional carbon nanomaterial. With their excellent conductivity and chemical stability, they are widely used in lithium-ion batteries. Electrospinning is a method for large-scale production of carbon nanofibers, which has been widely used in the industrial field. The synthesis of carbon nanofibers for lithium-ion batteries using electrospinning technology has broad application prospects. In the electrospinning process, the addition of some heteroatoms or pore-forming agents, activators can synthesize heteroatom-doped, suitable porosity, defect-rich high-performance carbon nanofibers in one step. The Jiang team used electrospinning technology combined with carbothermal reduction to synthesize heteroatom-doped carbon nanofibers (Fig. 2.4a) [14]. In the synthesis process, the $Cu(CH_3COO)_2$ precursor is evenly distributed in the polymer nanofibers. During the subsequent high-temperature pyrolysis process, the copper salt decomposes in situ, the copper particles are evaporated, and carbon bubbles of nanometer size are formed on the fiber surface. Scanning electron microscope analysis results show that the fibers present an interlinked structure, which is very conducive to electron transmission and can effectively enhance the kinetics of lithium-ion storage (Fig. 2.4b). In addition, the nitrogen element in the polymer spontaneously dopes the carbon fiber, and the spontaneous introduction of nitrogen improves the conductivity and activity of the electrode material. Based on these advantages, this unique structure of carbon fiber still has a lithium-ion storage capacity of 480 mA h/g after 900 cycles at a current density of 800 mA/g (Fig. 2.4c). Bacterial cellulose is a more widely sourced nanofiber network with high purity, high polymerization, and high mechanical performance (Fig. 2.4d). After high-temperature carbonization, bacterial cellulose presents an interlinked carbon fiber network (Fig. 2.4e). Carbon nanofibers synthesized from bacterial cellulose have multiple advantages, such as structural stability, mechanical flexibility, large specific surface area, open porous structure, etc. Another major advantage of bacterial cellulose is that its morphology, porosity, crystallinity, and orientation have great adjustability. Adding additives during pyrolysis can change the electrochemical properties of carbon nanofibers after carbonization. The Cao team used KOH to activate bacterial cellulose, prepared a layered porous structure and large surface area nanocarbon fiber network, and showed excellent performance when used as a negative electrode for lithium-ion batteries [15]. In addition, the lithium-ion storage performance of bacterial cellulose-derived carbon can be further improved by heteroatom doping. It is worth noting that bacterial cellulose also has broad application prospects in the field of self-supporting electrodes for lithium-ion batteries.

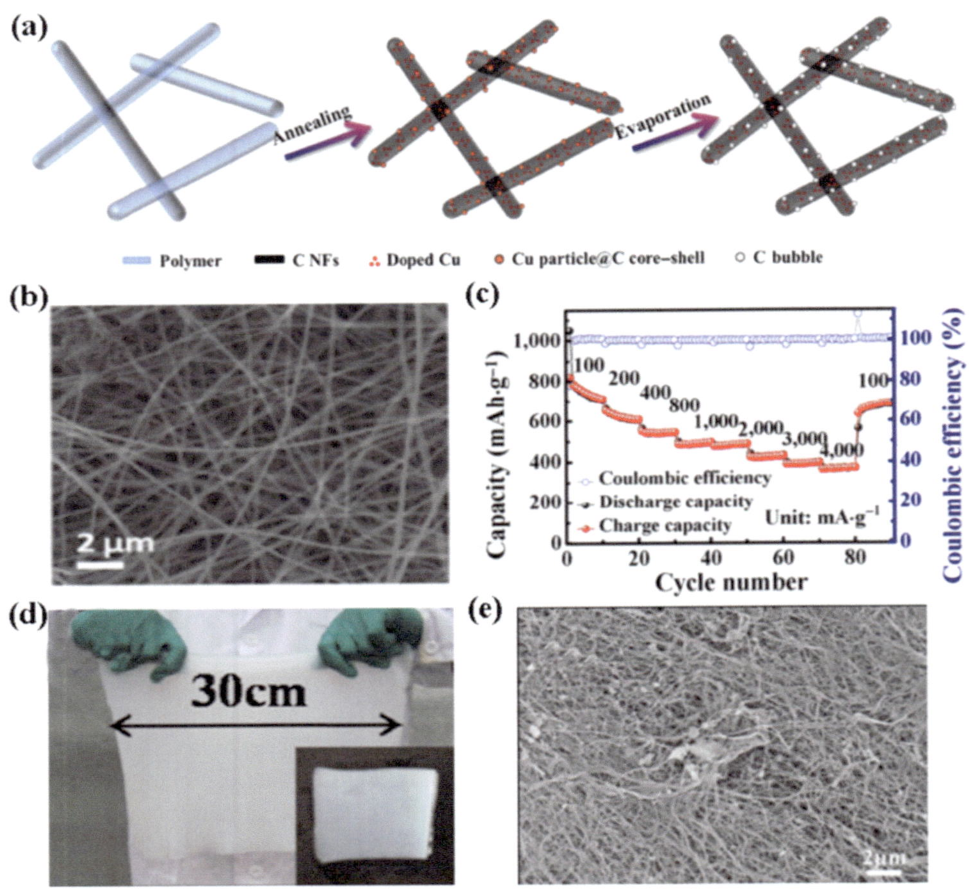

Fig. 2.4 **a** Schematic diagram of heteroatom-doped carbon nanofiber synthesis; **b** Scanning electron microscope image of carbon nanofibers and magnification image when used as a negative electrode for lithium-ion batteries (**c**) [14]. **d** Physical image of bacterial cellulose; **e** scanning electron microscope image of carbon fibers obtained after high-temperature carbonization of bacterial cellulose [15]

2.4 Biomass-Derived Carbon

Over the years, various forms of carbon materials including graphene, carbon nanotubes, fullerenes, etc., have been developed for a wide range of industrial and non-industrial applications. The development of these complex carbon-based materials proves that carbon has infinite application possibilities. However, due to the current level of technology, the synthesis and regulation of materials such as graphene are relatively complex, limiting their large-scale application. Therefore, the development of low-cost, efficient carbon materials is of great significance for the development of lithium-ion batteries. Among all carbon precursors, plant and animal waste are widely used sources of biomass. Porous carbon synthesized through carbonization and activation technologies can be used in the field of energy storage. The common strategy for synthesizing biomass porous carbon is to increase its specific surface area and defects by various physical or chemical methods. In addition, introducing functional heteroatoms into porous carbon through doping or hybridization is another effective method to improve its electrochemical performance. Compared with biomass carbon obtained by direct carbonization of precursors, doped carbon materials have better characteristics, such as enhanced electron conductivity, high chemical and thermal stability, and adjustable surface oxidation-reduction sites or acidity and alkalinity. These features have unique advantages in lithium storage.

The exciting thing is that many biomasses contain abundant heteroatoms. Selecting the appropriate biomass precursor can yield excellent heteroatom-doped carbon materials after carbonization. The most common method of treating biomass is direct carbonization. Common carbonization methods include high-temperature pyrolysis and hydrothermal carbonization. The former involves high-temperature treatment in the absence of oxygen, while the latter usually takes place in a reaction kettle with water as the solvent at relatively low temperatures. Pyrolysis treatment conditions (temperature, heat preservation time, heating rate, atmosphere, etc.) and the reasonable choice of biomass materials are crucial for improving battery performance [18]. The Zhou team developed a simple method, combining hydrothermal and graphitization, to synthesize high-quality graphene sheets from wheat straw (Fig. 2.5a) [16]. The wheat straw-derived graphene sheets have an ultra-thin nanosheet framework (Fig. 2.5b), high graphitization (up to 90.7%), and a rich mesoporous structure. When used as an anode material for lithium-ion batteries, this type of graphene sheet has a reversible capacity of up to 463.5 mA h/g at 1 C, and even at a large current density of 5C, the reversible capacity is still 306.8 mA h/g, demonstrating excellent rate performance (Fig. 2.5c). This method of preparing high-quality graphene from biomass through environmentally friendly, catalyst-free pathways is very suitable for large-scale industrial applications. Although hydrothermal carbonization is convenient and fast, the directly carbonized biomass carbon materials often have undeveloped pores, and it is easy to encounter bottlenecks when used to increase the capacity of electrode materials. Therefore, a series of measures are taken to increase their specific surface area and porosity. Common methods to increase the surface area include templating and chemical activation. Zhu's team used sustainable biomass raw material cotton cellulose as the raw material, absorbed $Mg(NO_3)_2$ solution in cellulose fibers, then underwent drying and carbonization processes, and finally removed the MgO template with acid to obtain high specific surface area porous carbon [19]. This porous carbon, when used as a negative electrode for lithium-ion batteries, still has a reversible capacity of 793 mA h/g after 500 cycles at a current density of 0.5 A/g. Xiao's team adopted a more convenient activation method, using wheat straw as the raw material and KOH activation to synthesize nitrogen-rich porous carbon with a layered porous structure [17]. The wheat straw-derived carbon after KOH activation has a large specific surface area and a rich porous structure (Fig. 2.5d). When used as a negative electrode for lithium-ion batteries, it still maintains a reversible capacity of 344 mA h/g at a current density of 18.5 A/g (Fig. 2.5e). KOH activation is the most common and effective method. The reaction of biomass carbon with KOH will significantly etch the carbon matrix, resulting in a large number of porous structures. In addition, the intermediate metal K is embedded in the carbon matrix, causing the carbon lattice to expand. Removing the intercalated metal K can further enhance the porosity. There are many reports on KOH activation of biomass. In addition to KOH, $ZnCl_2$, H_3PO_4, $FeCl_3$, $MgCl_2$ are also often used to activate biomass carbon materials. At present, activation is still the preferred method for producing porous carbon with extremely high surface area and other adjustable characteristics.

In addition to plant biomass, animal biomass is also widely used to synthesize porous carbon materials for use as negative electrodes for lithium-ion batteries. Ma's team used natural silk (Fig. 2.6a) as the raw material, and adopted a strategy of simultaneous activation and graphitization to synthesize layered porous carbon materials [20]. The novel carbon material has a typical flake structure (Fig. 2.6b), and after activation by metal salts $ZnCl_2$ and $FeCl_3$, it exhibits a rich porous structure, with a specific surface area of up to 2494 m^2/g and a pore volume of 2.28 cm^3/g (Fig. 2.6c). The high specific surface area provides abundant active sites, and the rich porous structure can effectively promote the penetration of the electrolyte and the transfer of charge, significantly improving the

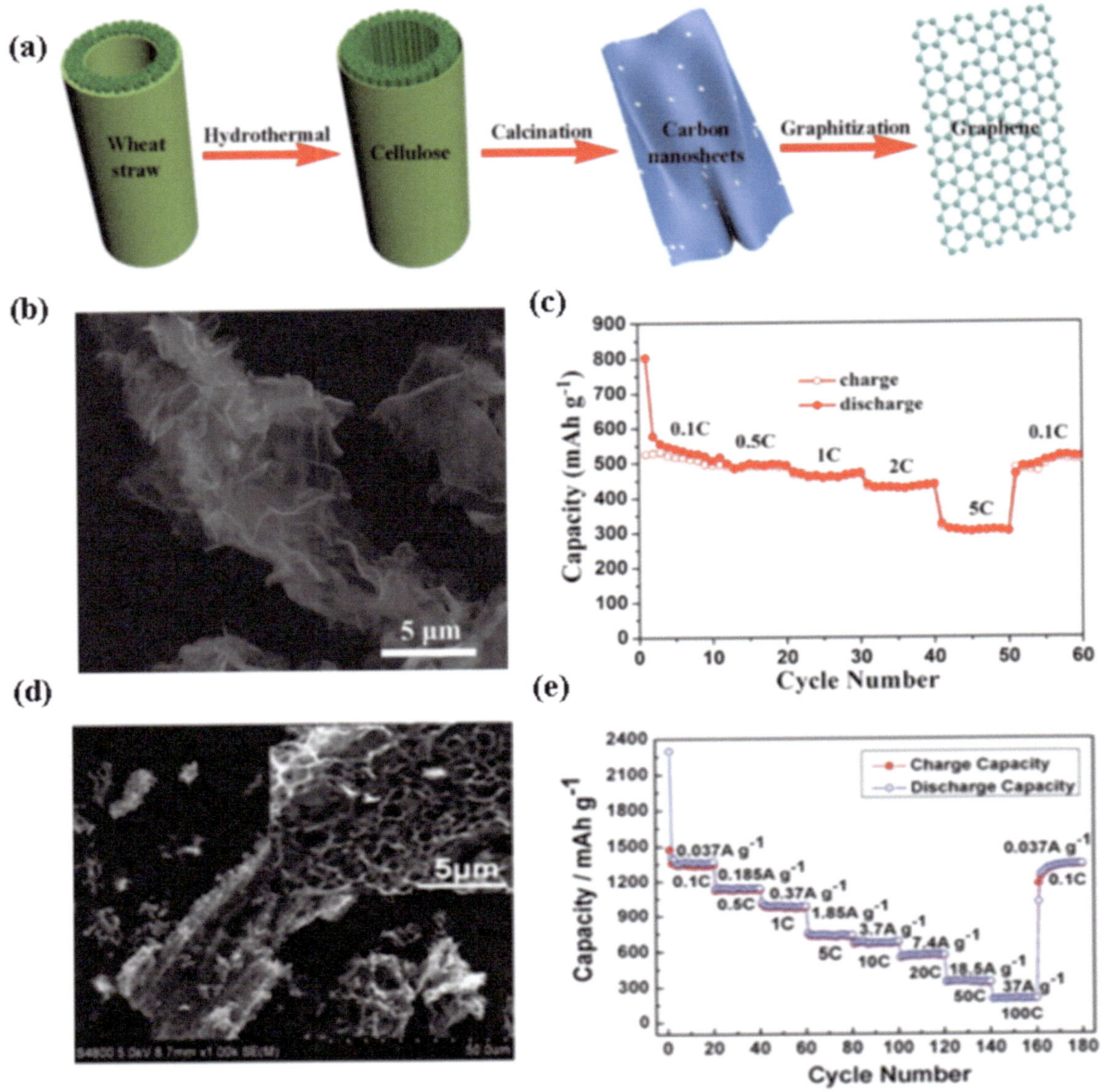

Fig. 2.5 **a** Pathway diagram for synthesizing graphene nanosheets from wheat straw; **b** scanning electron microscope image of wheat straw-derived graphene sheets and their rate performance as a lithium-ion battery anode (**c**) [16]. Scanning electron microscope image of porous carbon obtained by KOH activation of wheat straw (**d**) and its rate performance as a lithium-ion battery anode (**e**) [17]

electrode kinetics while reducing the polarization effect. In addition, thanks to the rich nitrogen content in silk, the porous carbon derived from silk has a high nitrogen content (4.7%), and the rich nitrogen content can significantly enhance the electrochemical activity and conductivity of the carbon material. Based on these advantages, the porous carbon derived from silk shows excellent electrochemical performance as a lithium-ion battery anode, with reversible capacities of 1604 mA h/g and 261 mA h/g at current densities of 0.2 A/g and 20 A/g, respectively. Even when the current density increases to 37.2 A/g, there is still a reversible capacity of 212 mA h/g (Fig. 2.6d), and after 300 cycles at a current density of 3720 mA/g, the capacity is as high as 714 mA h/g (Fig. 2.6e), showing excellent cycle stability. Gelatin is another common animal-based biomass, often derived from animal fur, and is a renewable biomass. The Wang team used fish scales as raw materials, extracted gelatin as a carbon source, and successfully synthesized porous carbon with a rich structure (Fig. 2.6f) with the help of KCL and ice templates through freeze-drying, carbonization, and aqueous steps. This carbon material derived from fish scales is an interconnected porous nanoplate structure (Fig. 2.6g) [21], with a carbon plate thickness of about 1.4 nm. In addition, the abundant heteroatoms in fish scales also doped the carbon material during the high-temperature annealing process. Thanks to the unique porous nanoplate

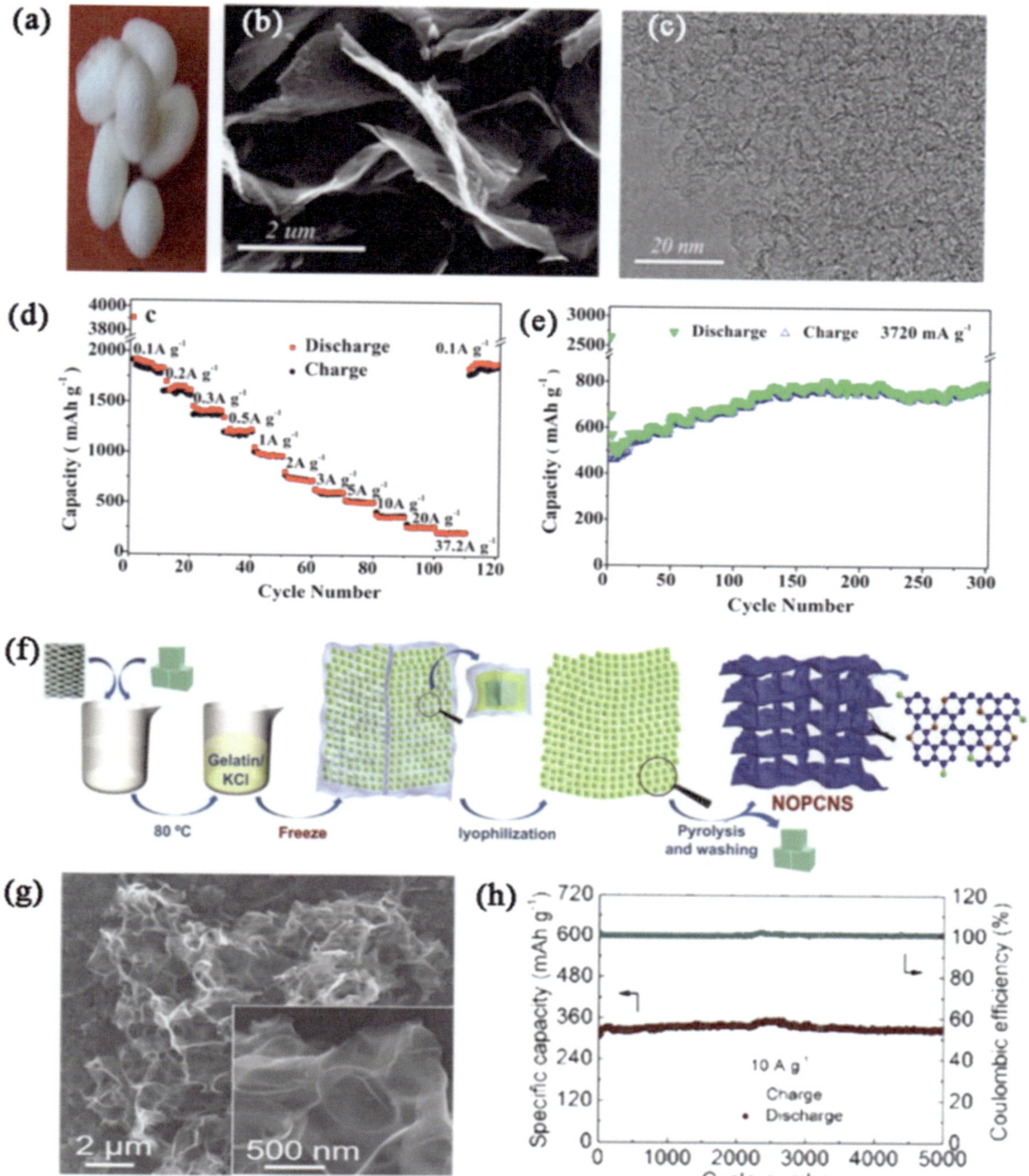

Fig. 2.6 **a** Scanning electron microscope image of natural silk, scanning electron microscope image of porous carbon derived from silk (**b**), transmission electron microscope image (**c**), and rate performance graph as a negative electrode for lithium-ion batteries (**d**) and cycle performance graph at a current density of 3720 mA/g (**e**) [20]. Schematic diagram of the synthesis of porous carbon prepared from fish scales (**f**), scanning electron microscope image of interlinked porous carbon derived from fish scales (**g**) and long cycle performance graph as a negative electrode for lithium-ion batteries at a current density of 10 A/g (**h**) [21]

structure and rich heteroatom doping (N:8.2 at.%, O:8.4 at.%), the fish scale-derived carbon shows excellent rate performance and cycle stability as a lithium-ion battery, and after 5000 cycles at a high current density (10 A/g), the capacity does not show obvious decay (Fig. 2.6h). In addition to natural silk and fish scales, other animal biomasses, such as shrimp shells, discarded bones, feathers, and expired milk, have also been studied for use in lithium-ion battery anodes.

In addition to the commonly used direct carbonization method and the post-activation carbonization method for preparing lithium-ion battery anodes using biomass materials, light atom doping is another effective way to improve the lithium-ion storage performance of biomass-based carbon materials. Common non-carbon doping elements include N, S, B, etc. Heteroatom doping can change the type and content of functional groups on the surface of carbon materials, significantly improving the conductivity of carbon materials. Among various heteroatom doping strategies, N doping is the most common. Nitrogen doping can provide abundant attachment points for lithium ions, change the density of surrounding electron clouds, enhance the adsorption performance of lithium ions, and thus improve the energy density and power density of carbon materials.

The Hu team used sugarcane bagasse as raw material, melamine as a nitrogen source, and synthesized nitrogen-rich carbon material with a multi-level porous structure through KOH activation and nitrogen doping (Fig. 2.7a) [22]. The lithium-ion storage performance of porous carbon with different N doping contents was systematically studied (Fig. 2.7b). The appropriate N-doped carbon material and undoped carbon material at 5 The reversible capacity is 142.7 and 60.8 mA h/g under a current density of A/g. The research results show that N doping can significantly improve the capacity and rate performance of carbon materials, which is mainly due to N doping promoting the insertion and removal of lithium ions, enhancing the kinetics of electrode reactions. In addition to cyanamide, commonly used N sources for doping biomass include dicyandiamide, urea, and ammonia. S doping is another effective way to improve the lithium-ion storage performance of carbon materials. The electronegativity of S atoms is significantly

Fig. 2.7 **a** Schematic diagram of nitrogen-doped carbon prepared from sugarcane bagasse, **b** lithium–ion storage rate performance of carbon materials with different nitrogen contents prepared from sugarcane bagasse [22] (**c**) Scanning electron microscope image of S-doped porous carbon prepared from pharmaceutical residue, rate performance graph (**d**) and cycle performance graph under a current density of 5 A/g [23]

different from that of carbon atoms, so S doping can to provide more active sites, in addition, S doping can also increase the interlayer spacing of carbon, effectively improving the rate performance of carbon materials [24]. Based on the above theory, Du's team used pharmaceutical residue as raw material and thioacetamide as sulfur source for doping, successfully synthesizing interconnected porous sponge structure carbon materials (Fig. 2.7c) [23]. After S doping, the lithium-ion storage performance of carbon materials has been significantly improved. By adjusting the amount of S doping, the carbon material with the optimal amount of S doping shows excellent rate performance and cycle stability. The reversible capacity under a current density of 5A/g is still 388 mA h/g (Fig. 2.7d), and the capacity remains at 364 mA h/g after 500 cycles (Fig. 2.7e), which is significantly improved compared to the rate performance and capacity of carbon materials without S doping. In addition to thioacetamide, thiourea, sulfur powder, and dimethyl sulfide are also used for doping carbon materials.

Compared to single-atom doping, dual-atom doping is more efficient in enhancing the lithium-ion storage performance of carbon materials. The synergistic effect of dual atoms can alter the electronic transport properties of carbon materials, provide additional active sites, introduce more defects and disorder, and is more conducive to improving the lithium-ion storage performance of carbon materials. Common dual-atom doping includes N,S co-doping, N,P co-doping, N,B co-doping, etc. Zhao's team used expired egg yolk as raw material to prepare capsule-like N,P co-doped porous carbon (Fig. 2.8a) [25]. The proteins and phospholipids in the egg yolk provide a rich source of nitrogen and phosphorus, and the synergistic effect of N,P

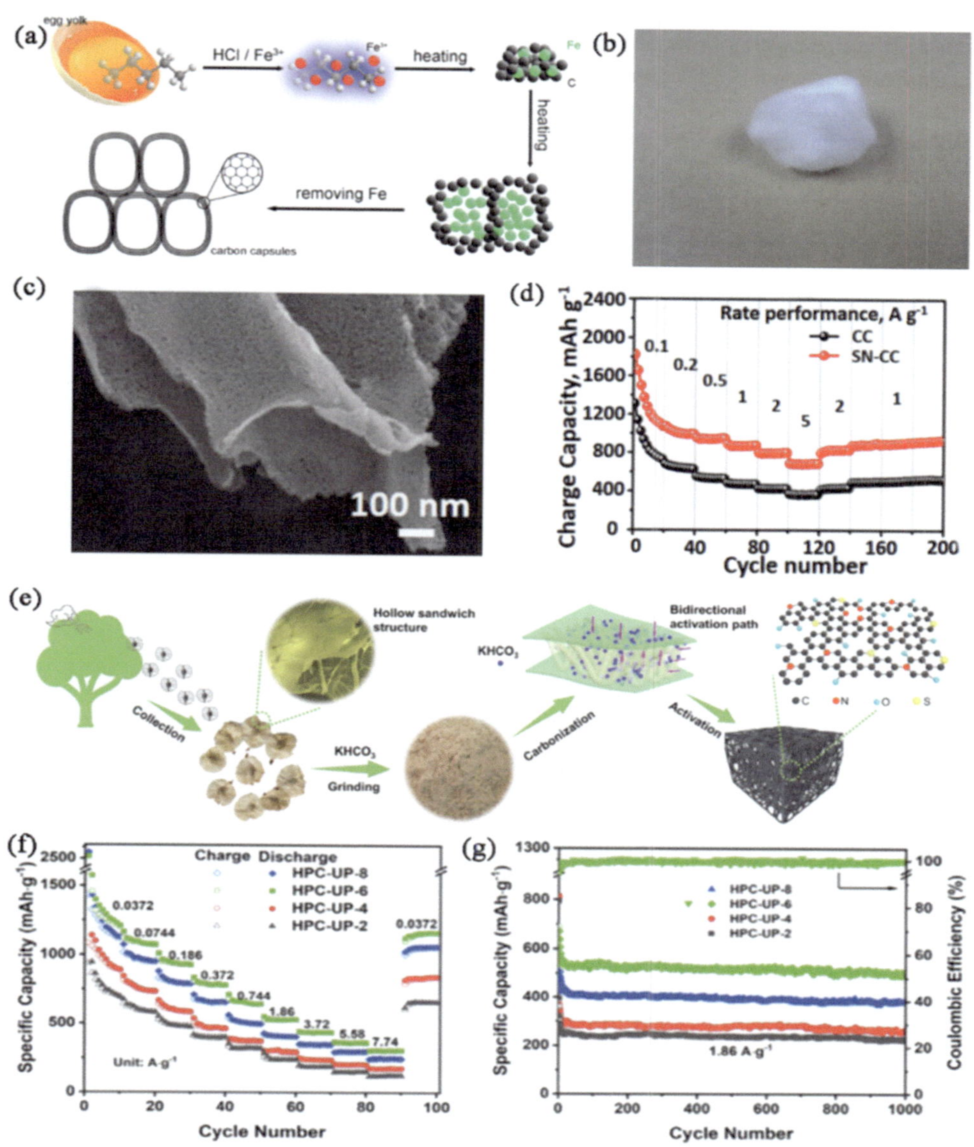

Fig. 2.8 **a** Synthesis pathway diagram of N,P co-doped porous carbon prepared from egg yolk [25]; Photo of cotton (**b**) and the scanning electron microscope image of its derived N,S co-doped porous carbon (**c**) and its rate performance graph as a lithium-ion battery anode (**d**) [26]; **e** synthesis pathway diagram of N, O, S ternary doped porous carbon prepared from tree fruit, and its rate (**f**) and cycle performance graph (**g**) as a lithium-ion battery [27]

co-doping endows the egg yolk-derived porous carbon with excellent lithium-ion storage performance. Yang's team used discarded cotton as a carbon source and thiourea as a source of phosphorus and nitrogen to prepare N,S co-doped porous carbon (NS-CC) (Fig. 2.8b, c) [26]. This dual-doped carbon material has a typical flake structure and a rich pore structure. N,S dual-atom doping introduces a wealth of defects into the carbon material, effectively increasing the interlayer spacing of carbon. The rich pore structure provides abundant active sites for lithium-ion storage, shortens the ion transport path, and enhances the electrode reaction kinetics. Based on these advantages, NS-CC exhibits excellent electrochemical performance as a negative electrode for lithium-ion batteries, maintaining a reversible capacity of 600 mA h/g even at a high current density of 5 A/g (Fig. 2.8d). In addition to adding nitrogen and sulfur sources to prepare heteroatom-doped carbon materials, cleverly using the abundant heteroatoms in plants themselves to prepare multi-atom doped carbon materials is also an efficient approach. Sun's team used fruit seeds to prepare N,O, and S tri-element co-doped porous carbon with abundant defects and high specific surface area through a simple activation strategy (Fig. 2.8e) [27]. By adjusting the ratio of the activator $KHCO_3$ to biomass, a variety of structurally rich porous carbons were prepared. The optimized porous carbon as a negative electrode for lithium-ion batteries has a capacity of 361.3 mA h/g at an energy density of 7.44 A/g, and a capacity retention rate of up to 93.4% after 1000 cycles at a current density of 1.86 A/g (Fig. 2.8f, g). The excellent electrochemical performance is mainly attributed to the unique porous structure, abundant defects, and larger interlayer spacing of carbon.

2.5 Conclusion

This chapter introduces the application of carbon materials (graphite, graphene, carbon nanotubes/carbon fibers, biomass-derived carbon) in the anode of lithium-ion batteries. It specifically includes an introduction to the synthesis methods of carbon materials, as well as a summary of common modification strategies. In summary, the widespread application of carbon materials in lithium-ion batteries has promoted the development of lithium-ion batteries. To date, graphite has become the most widely used carbon material, used in the anode of commercial lithium-ion batteries. Graphene, with its excellent conductivity and large specific surface area, has become the best performing carbon material. In the future, with the breakthrough in large-scale production technology of graphene, it will be widely used in lithium-ion batteries. Carbon nanotubes/carbon fibers have high strength, low density, and excellent conductivity. As anode materials for lithium-ion batteries, they can enhance lithium embedding capacity and ion transmission rate. In addition, carbon fibers and carbon nanotubes can also serve as battery current collectors, and are expected to be applied in the field of flexible lithium-ion batteries and wearable lithium-ion batteries in the future. Biomass-derived carbon is a carbon material made through the pyrolysis, carbonization, and activation process of biomass. It has sustainability, low cost, and abundant resources. The application of biomass-derived carbon in the anode of lithium-ion batteries is gradually gaining attention. The large-scale application of biomass-derived carbon will further reduce the cost of lithium-ion batteries. In summary, carbon materials have a wide range of application prospects in the anode of lithium-ion batteries. The preparation technology of high-performance carbon materials can significantly improve the energy density, power density, and cycle life of the battery, promoting the development of lithium-ion batteries in the field of energy storage.

References

1. Xie L, Tang C, Bi Z, Song M, Fan Y, Yan C, Li X, Su F, Zhang Q, Chen C (2021) Adv Energy Mater 11:2101650
2. Liu H, Liu X, Li W, Guo X, Wang Y, Wang G, Zhao D (2017) Adv Energy Mater 7:1700283
3. Cheng Q, Zhang Y (2018) J Electrochem Soc 165:A1104–A1109
4. Cheng Q, Yuge R, Nakahara K, Tamura N, Miyamoto S (2015) J Power Sour 284:258–263
5. Ahn W, Song HS, Park S-H, Kim K-B, Shin K-H, Lim SN, Yeon S-H (2014) Electrochim Acta 132:172–179
6. Fan Z, Yan J, Ning G, Wei T, Zhi L, Wei F (2013) Carbon 60:558–561
7. Ting P-M, Huang J-Y, Muruganantham R, Liu W-R (2022) Mater Today Commun 31:103498
8. Shan H, Li X, Cui Y, Xiong D, Yan B, Li D, Lushington A, Sun X (2016) Electrochim Acta 205:188–197
9. Kang C, Patel M, Rangasamy B, Jung K-N, Xia C, Shi S, Choi W (2015) J Power Sour 299:465–471
10. Kang C, Cha E, Baskaran R, Choi W (2016) Nanotechnology 27:105402
11. Ren J, Zhang Y, Bai W, Chen X, Zhang Z, Fang X, Weng W, Wang Y, Peng H (2014) Angew Chem Int Ed 53:7864–7869
12. Li X, Xue C, Liu Y, Zhao J, Zhang J, Zhang J (2023) Electrochim Acta 440:141697
13. Yoon S, Lee S, Kim S, Park K-W, Cho D, Jeong Y (2015) J Power Sour 279:495–501
14. Zhang H, Zhang G, Li Z, Qu K, Shi H, Zhang Q, Duan H, Jiang J (2018) Nano Res 11:3791–3801
15. Wang W, Sun Y, Liu B, Wang S, Cao M (2015) Carbon 91:56–65
16. Chen F, Yang J, Bai T, Long B, Zhou X (2016) J Electroanal Chem 768:18–26
17. Chen L, Zhang Y, Lin C, Yang W, Meng Y, Guo Y, Li M, Xiao D (2014) J. Mater. Chem. A 2:9684–9690
18. Baskar AV, Singh G, Ruban AM, Davidraj JM, Bahadur R, Sooriyakumar P, Kumar P, Karakoti A, Yi J, Vinu A (2023) Adv Funct Mater 33:2208349
19. Zhu C, Akiyama T (2016) Green Chem 18:2106–2114
20. Hou J, Cao C, Idrees F, Ma X (2015) ACS Nano 9:2556–2564
21. Liu M, Zhang Z, Dou M, Li Z, Wang F (2019) Carbon 151:28–35

22. Wan H, Hu X (2019) Solid State Ionics 341:115030
23. Wan H, Ju X, He T, Chen T, Zhou Y, Zhang C, Wang J, Xu Y, Yao B, Zhuang W, Du X (2021) J Alloy Compd 863:158078
24. Chen F, Ren J, Ma L, Luo X, Wu N, Ma S, Li B, Song Z, Zhou X (2020) Int J Electrochem Sci 15:5803–5820
25. Zhao H, Gao Y, Wang J, Chen C, Chen D, Wang C, Ciucci F (2016) Mater Lett 167:93–97
26. Xiong J, Pan Q, Zheng F, Xiong X, Yang C, Hu D, Huang C (2018) Front Chem 6:78
27. Zhang J, Zhang D, Li K, Tian Y, Wang Y, Sun T (2021) J Colloid Interface Sci 598:250–259

3 Silicon Anode

In addition to carbon materials, silicon anode materials of the embedded type have also received widespread attention from researchers. Because silicon anode materials have a relatively high theoretical capacity for storing lithium ions, this means they can store more capacity, giving the battery a higher energy density. However, the application of silicon in lithium-ion batteries also faces some challenges, as silicon undergoes significant volume expansion during lithiation, leading to material fracture and capacity decay. Finding suitable synthesis methods for silicon anode materials is of great significance for improving battery performance.

3.1 Silicon

Silicon is a non-metallic element that is widely present in nature. It is the second most abundant element in the earth's crust, accounting for about 27.7% of the crust's mass, and its chemical symbol is Si. Silicon has a variety of crystal structures, the most common of which are the diamond-cubic crystal structure and the hexagonal close-packed crystal structure.

3.1.1 Structure of Silicon

Diamond-cubic crystal structure: The diamond-cubic crystal structure, also known as the rhombohedral lattice, is one of the most common crystal structures of silicon. As shown in Fig. 3.1, in this structure, silicon atoms form a face-centered cubic lattice, with each silicon atom connected to four surrounding silicon atoms through covalent bonds, forming a three-dimensional solid network. This structure gives diamond-type silicon extremely high hardness and good thermal conductivity.

Hexagonal close-packed crystal structure: The hexagonal close-packed crystal structure, also known as the hexagonal close-packed lattice, is a crystal structure of silicon under high temperature and high pressure conditions. As shown in Fig. 3.2, in this structure, silicon atoms are arranged in a hexagonal close-packed manner, forming hexagonal layers. Each silicon atom is connected to the atoms in the bottom and top layers, forming a three-dimensional structure. This structure is more common in natural silicon minerals.

3.1.2 Electrochemical Characteristics of Silicon Anode During Lithium Storage Process

The lithium storage process of silicon as an anode material in lithium-ion batteries involves complex electrochemical reactions, including the intercalation and deintercalation of lithium ions. Here are some electrochemical characteristics of the silicon anode during the lithium storage process.

Lithium intercalation reaction: During the charging process, lithium ions enter the silicon anode from the electrolyte and are intercalated into the lattice structure of silicon. This causes the silicon lattice to expand, resulting in a volume expansion effect. This reaction can be represented as:

$$Si + xLi^{+} + xe^{-} \rightarrow Li_{x}Si$$

Lithium deintercalation reaction: During the discharge process, the lithium ions intercalated in silicon are deintercalated and return to the electrolyte. This causes the silicon lattice to shrink, resulting in a volume contraction effect. The deintercalation reaction can be represented as:

$$Li_{x}Si \rightarrow Si + xLi^{+} + xe^{-}$$

where, x represents the number of intercalated lithium ions, and e^{-} represents electrons.

During the lithium storage process, the solvent and salt in the electrolyte react to form a Solid Electrolyte Interphase (SEI) layer. This thin film can protect the electrode material from corrosion by the electrolyte, but it may also affect the transmission rate of lithium ions, this process is the formation of the SEI layer.

X. Li et al., *Electrode Development for Lithium-ion Batteries*, https://doi.org/10.1007/978-981-96-3021-9_3

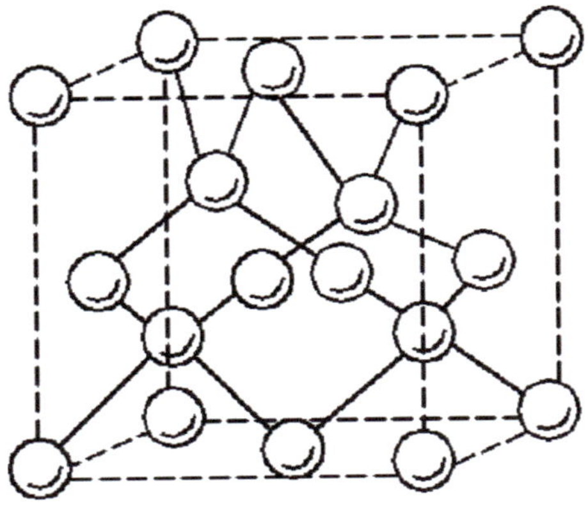

Fig. 3.1 Diamond-cubic crystal structure diagram

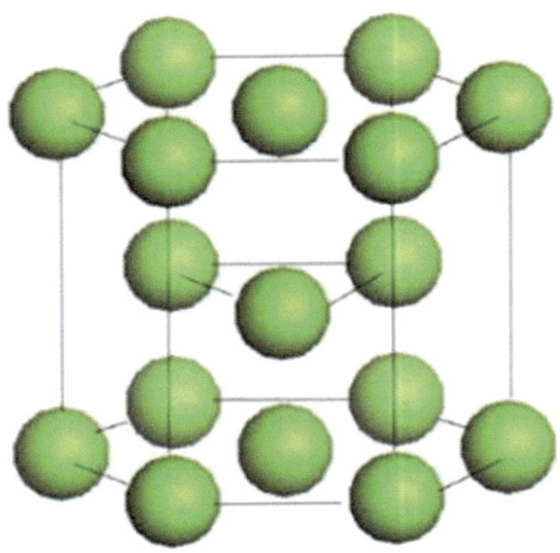

Fig. 3.2 Hexagonal close-packed crystal structure diagram

The lithium intercalation reaction of silicon causes the silicon lattice to expand, which results in a volume expansion effect, leading to the expansion and contraction of the electrode material. This volume change may cause the electrode particles to fracture, affecting the stability and cycle life of the electrode. On the other hand, due to volume expansion and structural damage, the silicon anode may experience capacity decay in long-term charge–discharge cycles, reducing the life of the battery. Therefore, there are serious issues affecting energy storage performance, such as volume expansion effect and capacity decay, in the lithium storage process of the silicon anode.

3.1.3 Advantages and Disadvantages of Silicon Anode During Lithium Storage Process

The application of silicon as an anode material in lithium-ion batteries has received extensive research and attention [1, 2]. However, there are also some advantages and challenges in the lithium storage process of silicon electrodes. Here are the advantages and disadvantages of silicon electrodes during the lithium storage process.

Silicon is abundant in the earth's crust, relatively rich in resources, and has a low working voltage (<0.5 V vs. Li/Li^+). This makes silicon electrodes sustainable and reliable as anode materials for lithium-ion batteries. Compared with traditional carbon anode materials, silicon electrodes can store more lithium ions, thereby achieving higher energy density. Therefore, silicon as a battery anode material has advantages such as high capacity and high energy density.

However, silicon anodes undergo volume expansion during lithium storage, leading to the expansion and contraction of the electrode material. This may cause the electrode particles to fracture, the electrode material to pulverize, and a decrease in conductivity. Due to the structural damage caused by volume expansion, silicon electrodes may gradually lose activity during the cycle of charging and discharging, leading to a decrease in battery capacity and a shortened lifespan. Additionally, the high cost of producing high-purity silicon electrode materials and combining silicon with conductive additives may affect the overall cost of the battery.

Despite the high capacity and energy density of silicon anodes in lithium-ion batteries, challenges such as volume expansion and capacity decay need to be overcome to realize their commercial application. Researchers are constantly exploring methods such as nanostructure, new composite material design, and electrode engineering to improve the performance of silicon electrodes and promote the development of silicon electrodes in lithium-ion batteries.

3.2 Strategies to Improve the Lithium Storage Performance of Silicon Anodes

3.2.1 Structural Design

In the field of lithium-ion batteries, nanostructure design is often used to improve the performance of electrode materials, especially high-capacity anode materials like silicon. Silicon, as an anode material, undergoes volume changes during charging and discharging, leading to the expansion and contraction of the electrode material. This volume change may cause the electrode to fracture, decrease battery performance, and reduce cycle life. Nanostructure design, by preparing silicon materials into nanoparticles or nanostructures, can to some extent alleviate the mechanical stress caused by volume changes, thereby improving the stability and lifespan of the electrode [3].

3.2.1.1 Porous Silicon

Zheng and his collaborators [4] used a ball milling-etching-thermal decomposition process to successfully prepare a nano-porous silicon/graphite/carbon composite material. At a current density of 0.2 mA/cm^2, the reversible capacity of this composite material is about 700 mAh/g. Even after 120 cycles, the capacity does not significantly decrease. This is due to the nanopore structure in the silicon particles, which provides a certain buffering effect during volume expansion. In addition, this material has good electronic and ionic conductivity, making the composite material have outstanding electrochemical performance.

On the other hand, Lu and his team [5] a non-filled carbon coating structure was designed and applied to porous silicon microparticles. Unlike the conventional carbon coating of individual silicon particles, they coated the carbon shell on the outer surface of the secondary porous silicon microparticles composed of silicon nanoparticles, as shown in Fig. 3.3. This unique non-filled carbon coating can effectively prevent the electrolyte from penetrating into the silicon particles, making the solid electrolyte interface (SEI) mainly formed on the surface of the carbon shell. In addition, because there is a certain space inside the carbon shell, it can be used to buffer the volume change of silicon particles, thereby improving the stability of the microparticles and the SEI film. The results show that this composite structure has excellent cycle stability. After 1000 cycles at a C/4 rate, it still has a reversible specific capacity of 1500 mAh/g. The areal capacity of this structure can be as high as over 3 mAh/cm^2, and there is no obvious capacity decay after 100 cycles.

3.2.1.2 Silicon Nanowires

Constructing silicon nanowire structures is considered one of the effective methods to solve the volume effect of silicon anodes. Experiments have found that when the Si nanowires oriented along <112> are charged, the nanowires expand instead of elongating, and it can be found that the lithiated silicon nanowires have highly anisotropic volume expansion, resulting in a surprising dumbbell-shaped cross-section. Under the drive of progressive charging, the stress concentration in the neck region causes a single nanowire to break. In addition, the fully lithiated phase of the silicon anode nanowire is $Li_{15}Si_4$, not the commonly believed $Li_{22}Si_5$, indicating that the maximum capacity of Si at room temperature is 3579 mAh/g.

In addition, Chan and his team [6] directly grew silicon nanowire arrays on the current collector. During the cycling process, these silicon nanowires will expand radially and longitudinally, successfully avoiding the pulverization problem of silicon materials. The scanning electron microscope images before and after the cycle at the same magnification are shown in Fig. 3.4. Because the nanowires are directly grown on the current collector, electrons can be transmitted along the one-dimensional nanowires. The battery constructed with this one-dimensional nanowire array has a first coulombic efficiency of 73%, a specific capacity of 4277 mAh/g, and exhibits excellent electrochemical performance. However, this method is costly and not suitable for large-scale production applications. Similarly, Huang and his team [7] prepared a thin film of silicon nanowire array with carbon coating by corroding silicon wafers with metal catalysts and pyrolyzing carbon aerogel on it. Within the voltage range of 0.02–2.0 V, and at a rate of 150 mAh/g, after 40 cycles, it still maintains a reversible capacity of 1326 mAh/g, and the coulombic efficiency is 84%. These excellent performances are attributed to the uniformity and continuity of the carbon coating, which not only enhances the conductivity of the silicon anode nanowire material but also alleviates the volume change during the lithium-ion deintercalation process.

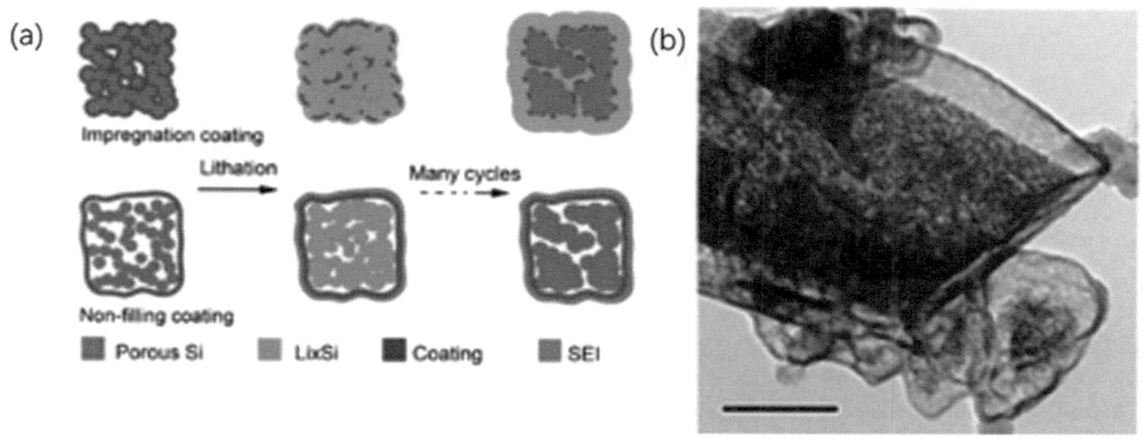

Fig. 3.3 Schematic diagram of the non-filled carbon coating—porous silicon microparticle structure [5]

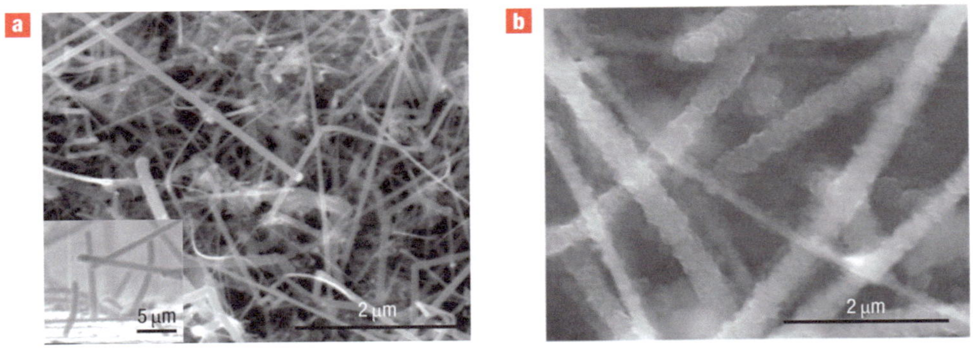

Fig. 3.4 SEM images of silicon nanowires before (**a**) and after (**b**) electrochemical cycling [7]

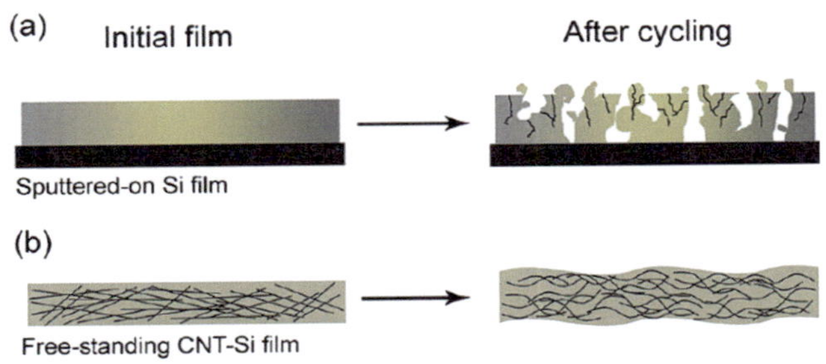

Fig. 3.5 **a** Silicon thin film and **b** Schematic diagram of carbon nanotubes filled into silicon thin film before and after charge and discharge cycling [8]

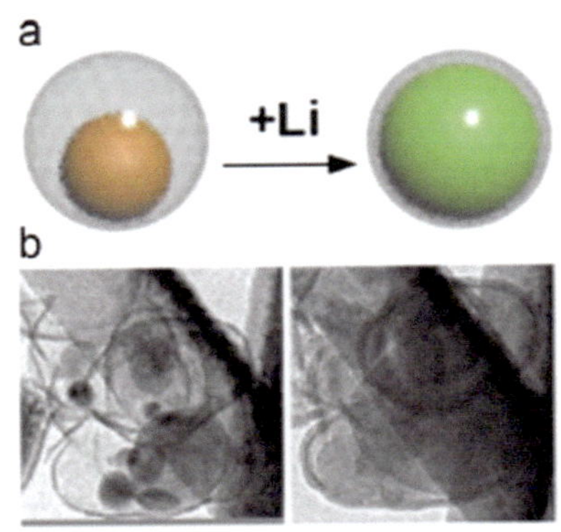

Fig. 3.6 A magnified schematic of a single Si@void@C particle [10]

3.2.1.3 Nano Silicon Thin Film

Cui and his team [8] successfully constructed a composite thin film structure of freestanding carbon nanotubes (CNT) and silicon (Fig. 3.5). They filled the silicon thin film with carbon nanotubes, forming a structure similar to a "reinforced concrete" network. This structure allows the electrode to perform excellently in terms of mechanical and conductive properties. The research results show that after a long period of cycling, the thin film with added carbon nanotubes still maintains its integrity. Even if small cracks appear in some areas, the carbon nanotube network remains connected. During the lithium embedding process, this composite thin film can also release stress caused by volume changes by forming a "ripple" structure. The composite material of carbon nanotubes and nano silicon thin film has a high specific capacity (2000 mAh/g) and stable cycling performance. In addition, this composite thin film does not need to be coated on the current collector and can be directly used in the battery, thereby significantly reducing the weight of the battery.

3.2.1.4 Egg Yolk-Shell Structure Design

Another clever method to solve the volume expansion of Si during lithiation is the design of egg yolk-shell Si–C composite material [9, 10]. The yolk is silicon nanoparticles, and the shell is composed of a thin layer of carbon. Most importantly, there is enough space between the silicon yolk and the carbon shell for the silicon yolk to expand freely without cracking or shattering. As shown in Fig. 3.6, during the lithiation process of the silicon yolk, the gap between the silicon yolk and the C shell is filled. The carbon shell maintains the stability and integrity of the structure, with minimal volume change during the lithiation process. At the same time, the conductivity of the silicon yolk is always maintained through contact with the carbon shell, forming the conductive network of the electrode. Since the carbon shell is in contact with the liquid electrolyte, the SEI layer formed on the carbon shell will not break during the cycling process. The composite anode structure maintains a capacity retention rate of 100% for the first 300 cycles, and a capacity retention rate of 74% after 1000 cycles [9]. This clever design can stabilize the material structure, improve the cycling stability performance, without significant capacity decay.

3.2.1.5 Double-Wall Silicon Nanotube Design

So far, the performance of the double-wall silicon nanotube anode surpasses all other silicon anode designs, as shown in Fig. 3.7. The Li^+ on the shell of the double-wall silicon nanotube. The silicon oxide layer that can penetrate to the surface forms a limiting effect, effectively preventing the silicon tube from expanding outward, rather than inward.

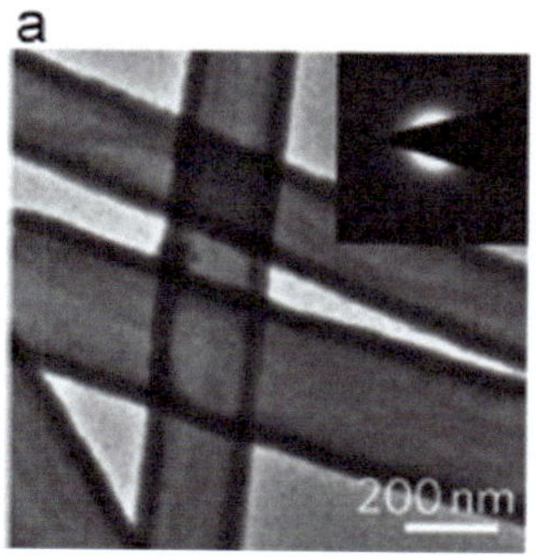

Fig. 3.7 Transmission electron microscope image of DW SiNTs [11]

Since the inner wall of the silicon nanotube is not exposed to the liquid electrolyte, a stable solid electrolyte interphase layer is formed on the outside of the tube. Batteries assembled with this design material have been proven to have a capacity about 8 times that of traditional graphite anodes and can retain 85% of the capacity after up to 6000 cycles [11]. In addition, due to the high surface area of silicon nanotubes, the charging rate of this silicon nanotube battery can be as high as 20 C. This design alleviates the chemical–mechanical effects related to silicon volume expansion during lithiation.

3.2.2 Silicon/Carbon Composite Materials

Carbon, as a conductive and stable material, can be used to improve the conductivity of silicon and alleviate the mechanical stress caused by volume expansion. In addition, carbon materials can stabilize the interface between silicon and the electrolyte to a certain extent, reduce the formation rate of the SEI layer, and help improve the stability of the electrode, thereby improving the life of the battery. Studies have found that adding an appropriate amount of carbon material can not only provide a channel for lithium-ion transport but also increase the insertion points for lithium ions. In addition, the lithium insertion potential of silicon is close to that of carbon materials, and the volume effect of silicon can be well alleviated by compounding with silicon carbon materials, improving the electrochemical stability of silicon materials.

3.2.2.1 Graphite

Adding a small amount of graphite to the silicon anode can effectively improve the electrochemical performance of the electrode. Research has found that in composite materials composed of silicon nanoparticles and graphite sheets and intercalation layer structures, graphite not only serves as an effective substrate for stabilizing the solid electrolyte interface (SEI) layer but also prevents the agglomeration of silicon nanoparticles and promotes electron transport in the anode. This composite material can usually be prepared by directly ball milling a mixture of silicon and graphite powder.

The team of Cili Jie from Shandong University [12] successfully prepared micron-sized walnut-shaped porous silicon/reduced graphene oxide (P-Si/rGO) core-shell composite materials using silicon metal powder and graphene oxide through in-situ reduction and dealloying processes. The unique core-shell porous structure feature has excellent charge transfer pathways, which can produce ideal electrochemical performance, with good reversible capacity, rate performance, and cycle performance. The synthesis schematic process is shown in Fig. 3.8. Kim and his team [13] used a high-energy ball milling method to disperse nano-silicon powder in the graphite structure, effectively slowing down the drastic volume changes of silicon materials during charging and discharging. In addition, graphite can also block some silicon nanoparticles from direct contact with the electrolyte. Compared with pure nano-silicon, silicon/graphite composite materials have better cycle performance, and the capacity retention rate is close to 90% after 15 cycles.

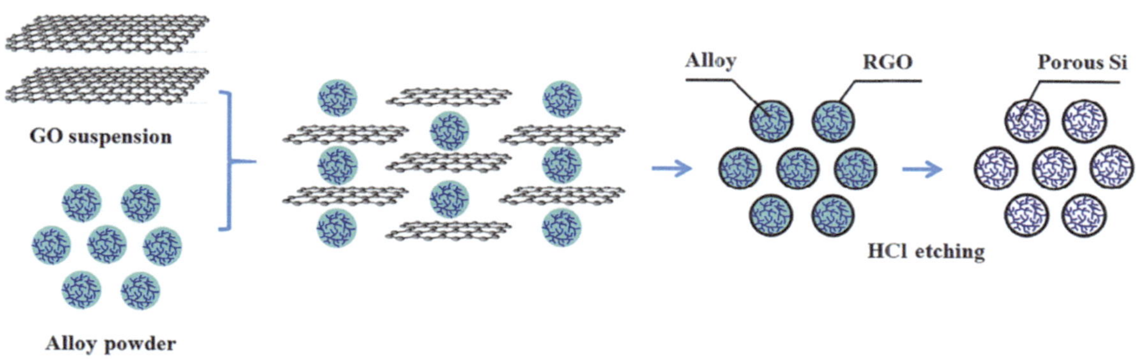

Fig. 3.8 Schematic diagram of the synthesis of walnut-like porous silicon/reduced graphene oxide (P-Si/rGO) core-shell composite material [12]

3.2.2.2 Carbon Nanotubes/Carbon Fibers

One-dimensional materials such as carbon nanotubes (CNTs) and carbon fibers (CNFs) are often used in composites with silicon materials [14, 15]. Compared with graphite, carbon nanotubes and carbon fibers have better conductivity, excellent mechanical strength, and a larger aspect ratio, which can improve the conductivity and mechanical elasticity of the electrode. Studies have found that when carbon nanotubes or carbon nanofibers are combined with silicon, these one-dimensional carbon materials can provide a buffer space for silicon volume expansion, thereby reducing the adverse effects of volume changes on the electrode. At the same time, the three-dimensional conductive network constructed can also improve the overall electron conductivity of the electrode.

Zhang and his team [16] added an appropriate amount of nano-silicon and citric acid to a mixture of nickel acetate and ethanol, and prepared silicon@carbon particles (Si@C) through spray drying. Then, under the action of a nickel oxide catalyst, using the method of chemical vapor deposition, the silicon@carbon particles were further grown into silicon/carbon composite materials co-coated with carbon nanotubes and nano carbon fibers. The process of preparing mesoporous silicon/carbon nanocomposites by the magnesiothermic reduction method is shown in Fig. 3.9. By comparing pure silicon with two different stages of composite materials, it was found that the co-coated composite material performed best in terms of cycle stability. After 50 cycles at a current density of 300 mA/g, the capacity still remained at 1195 mAh/g. The capacity retention of silicon@carbon particles without carbon nanotube and nano carbon fiber coating was about 601 mAh/g, while the capacity of pure silicon nanoparticles was almost zero after 15 cycles. The excellent performance is due to the fact that carbon nanotubes and nano carbon fibers can not only provide good electron conduction channels for silicon@carbon particles, but also provide a certain space interwoven with silicon@carbon particles, thereby alleviating the volume expansion of silicon and improving the integrity and stability of the electrode during the cycle.

3.2.2.3 Carbon Gel

Carbon gel prepared by the sol-gel method has a large number of nanopores inside, forming a continuous and uniform conductive network structure, which can effectively alleviate the volume expansion of silicon materials and enhance electron conductivity [17]. In Hasegawa and others [18] in the study, carbon aerogel microspheres containing silicon powder were successfully prepared by adding silicon powder to the reverse emulsion polymerization process of resorcinol and formaldehyde, followed by freeze-drying and carbonization in an inert atmosphere. Scanning electron microscopy and X-ray diffraction results confirmed the formation of a composite structure of silicon materials encapsulated by carbon aerogel microspheres. It was found that the mesoporous structure of the carbon aerogel microspheres could buffer the volume expansion of silicon, thereby improving the cycle performance of the battery. Despite a lower initial Coulombic efficiency, this is due to the large surface area of both the nanoporous structure of the carbon aerogel and the nanosilicon. During the initial charging process, a large area of solid electrolyte interface (SEI) film forms on the surface, leading to excessive loss of lithium ions. Meanwhile, during the cycling process, the carbon aerogel shell gradually develops cracks and shows a trend of expansion, indicating that carbon aerogel is not as stable as other carbon materials.

3.2.2.4 Silicon–Carbon Multicomponent Composite

In addition to the combination of silicon–carbon materials, many studies have explored the composites of silicon with carbon materials and other nanomaterials, as well as doping methods in silicon–carbon composites, also achieving excellent electrochemical performance. Morita et al. [19]

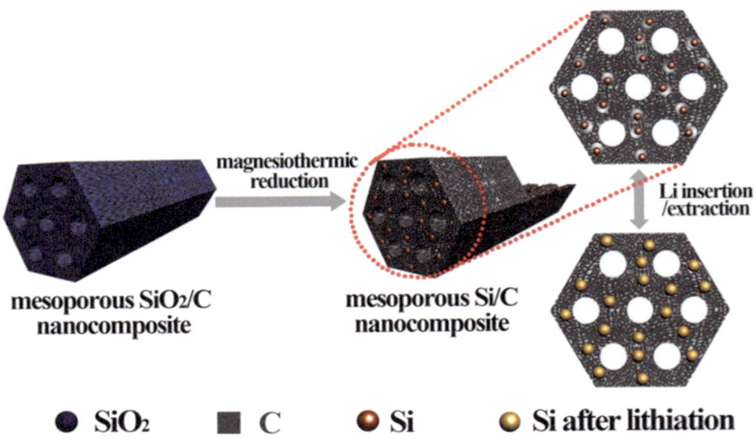

Fig. 3.9 Schematic diagram of the preparation of mesoporous silicon/carbon nanocomposite by magnesiothermic reduction method [16]

prepared a nanosilicon cluster-SiO_x–carbon composite material using the disproportionation of silicon monoxide and the polymerization of furfuryl alcohol, which showed good cycle life (up to 200 cycles) and relatively high capacity (about 700 mAh/g). This composite structure benefits from the tight combination of SiO_x and nanosilicon clusters, forming a stable contact that can be maintained even during the charge–discharge cycling process. In addition, nanosilicon particles and silica phase are evenly dispersed, effectively suppressing the pulverization of silicon particles caused by drastic volume changes during the lithium ion insertion/extraction process. Lee et al. [20] prepared a carbon-coated nanosilicon dispersed oxide/graphite composite material, and the scanning electron microscope images of different nanosilicon, oxide, and carbon composite materials are shown in Fig. 3.10. They obtained the carbon-coated nanosilicon dispersed oxide/graphite composite material by mixing the nanosilicon dispersed oxide obtained by mechanochemical reduction with graphite, and then performing asphalt pyrolysis in an argon atmosphere at 900 °C. This composite material exhibits excellent electrochemical performance. The results of these methods show that by combining different nanomaterials, or by doping appropriately during the composite process, the performance of electrode materials can be significantly improved.

3.2.3 Coating Technology

The method of coating or applying an external material on the surface of silicon electrode materials using coating technology can, to a certain extent, absorb the mechanical stress caused by volume expansion, reduce the risk of electrode material fracture, and improve electrode stability. At the same time, it can form a stable interface, reduce the direct contact between the electrode material and the electrolyte, reduce the corrosion and damage of the electrode, lower the SEI layer formation rate, and help improve the stability and lifespan of the electrode. In addition, the coating material can enhance the interface affinity between the electrode material and the electrolyte, improving the stability of the electrode–electrolyte interface.

3.2.3.1 Alucone Coating

He and others [21] also used lattice-resolution in-situ TEM to study the mechanism of Alucone-coated silicon particles after electrochemical cycling [21]. They found that the Alucone coating process chemically corroded the thin SiO_x layer and replaced the silicon oxide layer with a mechanically elastic metal–carbon–oxygen network. After lithiation, the Alucone coating becomes more conductive. Due to the good conductivity and mechanical resilience of the Alucone

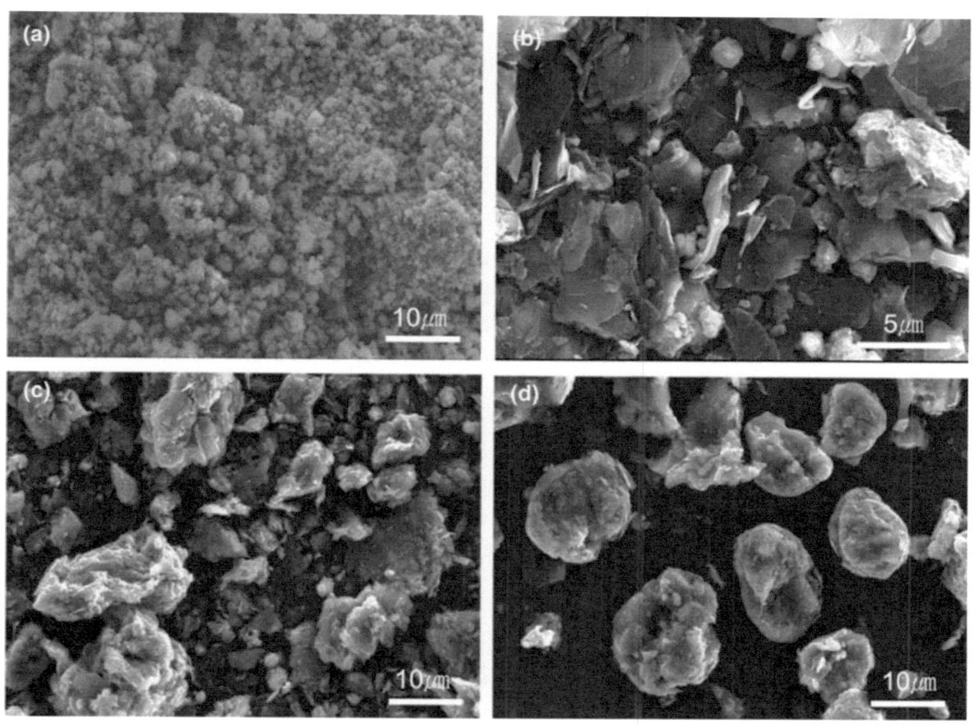

Fig. 3.10 Scanning electron microscope images of nanosilicon with different oxides and carbon composite materials [20]

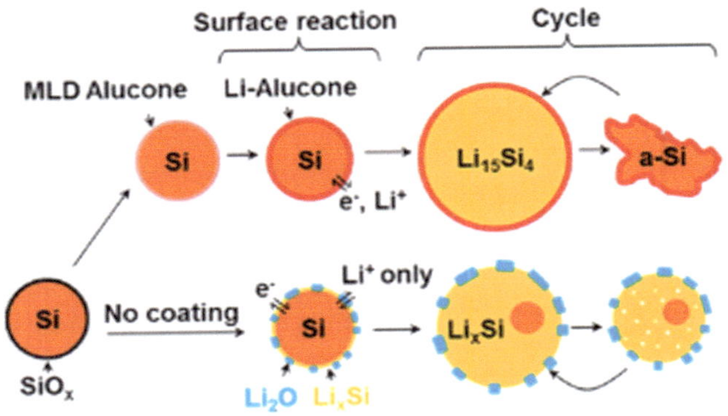

Fig. 3.11 Schematic diagram of the reaction and cycling behavior of nano-silicon surfaces under different coating conditions [21]

coating for Li^+ and e^-, the silicon particles exhibit a "balloon-like" behavior, which can quickly undergo reversible lithiation behavior. In contrast, in SiO_x coated Si particles, there is not enough delithiation, the coated Si particles do not form pores, and irregularly shaped amorphous silicon is formed after the charge and discharge process (Fig. 3.11). Therefore, the Alucone coating can improve the electrical and mechanical properties of the silicon anode, thereby improving the cycle performance and coulombic efficiency of the battery.

3.2.3.2 Metal Coating

Metal coating is another method to increase electron conduction in nanostructured Si materials. Metal coatings do not react with lithium and electrolyte, but can transmit electrons of lithium ions. McDowell and others reported [22] that Cu metal coating applied to Si nanowires can improve the rate performance and cycle life of the battery. As shown in Fig. 3.12, in-situ TEM observations indicate that the Cu coating helps to suppress volume expansion during lithiation. The Cu coating is only deposited on one side of the nanowire, avoiding the closure of the Li^+ inlet and outlet. During the lithiation process, although the length and width of the silicon nanowire increase, the Cu-coated Si nanowire bends without breaking due to the constant copper carrier. TEM images show that Cu and Li do not undergo lithiation during the entire process. In the composite material, the Cu coating not only maintains the integrity of the structure but also maintains high conductivity.

3.2.3.3 Conductive Polymer Coatings

The preparation of silicon–carbon composite materials is one of the important ways to solve the poor conductivity of Si and improve the output power density of Si-based anodes. Traditionally, people use a mixture of carbon materials and silicon nanoparticles, but this design has insufficient conductivity, and not all silicon nanoparticles

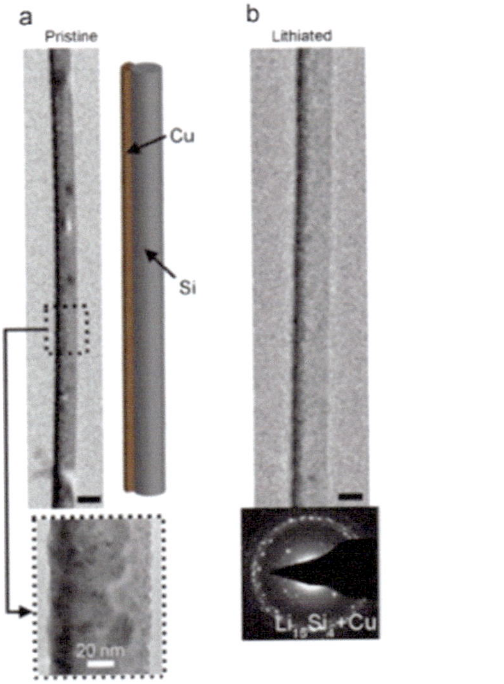

Fig. 3.12 TEM images and schematic of the original single-crystal Si nanowire with copper (dark contrast) on the left wall [22]

can react with Li^+ during the electrochemical cycle. This is because only Si materials that are in close contact with carbon materials are effective in the battery cycle, and Si materials far from the carbon material network will not react with lithium ions. Compared with the traditional design, Si particles that are not adjacent to carbon materials lose their activity during the volume change process, leading to the decay of battery capacity. Liu et al. [23] showed that using conductive polymer binders to encapsulate silicon anode materials is a more effective way, as shown in Fig. 3.13. The conductive polymer has elasticity during the silicon anode cycle, effectively providing a conductive channel for all silicon anode particles. In addition, this method does not require conductive carbon additives, which can enhance the effective load of silicon materials, thereby improving the energy density of lithium-ion batteries.

3.2.4 Alloying

By using alloying, silicon is alloyed with other metal elements to alleviate the volume expansion of silicon, thereby improving the performance of silicon anode materials. Common methods include silicon-lithium alloy, silicon-tin alloy, silicon-aluminum alloy, silicon-titanium alloy, silicon-magnesium alloy, etc.

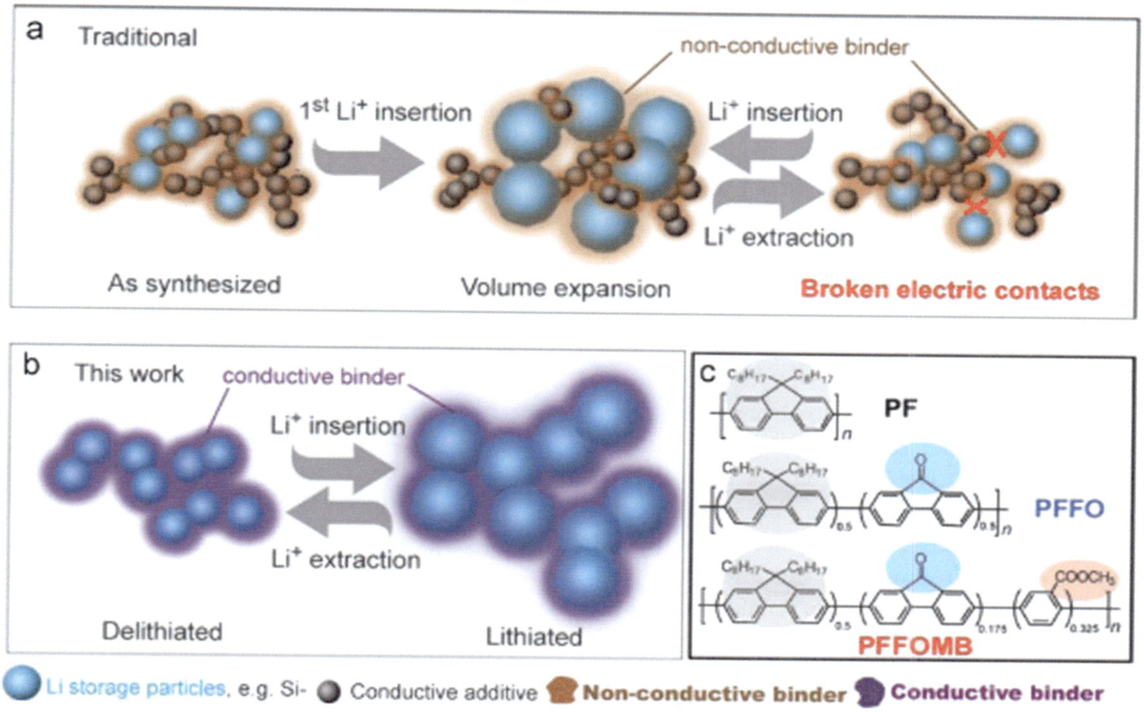

Fig. 3.13 Schematic of the technical approach to solving the problem of battery material volume change with conductive polymer coating and traditional carbon composite [23]

3.2.4.1 Silicon–Lithium Alloy

Silicon-lithium alloy as a lithium-ion battery anode material can improve the performance of the battery to a certain extent, especially for the possible volume expansion problem of silicon anode materials [24]. In the 1970s, lithium-silicon alloy was first reported as the anode material for high-temperature molten salt lithium batteries. Further research was conducted on Li–Si alloys, and a Li–Si equilibrium phase diagram was established. With the continuous development of in-situ technology, the mechanism of electrochemical lithiation of carbon-silicon anodes was studied using in-situ scanning electron microscopy (SEM), transmission electron microscopy (TEM), nuclear magnetic resonance (NMR) and X-ray diffraction (XRD). Unlike intercalation electrodes, alloy composite electrodes undergo large volume changes during the delithiation process and may undergo multiple crystal phase changes, which shortens the cycle life and leads to battery failure. It was found that after the lithium-silicon alloy, it can alleviate pulverization and alleviate capacity decay.

3.2.4.2 Silicon–Tin Alloy

Silicon anode materials undergo volume expansion during the lithium storage process, leading to electrode rupture and damage. Alloying silicon with tin can alleviate the volume expansion effect of silicon. Because the volume expansion of tin is relatively small, this can reduce the mechanical stress of the electrode, improve the stability and cycle life of the electrode. Moreover, tin has a high lithium storage capacity, alloying silicon with tin can increase the total lithium storage capacity of the electrode. This means that the battery can store more lithium ions, thereby improving the energy density and energy storage capacity of the battery. The materials chosen for the study are silicon and tin [25], and using advanced co-deposition technology, using tin metal catalyst, the uniform mixed growth of crystalline-amorphous silicon nanowires was achieved through the gas-liquid-solid mechanism. Silicon nanowires with a diameter of less than 300 nm are not easy to break, and when grown in this way, the current collector electrode has direct conductivity, and its electrochemical performance shows stable cycling performance. The electrode system has good electrochemical performance, can grow on a large scale, and has been successfully grown on flexible carbon paper current collectors. These findings will have an impact on the development of flexible batteries and wearable energy storage.

3.2.4.3 Silicon–Aluminum Alloy

Silicon–aluminum alloy has a high lithium-ion storage capacity. Compared with traditional carbon anode materials, silicon–aluminum alloy can store more lithium ions, thereby improving the energy density of the battery.

Hwang et al. [18] synthesized micrometer-sized Al–Si nanoparticles by selectively etching Al–Si alloy powder. Subsequently, through the selective thermal wet oxidation process of the etched Al–Si particles, a thin Al_2O_3 layer was introduced onto the silicon foam surface. The obtained Si/Al_2O_3 foam anode has excellent cycle stability (the capacity retention rate is 78% after 300 cycles at a C/5 rate) and excellent rate performance.

3.3 Summary

Silicon anodes have significant electrochemical characteristics and advantages and disadvantages during the lithium storage process. The structure of silicon has a high lithium storage capacity, can achieve a higher theoretical specific capacity, but volume expansion occurs during the lithium storage process, causing severe expansion stress and particle rupture, thereby damaging the cycle stability of the electrode. Although the characteristics of silicon anodes include high capacity and low cost, solving its volume expansion problem is still a challenge, and innovative design and material engineering are needed to fully exploit its lithium storage potential to achieve reliable battery performance.

To improve the lithium storage performance of silicon anodes, a series of strategies can be adopted. Firstly, by using a nanostructure design, silicon materials can be dispersed into nanoparticles or nanowires to reduce volume expansion differences, lower expansion stress, enhance the stability of the electrode, and increase surface area, which aids in the rapid diffusion of lithium ions. Secondly, the application of a protective coating layer can isolate silicon from the electrolyte, reducing the reaction at the solid–liquid interface, thereby mitigating the damage to the electrode from volume expansion. In addition, alloying doping is another important strategy. By alloying or doping silicon with other elements, its lithium insertion/extraction kinetics can be improved, expansion can be reduced, and the capacity and stability of the electrode can be enhanced. A porous structure design can also be used to prepare porous silicon structures, such as nanopores or porous particles, to accommodate volume expansion and provide more lithium insertion sites, while reducing the interface area between the solid and the electrolyte, improving electrode performance. Finally, adding high conductivity carbon materials (such as carbon nanotubes or graphene) to the silicon anode can enhance the conductivity of the electrode, promote uniform current distribution, and alleviate electrode expansion issues. The combined application of these strategies can significantly improve the lithium storage performance of silicon anodes, making them more suitable for high-performance and long-life lithium-ion battery applications.

References

1. Wang J, Shen HP, Cheng MX et al (2023) Research progress on high energy density silicon anode lithium-ion batteries. Power Technol 47(3):282–286
2. Han MY, Zhao LN, Sun J (2021) Research progress on silicon and silicon-based anode materials. J Higher Educ Chem 042(012):3547–3560
3. Gu M, He Y, Zheng J et al (2015) Nanoscale silicon as anode for Li-ion batteries: the fundamentals, promises, and challenges. Nano Energy 17:366–383
4. Zheng Y, Yang J, Wang J et al (2007) Nano-porous Si/C composites for anode material of lithium-ion batteries. Electrochim Acta 52(19):5863–5867
5. Lu Z, Liu N, Lee HW et al (2015) Nonfilling carbon coating of porous silicon micrometer-sized particles for high-performance lithium battery anodes. ACS Nano 9(3):2540–2547
6. Chan CK, Peng H, Liu G et al (2008) High-performance lithium battery anodes using silicon nanowires. Nat Nanotechnol 3(1):31–35
7. Huang R, Fan X, Shen W et al (2009) Carbon-coated silicon nanowire array films for high-performance lithium-ion battery anodes. Appl Phys Lett 95(13)
8. Cui LF, Hu L, Choi JW et al (2010) Light-weight free-standing carbon nanotube-silicon films for anodes of lithium ion batteries. ACS Nano 4(7):3671–3678
9. Liu N, Lu Z, Zhao J et al (2014) A pomegranate-inspired nanoscale design for large-volume-change lithium battery anodes. Nat Nanotechnol 9(3):187–192
10. Favors Z, Wang W, Bay HH et al (2014) Scalable synthesis of nano-silicon from beach sand for long cycle life Li-ion batteries. Sci Rep 4(1):5623
11. Wu H, Chan G, Choi JW et al (2012) Stable cycling of double-walled silicon nanotube battery anodes through solid-electrolyte interphase control. Nat Nanotechnol 7(5):310–315
12. Zhai W, Ai Q, Chen L et al (2017) Walnut-inspired microsized porous silicon/graphene core–shell composites for high-performance lithium-ion battery anodes. Nano Res 10:4274–4283
13. Kim BC, Uono H, Sato T et al (2004) Li-ion battery anode properties of Si-carbon nanocomposites fabricated by high energy multiring-type mill. Solid State Ionics 172(1–4):33–37
14. Kim T, Mo YH, Nahm KS et al (2006) Carbon nanotubes (CNTs) as a buffer layer in silicon/CNTs composite electrodes for lithium secondary batteries. J Power Sour 162(2):1275–1281
15. Zhang M, Hou X, Wang J et al (2014) Interweaved Si@C/CNTs&CNFs composites as anode materials for Li-ion batteries. J Alloy Compd 588:206–211
16. Zhang R, Du Y, Li D et al (2014) Highly reversible and large lithium storage in mesoporous Si/C nanocomposite anodes with silicon nanoparticles embedded in a carbon framework. Adv Mater 26(39):6749–6755
17. Shen J, Xue H, Wu G (2007) Microstructure control of carbon aerogels. J Tongji Univ 35(6):779
18. Hasegawa T, Mukai SR, Shirato Y et al (2004) Preparation of carbon gel microspheres containing silicon powder for lithium ion battery anodes. Carbon 42(12–13):2573–2579
19. Hwang G, Park H, Bok T et al (2015) A high-performance nanoporous Si/Al_2O_3 foam lithium-ion battery anode fabricated by selective chemical etching of the Al-Si alloy and subsequent thermal oxidation. Chem Commun 51(21):4429–4432
20. Lee HY, Lee SM (2004) Carbon-coated nano-Si dispersed oxides/graphite composites as anode material for lithium ion batteries. Electrochem Commun 6(5):465–469
21. He Y, Piper DM, Gu M et al (2014) In situ transmission electron microscopy probing of native oxide and artificial layers

on silicon nanoparticles for lithium ion batteries. ACS Nano 8(11):11816–11823
22. McDowell MT, Lee SW, Wang C et al (2012) The effect of metallic coatings and crystallinity on the volume expansion of silicon during electrochemical lithiation/delithiation. Nano Energy 1(3):401–410
23. Gu M, Xiao XC, Liu G et al (2014) Mesoscale origin of the enhanced cycling-stability of the Si-conductive polymer anode for Li-ion batteries. Sci Rep 4(1):3684
24. Zhao H, Yuan W, Liu G (2015) Hierarchical electrode design of high-capacity alloy nanomaterials for lithium-ion batteries. Nano Today 10(2):193–212
25. Loveridge MJ, Malik R, Paul S et al (2018) Binder-free Sn–Si heterostructure films for high capacity Li-ion batteries. RSC Adv 8(30):16726–16737

4 Lithium Titanate Anode

Advanced energy storage technology is crucial for clean, sustainable, and safe energy, as the effective utilization of new energy depends on cost-effectiveness and efficient energy storage. Many new technologies rely on breakthroughs in energy storage technology, including electric vehicles, smart grids, and portable power sources. The widely used anode material, graphite, cannot meet the strict requirements of lithium-ion batteries for safety, cycle stability, and rate performance. The development of advanced anode materials, which can overcome the shortcomings of graphite anodes (such as the formation of lithium dendrites and poor electrode–electrolyte interfaces during charging), is crucial for improving the cycle stability and operational safety of lithium-ion batteries. Spinel $Li_4Ti_5O_{12}$ (LTO) has received widespread attention due to its long cycle stability, zero strain characteristics, and high thermal stability. However, LTO has a low theoretical capacity and its low electronic conductivity limits its rate performance. Over the past few decades, researchers have conducted extensive research to improve the performance of LTO. This chapter summarizes the structure, preparation methods, existing problems, and modification strategies of LTO, and makes suggestions for future research.

4.1 Structure of Lithium Titanate Anode Materials

The composite oxide lithium titanate $Li_4Ti_5O_{12}$ is composed of metallic lithium and transition metal titanium, and is a type of spinel solid solution. Lithium titanate oxides, in addition to $Li_4Ti_5O_{12}$, also include Li_2TiO_3, Li_4TiO_4 and $Li_2Ti_3O_7$. Among them, Li_2TiO_3, Li_4TiO_4 and $Li_2Ti_3O_7$ are limited in practical applications because it is difficult to obtain pure phase powder during the preparation process. Spinel LTO, with its excellent thermal stability, cycle stability, capacity retention rate, high safety, and long service life, has received much attention in the field of new energy and is one of the currently industrialized anode materials [1–3].

The structure of LTO is similar to that of inverse spinel: in a unit cell, $32O^{2-}$ are arranged in a cubic close-packed arrangement occupying 32e sites, and 3/4 of the total Li^+ is surrounded by four O^{2-}. Adjacent to the regular tetrahedral ligand embedded in the gap, occupying the 8a site, the remaining 1/4 of Li^+ and all Ti^{4+} (atomic number ratio 1:5) are embedded in the gap by six O^{2-} adjacent to the regular octahedral ligand, occupying the 16d site. Therefore, its structure can be represented as $Li[Li_{1/3}Ti_{5/3}]O_4$. During the lithiation process, Li^+ at the 8a site and Li^+ at the 16d site move to the 16c site simultaneously, and the opposite occurs during delithiation, with Li^+ at the 16c site returning to the 8a and 16d sites. As shown in Fig. 4.1a and b, there are three occupied sites for lithium, which are the tetrahedral 8a site (yellow), the octahedral 16c site (green), and the 16d site (blue). Titanium atoms and oxygen atoms are represented by small gray balls and large red balls, respectively; in $Li_4Ti_5O_{12}$ atoms, the distance between 16d and 16c is 2.95 Å, between 16c and 8a is 1.81 Å, and between 8a and 16d is 3.46 Å [4, 5].

4.2 Preparation Methods of Lithium Titanate Anode Materials

There are various preparation methods for spinel LTO, and different preparation methods can yield products of different morphologies. The morphology of the electrode material will exhibit different electrochemical properties, each with its own advantages and disadvantages. As showing in Fig. 4.2a–d, there are many methods to synthesize LTO, among which the common ones are high-temperature solid-phase method, hydrothermal method, sol-gel method, and co-precipitation method. The LTO samples synthesized

X. Li et al., *Electrode Development for Lithium-ion Batteries*, https://doi.org/10.1007/978-981-96-3021-9_4

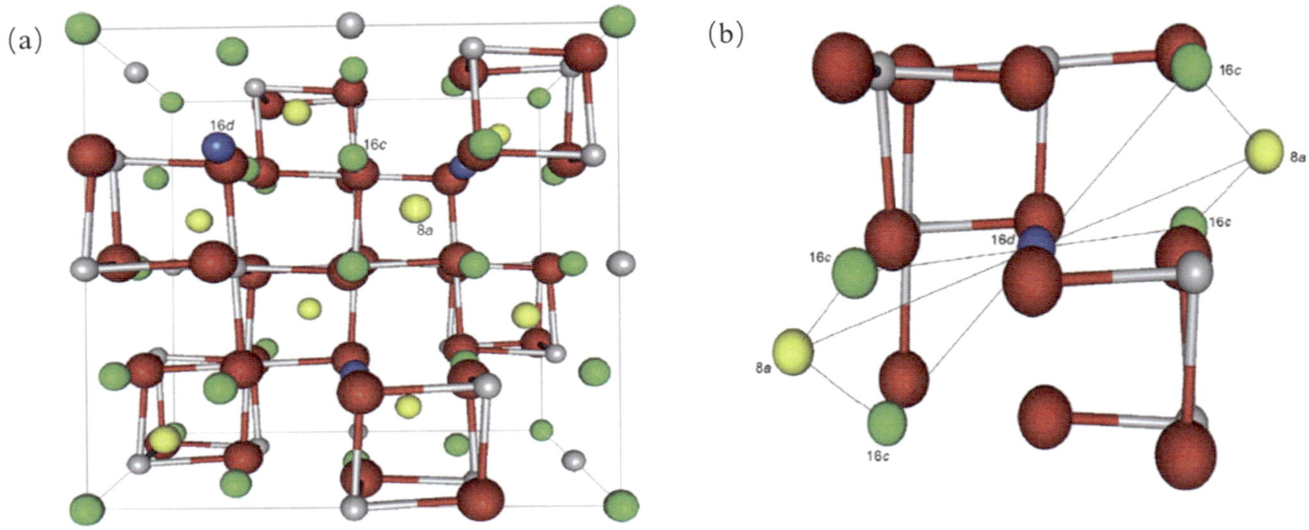

Fig. 4.1 **a** Crystal structure of $Li_4Ti_5O_{12}$; **b** atomic spacing of $Li_4Ti_5O_{12}$ [4]

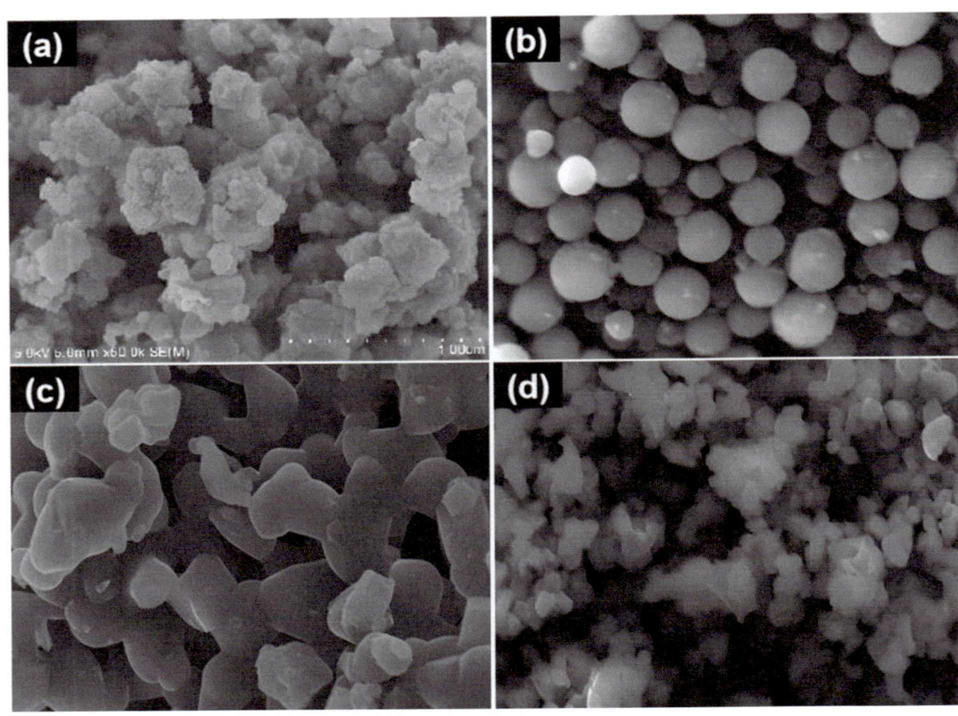

Fig. 4.2 SEM images of $Li_4Ti_5O_{12}$ negative electrode materials prepared by different methods: **a** high-temperature solid-phase method [6], **b** hydrothermal method [7], **c** sol-gel method [8], **d** co-precipitation method [9]

by these methods have good crystallinity, uniform particles, high purity, etc., and their application range is greatly improved.

4.2.1 High-Temperature Solid-Phase Method

The high-temperature solid-phase method is a traditional preparation process, which has the advantages of non-agglomerated powder particles, good filling, low cost, large output, and simple preparation process, but it also has the disadvantages of high energy consumption, low efficiency, coarse powder, and easy impurity mixing. The solid-phase method usually uses Li_2CO_3 or LiOH as the lithium source, and TiO_2 as the titanium source, mixed evenly with ball milling according to a certain stoichiometric ratio, and LTO material is obtained by calcining at high temperature from 700 to 1000 °C [6].

4.2.2 Hydrothermal Method

The hydrothermal method refers to a method of preparing electrode materials by dissolving and then crystallizing powder in a completely sealed pressure vessel, usually using pure water as the solvent. Compared with other powder preparation methods, the powder samples prepared by the hydrothermal method have the characteristics of small and uniform particle size, complete crystal growth, and small particle agglomeration. The hydrothermal synthesis of LTO usually uses tetrabutyl titanate as the titanium source, lithium hydroxide or lithium carbonate as the lithium source, reacts in a stainless steel reaction kettle lined with polytetrafluoroethylene at a certain temperature for several hours, and finally obtains nano-level LTO material through calcination [7].

4.2.3 Sol-Gel Method

The sol-gel method mainly uses compounds containing highly chemically active components as raw materials, uniformly mixes the raw materials under liquid conditions, undergoes chemical reactions such as hydrolysis and condensation, forms a stable and transparent sol system in the solution, and the sol slowly polymerizes between the gel particles after aging to form a gel with a three-dimensional network structure. The gel can be prepared into micrometer and even nanometer structure materials through drying and sintering solidification. The sol-gel method can not only synthesize LTO at a lower calcination temperature and a shorter calcination time, but also the product has a uniformly distributed, nearly cubic structure, narrow size distribution, nanometer level, thus having a higher initial discharge capacity and good cycle performance. The effect of the sol-gel method in synthesizing LTO is significantly better than the solid phase method, which has aroused widespread research interest [8].

4.2.4 Co-precipitation Method

The co-precipitation method mainly involves uniformly mixing the raw material solution in the dissolved state, adding a specific precipitant to the solution, and precipitating the components in the solution. First, prepare the precursor precipitate, and then calcine the precipitate to obtain LTO. This method is widely used due to its simple operation, simple preparation conditions, low cost, easy process control, and short synthesis cycle [9].

4.3 Problems with Lithium Titanate Negative Electrode Materials

$Li_4Ti_5O_{12}$ the material has become a focus of attention due to its excellent structural stability during the charge–discharge cycle. Compared with graphite anode materials, the lithium intercalation potential (1.55 V vs. Li^+/Li) is much higher than the lithium deposition potential, so it is not easy to generate lithium dendrites, puncture the separator, and cause battery self-ignition problems during the battery cycle. Although LTO has high safety, there are still the following problems.

4.3.1 Lower Theoretical Capacity

The theoretical capacity of LTO anode material is 175 mAh g^{-1}, which is lower compared to other anode materials. The capacity of commercial graphite anodes is about 1.5 times that of LTO, and the platform voltage is higher. The working voltage of the whole battery is generally around 2.2–2.5 V. When matching specific specifications of electric devices, the volumetric energy density is also lower. Therefore, to achieve the same specific energy as other materials, more lithium titanate batteries are needed, which will increase the cost, and the current market demand is for batteries with high volumetric energy. In improving capacity and overcoming the low energy density of LTO batteries, researchers have done a lot of work. Currently, relatively effective methods include element doping, preparing electrode materials with higher compaction density, increasing the proportion of electrode materials in the electrode sheet, finding electrode materials with higher redox potentials and combining them with LTO to form a full battery, thereby improving the specific capacity of LTO anode materials. However, these methods have limited capacity improvement and may also reduce the rate performance of the material. To improve capacity and meet rate performance, more in-depth research is needed [10].

4.3.2 Poor Rate Performance

In the electronic structure of LTO, there is an empty Ti3d state, and its higher band gap value (about 3.8 eV) makes the material itself insulating. In addition, the intrinsic conductivity and lithium ion diffusion coefficient of lithium titanate are both low, approximately 10^{-13} s cm^{-1} and 10^{-9}–10^{-13} cm^2 s^{-1}, under high rate conditions, the reversibility of the redox reaction is poor, the capacity fades quickly, and the rate performance is poor. Despite a lot of related research by scientists, such as preparing nano-sized lithium titanate anode materials, increasing surface conductive coatings and ion doping to improve the battery's rate performance, the improvement effect is not significant and it is still difficult to meet the needs of high-rate electric vehicles. The capacity and rate issues of LTO limit its wide range of applications, and we need to improve material performance from aspects such as preparation methods, structural modification, and soft battery packaging [11, 12]. As shown in Fig. 4.3a and b, Jin et al. [13] prepared mesoporous single-crystal lithium titanate (MSC-LTO). When traditional polycrystalline lithium titanate batteries work, Li^+ needs to diffuse over long distances and pass through grain boundaries to reach the electrolyte; in contrast, the electrolyte immersion in MSC-LTO greatly shortens the transmission distance of Li^+, giving the electrode material excellent rate performance.

4.3.3 Soft Pack Batteries are Prone to Gas Generation

LTO soft pack batteries are prone to generate CO, H_2 and CO_2 gases during use, causing battery swelling. The main reason is that the battery electrolyte contains trace amounts of water, which forms PF_5 during the electrode charging and discharging process, causing gas escape. The main cause of gas generation is that the Ti element on the electrode surface has a catalytic effect, accelerating the decomposition of the electrolyte and causing gas escape. The most effective solution to this problem is to coat a layer of carbon film on the synthesized electrode surface to isolate it from direct contact with the electrolyte, thereby avoiding the occurrence of swelling. As shown in Fig. 4.4a–e, He et al. [14] studied the gas generation mechanism of lithium titanate batteries. Lithium titanate soaked in electrolyte for more than 3 months will produce CO_2 gas, and an SEI film is

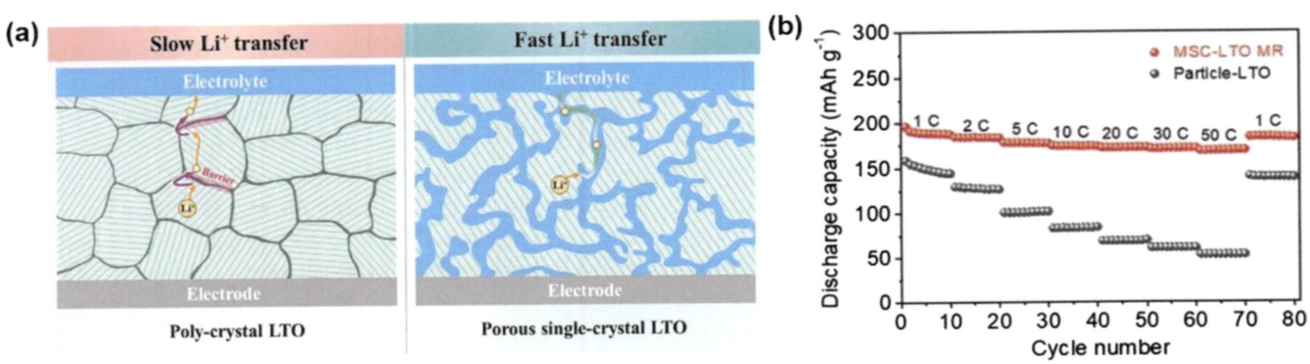

Fig. 4.3 **a** Schematic diagram of Li^+ transmission in polycrystalline LTO and MSC-LTO; **b** rate performance graph of polycrystalline LTO and MSC-LTO anode materials [13]

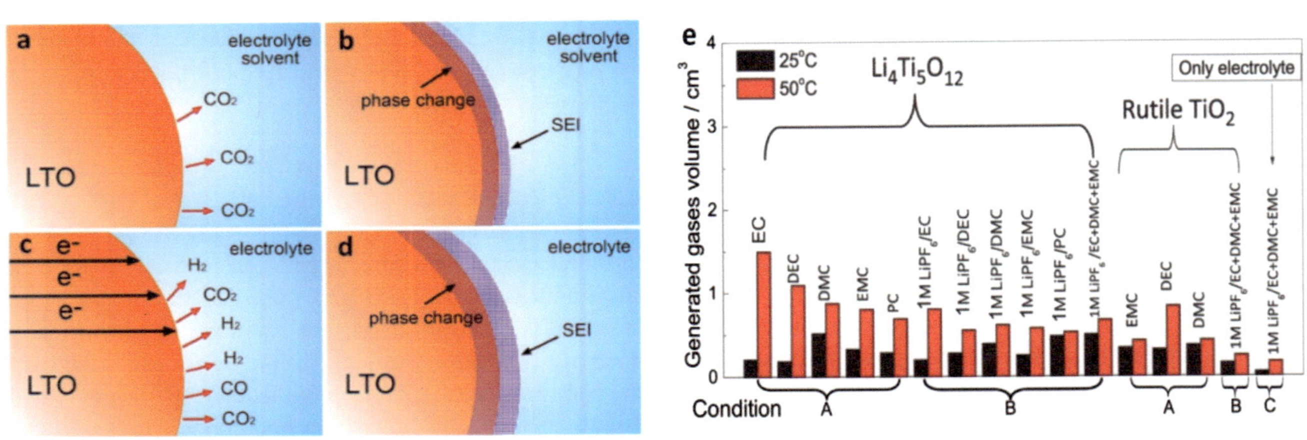

Fig. 4.4 **a**–**d** Gas generation reaction mechanism of $Li_4Ti_5O_{12}$ in different electrolytes; **e** the amount of gas produced by $Li_4Ti_5O_{12}$ and rutile TiO_2 in different electrolytes [14]

formed on the electrode surface. When $LiNi_{1/3}Co_{1/3}Mn_{1/3}O_2$ is used as the positive electrode, H_2, CO and CO_2 gases will be produced during storage or cycling, and an SEI film is also formed on the electrode surface.

4.4 Modification Methods of Lithium Titanate Anode Materials

$Li_4Ti_5O_{12}$ as an anode material for lithium-ion batteries, it has received much attention and has a good application prospect. However, the poor lithium-ion diffusion rate and low electronic conductivity of the material itself have affected the wide application in several aspects such as electrochemical energy storage and electric vehicles. The LTO battery material composed of large particle powder will have a low capacity retention rate during the long cycle process, and it is more obvious during high-rate charging and discharging. In order to obtain high specific capacity, high rate performance and good cycle stability, researchers in various industries have conducted a large number of experiments in the past few decades, including particle nanosization and surface coating, ion doping and synthesis of composite materials, which have significantly improved the performance of the material.

4.4.1 Particle Nanosization

Specific surface area, surface energy, and surface chemistry play a very important role in the charging and discharging process of lithium-ion batteries. Among them, smaller-sized nanomaterials can provide a larger electrode/electrolyte contact area and a shorter lithium-ion diffusion path length for Li^+ and electron transmission, there by achieving a higher charge and discharge rate, while suppressing the volume strain produced by the lithium insertion and extraction process, improving the electrochemical performance of the battery. Therefore, compared with traditional lithium-ion batteries, nanomaterials have a series of unique advantages and are very attractive to lithium-ion batteries [15, 16]. As shown in Fig. 4.5, Wang et al. [17] used oleic acid as a

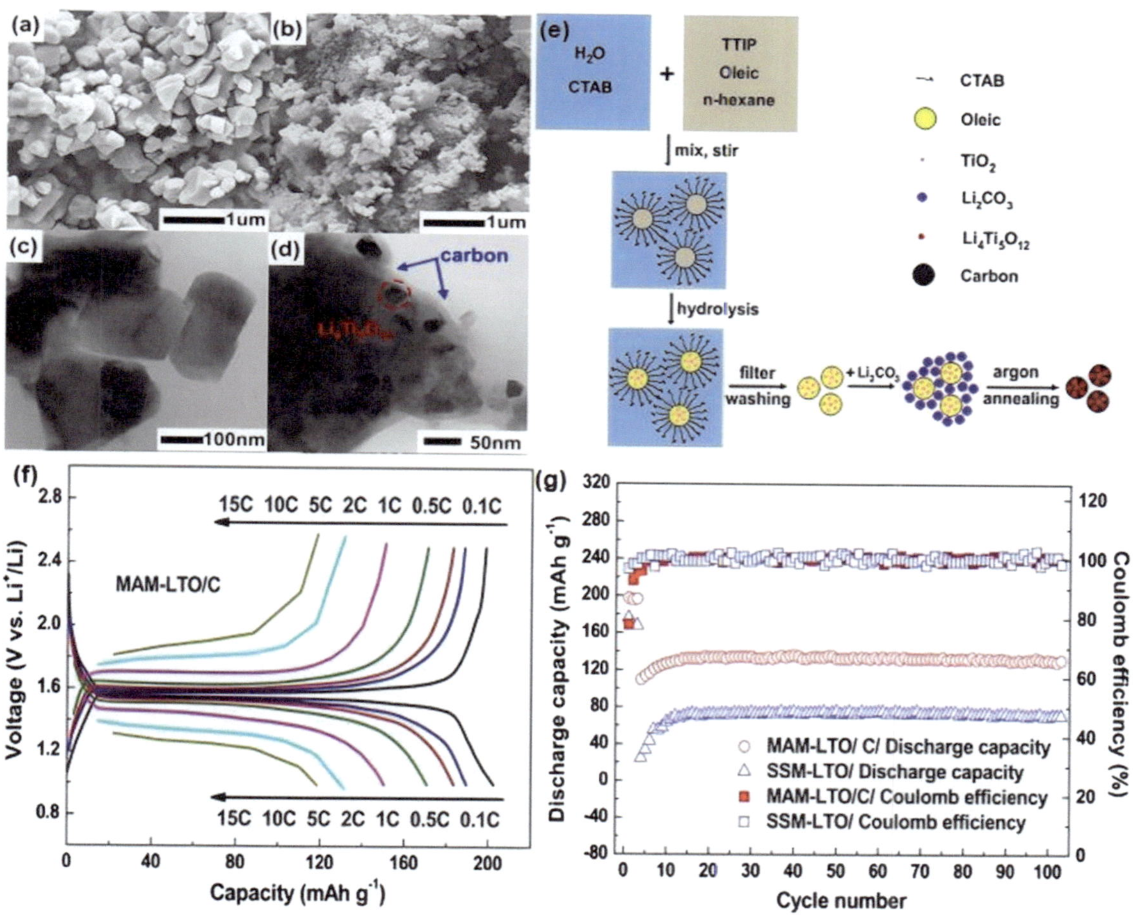

Fig. 4.5 **a** and **c** are the SEM and TEM images of the $Li_4Ti_5O_{12}$ anode material prepared by the one-step method, **b** and **d** are prepared by the two-step method; **e** preparation flow chart [17]. **f** Two-step method LTO 10 C rate performance chart, **g** 10 C 100 cycle performance chart [17]

carbon precursor to prepare different LTO materials by one-step solid-phase method and two-step solid-phase method. Among them, the material prepared by the two-step method has a particle size of about 25 nm. The charge and discharge test results show excellent rate performance and cycle stability. At a 10C rate, the discharge specific capacity is 136.3 mAh g^{-1}, and the capacity retention rate is 96.4% after 100 cycles. The improvement of material performance is attributed to the ultra-fine particles and uniform carbon coating layer.

4.4.2 Ion Doping

Ion doping is an important method to improve the conductivity of LTO anode materials. This method is mainly to dope ions into the crystal structure at the positions of lithium, titanium, and oxygen, and part of Ti^{4+} is converted to Ti^{3+}, Ti^{4+}/Ti^{3+}. As a charge compensation, the number of free electrons is increased, thereby improving the internal electronic conductivity of LTO. Various ions can be used as doping objects, and ion doping occurs at the vacancies of Li^+, Ti^{4+} and O^{2-}, and the related structural characteristics, ionic valence states, and conductivity have been widely studied. In addition to the improved electronic conductivity, the doped electrode materials cause lattice distortion due to the doped ions, thereby affecting the specific capacity and cycle performance of the material [18].

The elements doped at the Li site are mainly metal cations, such as Na^+, K^+, Mg^{2+} and Cu^{2+}, etc. The research on Ti site doping mainly focuses on the impact of transition metal cation and non-metal cation doping on LTO performance, including N^{3-}, Al^{3+}, Ga^{3+}, Co^{3+}, Fe^{3+}, Si^{4+}, etc. A large amount of research on O site ion doping focuses on halogen anions, such as F^-, Cl^-, Br^-, etc. [19]. As shown in Fig. 4.6a–c, Chen et al. [20] used the liquid phase method to prepare Cu-doped LTO anode materials. After Cu doping, the electronic conductivity of the material was improved, and the crystal structure of the material varied with different doping amounts. When the Cu doping amount is 0.05 ($Li_4Cu_{0.05}Ti_{4.95}O_{12}$), the material morphology is flaky, and at room temperature, the discharge specific capacity is 190 mAh g^{-1} at a 1 C rate, and the capacity retention rate is 92.2% after 500 cycles at 30 C.

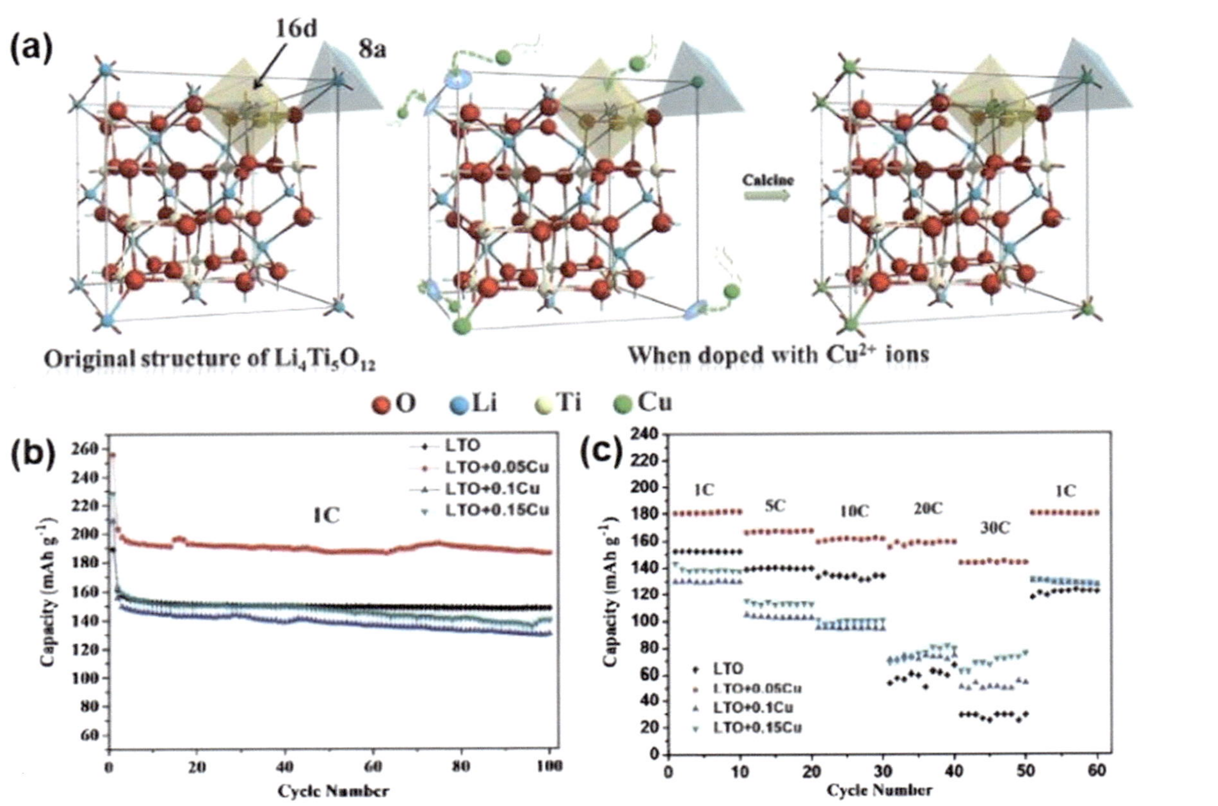

Fig. 4.6 **a** The cell structure of $Li_4Ti_5O_{12}$ and the doping position of Cu, **b** the cycle performance chart of different Cu doping amounts, **c** the rate performance chart of different Cu doping amounts [20]

4.4.3 Surface Coating

Coating modification is a commonly used modification method in industrial production, which can protect the core area of active materials from the influence of side reactions between the electrode and the electrolyte. Compared with particle nanosizing and ion doping, it is beneficial to improve the electrochemical performance of the material while maintaining the size, structure, and corresponding physicochemical advantages of the original material, and can act more directly on the material surface to optimize the charge transfer of the electrode/electrolyte interface. Depending on the coating material, it can be divided into carbon material coating, amorphous carbon coating, metal coating, etc., among which carbon material coating is common. As shown in Fig. 4.7a–d, Zhu and others prepared carbon-coated nanoporous microsphere LTO, the uniform carbon coating layer improved the rate performance of the material, and after 3000 cycles at 1 C rate in a cylindrical 18650 battery, the capacity hardly decayed [21].

Carbon is an excellent electron conductor, and the carbon coating layer can provide good conductivity for the electrode, accelerate charge transfer, and thus improve the electrochemical performance of the electrode material. Industrial production usually uses glucose and polyethylene glycol, which are uniformly mixed with the LTO precursor, and calcined in a high-temperature reaction furnace to form a carbon coating layer. The carbon source can inhibit the growth of LTO particles during calcination. However, a thick carbon coating layer may restrict the diffusion of Li^+. A uniform and appropriate carbon coating layer can provide good conductivity while retaining the transmission channel of Li^+. The carbon coating layer can also cover the catalytically active sites on the surface of LTO, preventing the electrode from reacting with the electrolyte to produce gas. During the formation of the SEI film, a continuous SEI film is formed on the surface of the carbon layer, which can inhibit continuous interface reactions during the cycle, helping to reduce the production of gas [22, 23].

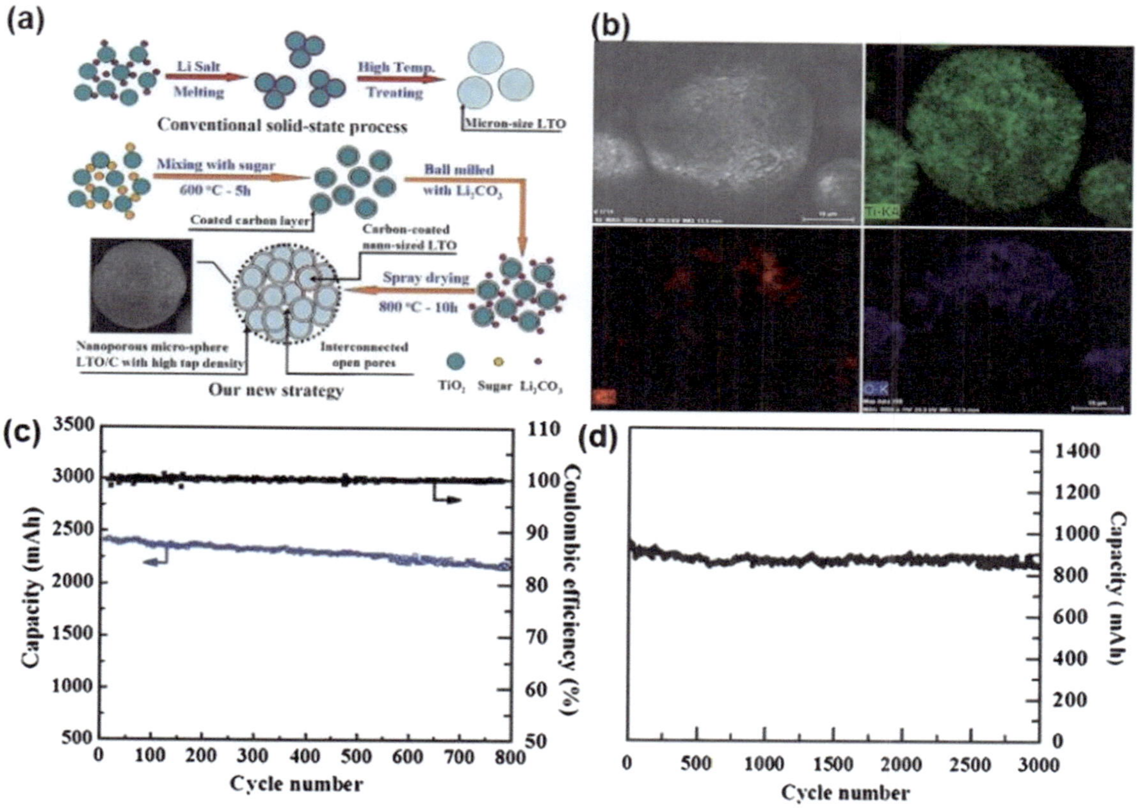

Fig. 4.7 **a** Traditional solid-phase synthesis and carbon-coated nanoporous $Li_4Ti_5O_{12}$ [15], **b** SEM image and corresponding titanium, carbon, oxygen EDX element distribution map, **c** cylindrical 26650 battery 1 C cycle performance chart, **d** cylindrical 18650 battery 1 C cycle performance chart [21]

4.5 Synthesis of Composite Materials

The electrochemical performance of lithium titanate negative electrode materials can be enhanced by forming composite materials with different materials. The main modification mechanism of composite materials is based on the design of the material interface, and interface design is a unique method to adjust the electronic/ionic conductivity of the material. This method has also received widespread attention because it can improve the performance of LTO battery electrode materials. For polycrystalline materials, the conduction of electrons and ions involves not only surface and bulk diffusion, but also grain boundary diffusion. For some materials, the diffusion rate of charges near the grain boundary may be higher than that of the body, so by increasing the grain boundary density, the surface conductivity of the material can be significantly improved. The grain boundary can also serve as a channel for Li^+ entering the particles can also serve as a lithium storage layer. By adjusting the properties of the grain boundaries, the electron and lithium ion transport performance of polycrystalline LTO can be improved, which may enhance its rate performance. By forming composite materials, not only can the grain boundary density be increased, but the grain boundary performance can also be improved. In addition, if multiple materials are combined into a composite material with an appropriate structure, its morphological stability can be further improved. Common composite materials include highly conductive carbon-based materials, silicon dioxide, titanium dioxide, and transition metal oxides [24, 25].

As shown in Fig. 4.8a–c, Liu et al. [26] prepared two different Si@$Li_4Ti_5O_{12}$ composite materials, Si microparticles are completely covered by porous LTO, with a thickness of about 30 nm. The composite material has excellent electrochemical performance, high capacity at high rates, and only 3.3% volume change after 150 charge–discharge cycles. The synthesis steps and electrode preparation process are simple, and this nanocomposite material may meet the needs of future high-capacity lithium-ion batteries.

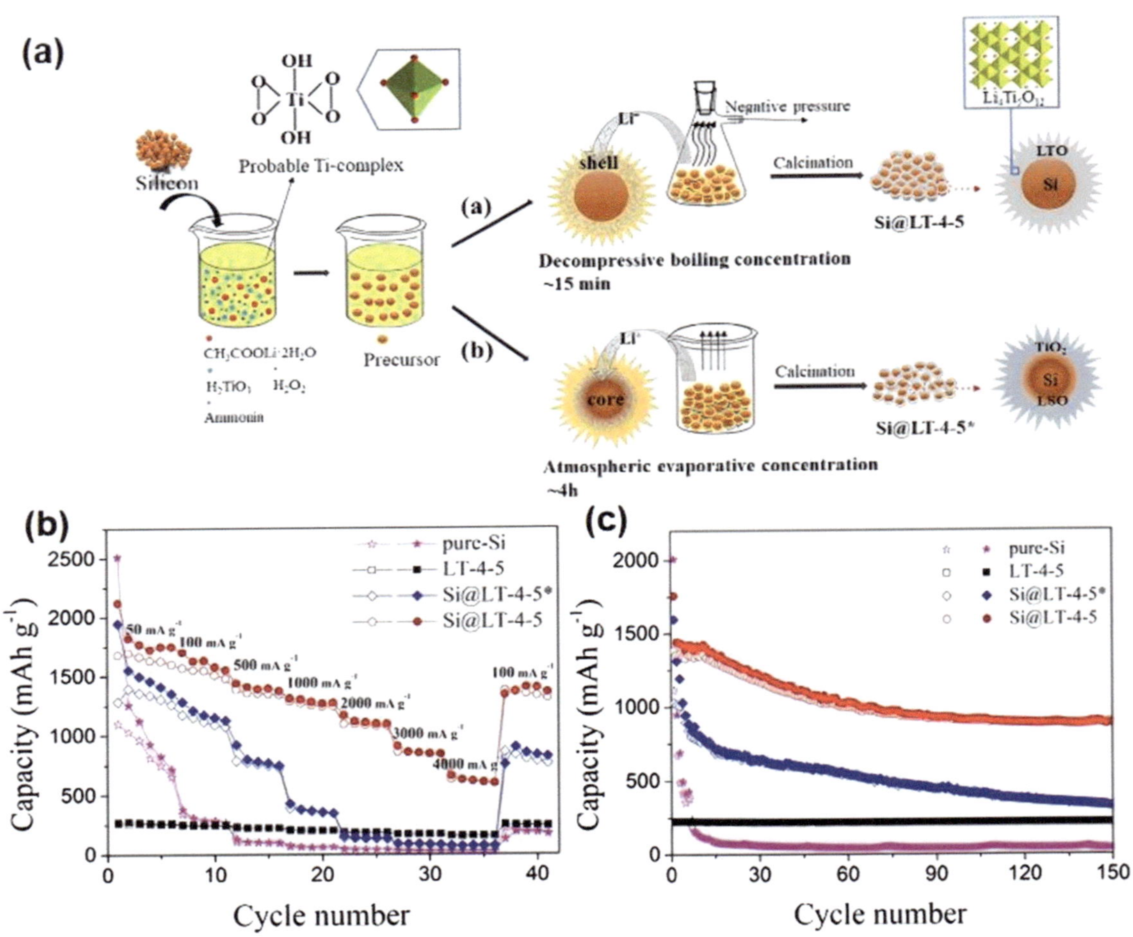

Fig. 4.8 **a** Schematic diagram of the preparation process of Si@$Li_4Ti_5O_{12}$ anode material with a core-shell structure; **b** rate performance graph of the material; **c** cycle performance graph of the material [26]

4.6 Conclusion

At present, the most widely used anode material is graphite. For two emerging fields, electrochemical energy storage and electric vehicles (including pure electric vehicles, hybrid electric vehicles, and micro hybrid electric vehicles), graphite anodes do not meet the safety and rate performance requirements. Spinel $Li_4Ti_5O_{12}$ is a promising material to replace graphite anodes due to its long cycle stability, zero strain characteristics, and high thermal stability. However, the poor conductivity of $Li_4Ti_5O_{12}$ limits its rate performance. To overcome this problem, many strategies have been developed and some progress has been made. However, to achieve large-scale commercial application of $Li_4Ti_5O_{12}$, more technological advances are still needed. In the past few decades, improving the rate performance and cycle stability of $Li_4Ti_5O_{12}$ has received widespread attention. Many of the developed technologies are too complex and not cost-effective. Nanostructures have broad application prospects and can effectively improve rate performance, but they may increase the reaction with the electrolyte, which has a negative impact on cycle stability. For $Li_4Ti_5O_{12}$ The future work of the negative pole, the combination of two or more effective strategies (such as nanostructures, carbon coating, doping, and electrode design, etc.) is the key to solving the problem, and it requires researchers to conduct more in-depth research.

References

1. Nitta N, Wu F, Lee J et al (2015) Li-ion battery materials: present and future. Mater Today 18(5):252–264
2. Yi T, Yang S, Xie Y (2015) Recent Advances of $Li_4Ti_5O_{12}$ as a promising next generation anode material for high power lithium-ion batteries. J Mater Chem A 3(11):5750–5777
3. Zhu G, Wang Y, Xia Y (2012) Ti-Based Compounds as anode materials for Li-ion batteries. Energy Environ Sci 5(5):6652–6667
4. Zhao B, Ran R, Liu M et al (2015) A comprehensive review of $Li_4Ti_5O_{12}$-based electrodes for lithium ion batteries: the latest advancements and future perspectives. Mater Sci Eng R 98:1–71
5. Vijayakumar M, Kerisit S, Rosso K et al (2011) Lithium diffusion in $Li_4Ti_5O_{12}$ at high temperatures. J Power Sour 196(4):2211–2220
6. Yuan MJ, Kan SR, Lu SG (2018) One-step synthesis of carbon-coated lithium titanate by high-temperature solid-phase method. Power Technol 42(3):333–342
7. Yan H, Zhang D, Guo G et al (2016) Hydrothermal synthesis of spherical $Li_4Ti_5O_{12}$ material for a novel durable $Li_4Ti_5O_{12}$/$LiMn_2O_4$ full lithium ion battery. Ceram Int 42(13):14855–14861
8. Zhang C, Zhang Y, Wang J (2013) $Li_4Ti_5O_{12}$ prepared by a modified citric acid sol-gel method for lithium-ion battery. J Power Sour 236:118–125
9. Li W, Wei A, Zhang L et al (2017) Structure and enhanced electrochemical performance of the CaF_2-modified $Li_4Ti_5O_{12}$ anode material. J Electroanal Chem 791:196–203
10. Yuan T, Tan Z, Ma C et al (2017) Challenges of spinel $Li_4Ti_5O_{12}$ for lithium-ion battery industrial applications. Adv Energy Mater 7(12):1–25
11. Chiu H, Lu X, Zhou J et al (2017) Annealing-regulated elimination of residual strain-induced structural relaxation for stable high-power $Li_4Ti_5O_{12}$ nanosheet anodes. Nano Energy 32:533–541
12. Dhaiveegan P, Peng H, Michalska M et al (2018) Investigation of carbon coating approach on electrochemical performance of $Li_4Ti_5O_{12}$/C composite anodes for high-rate lithium-ion batteries. J Solid State Electrochem 22:1851–1861
13. Jin X, Han Y, Zhang Z et al (2022) Mesoporous single-crystal lithium titanate enabling fast-charging Li-ion batteries. Adv Mater 34(18):1–11
14. He Y, Li B, Liu M et al (2012) Gassing in $Li_4Ti_5O_{12}$-based batteries and its remedy. Sci Rep 2:1–9
15. Kang J, Dong G, Li Z et al (2018) Preparation and electrochemical properties of nanorods and nanosheets structural $Li_4Ti_5O_{12}$ as anode for lithium ion batteries. J Mater Sci: Mater Electron 29:12615–12623
16. Li Y, Chen Q, Meng Q et al (2021) Synergy of a hierarchical porous morphology and anionic defects of nanosized $Li_4Ti_5O_{12}$ toward a high-rate and large-capacity lithium-ion battery. J Energy Chem 54:699–711
17. Wang Y, Rong H, Li B et al (2014) Microemulsion-assisted synthesis of ultrafine $Li_4Ti_5O_{12}$/C nanocomposite with oleic acid as carbon precursor and particle size controller. J Power Sour 246:213–218
18. Bai X, Li W, Wei A et al (2016) Preparation and electrochemical properties of Mg^{2+} and F^- co-doped $Li_4Ti_5O_{12}$ anode material for use in the lithium-ion batteries. Electrochim Acta 222:1045–1055
19. Hou L, Qin X, Gao X et al (2019) Zr-doped $Li_4Ti_5O_{12}$ anode materials with high specific capacity for lithium-ion batteries. J Alloy Compd 774:38–45
20. Chen C, Huang Y, An C et al (2015) Copper-doped dual phase $Li_4Ti_5O_{12}$–TiO_2 nanosheets as high-rate and long cycle life anodes for high-power lithium-ion batteries. Chemsuschem 8(1):114–122
21. Zhu G, Liu H, Zhuang J et al (2011) Carbon-coated nano-sized $Li_4Ti_5O_{12}$ nano porous micro-sphere as anode material for high-rate lithium-ion batteries. Energy Environ Sci 4(10):4016–4022
22. Pohjalainen E, Kallioinen J, Kallio T (2015) Comparative study of carbon free and carbon containing $Li_4Ti_5O_{12}$ electrodes. J Power Sour 279:481–486
23. Zhu W, Wang Y, Yu Y et al (2020) Core-shell structured a-Fe_2O_3@$Li_4Ti_5O_{12}$ composite as anode materials for high-performance lithium-ion batteries. J Alloy Compd 813:1–10
24. Zhang E, Zhang H (2019) Hydrothermal synthesis of $Li_4Ti_5O_{12}$–TiO_2 composites and $Li_4Ti_5O_{12}$ and their applications in lithium-ion batteries. Ceram Int 45(6):7419–7426
25. Wang R, Cao X, Zhao D et al (2020) Enhancing lithium storage performances of the $Li_4Ti_5O_{12}$ anode by introducing the CuV_2O_6 phase. ACS Appl Mater Interfaces 12(35):39170–39180
26. Liu M, Gao H, Hu G et al (2020) Facile preparation of core-shell Si@$Li_4Ti_5O_{12}$ nanocomposite as large-capacity lithium-ion battery anode. J Energy Chem 40:89–98

5 Sulfur Cathode

5.1 Research Background of Lithium–Sulfur Batteries

With the rampant exploitation and use of fossil fuels, environmental pollution is becoming increasingly severe, which has promoted the development and use of sustainable and clean energy. The utilization of sustainable energy such as wind and solar energy requires efficient and low-cost methods for storage and use. Rechargeable batteries are currently the most commonly used method. Lithium-ion batteries are widely used in various industries due to their high energy density, good cycle life, and high discharge voltage. With the continuous development of various portable devices and electric vehicles, people's requirements for power storage are getting higher and higher, and lighter, thinner, and higher-capacity lithium batteries are needed, which has promoted researchers to continuously study new battery systems. Lithium–sulfur batteries, due to their high theoretical energy density (2600 Wh kg^{-1}), are gradually coming into people's view. Sulfur is abundant on earth, has a low usage cost, and is environmentally friendly, so lithium–sulfur batteries are also considered one of the promising alternative batteries for the next generation of energy storage devices and electric vehicles.

5.2 Mechanism of Lithium–Sulfur Batteries

Lithium–sulfur batteries usually consist of sulfur as the active material of thecathode, lithium metal as theanode, a separator, and an organic liquid electrolyte. S_8 as an active substance involves a 16-electron transfer reaction ($S_8 + 16Li^+ + 16e^- \leftrightarrows 8Li_2S$), which has a high theoretical capacity of 1674 mAh g^{-1}, far higher than the theoretical capacity of traditional lithium batteries (273 mAh g^{-1}). Lithium–sulfur batteries have two discharge voltage platforms in the capacity-voltage diagram (Fig. 5.1). During the discharge process, S_8 is first reduced to a series of long-chain lithium polysulfides (Li_2S_x, $4 \leq x \leq 8$) intermediates, forming the first high voltage platform (~2.3 V vs. Li/Li^+). In the subsequent continuous lithiation process, soluble long-chain polysulfides are converted to solid short-chain sulfides Li_2S_2. Then, at about 2.1 V, the second discharge voltage platform is provided, and Li_2S_2 is converted to Li_2S. During the charging process, the reaction process is reversed. Therefore, the electrochemical reaction of S_8 is a multi-stage reaction process, and the produced polysulfides act as intermediates, involving the conversion between soluble long-chain polysulfides, insoluble S_8, and Li_2S_2/Li_2S. Therefore, the charge and discharge process of lithium–sulfur batteries is considered a multi-phase conversion reaction process.

5.3 Modification of the Cathode in Lithium–Sulfur Batteries

5.3.1 Sulfur–Carbon Composite Cathode

Currently, the practical application of lithium–sulfur batteries still faces many challenges. Firstly, the conductivity of sulfur is particularly poor, which limits the utilization rate of sulfur in the cathode and reduces the rate capability of lithium–sulfur batteries. The lithium polysulfide intermediates generated during the conversion reaction can also dissolve in ether electrolytes, causing a large loss of active material in the cathode, leading to a rapid reduction in battery capacity, a decrease in coulombic efficiency, passivation of the lithium metal surface, and the production of insoluble products. Sulfur will accompany an 80% volume expansion during the lithiation/delithiation process, which can damage the cathode structure of the lithium–sulfur battery and reduce the battery's cycle performance. Therefore, if we want high-performance lithium–sulfur batteries, we first need to optimize the sulfur cathode. An excellent sulfur cathode design should include high conductivity,

X. Li et al., *Electrode Development for Lithium-ion Batteries*, https://doi.org/10.1007/978-981-96-3021-9_5

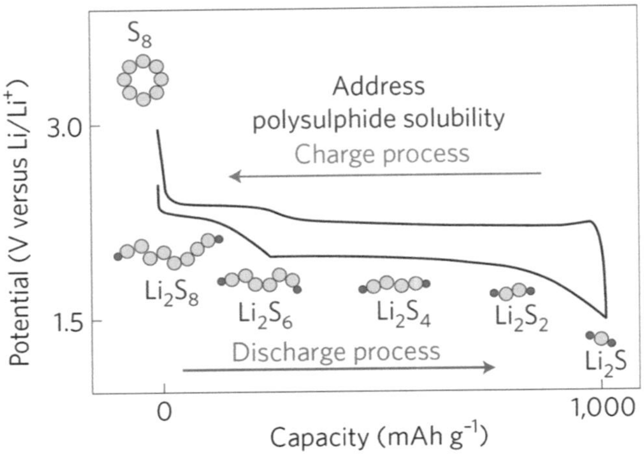

Fig. 5.1 Charge–discharge curve of lithium–sulfur battery [1]

outstanding anchoring ability for polysulfides, and a stable electrode structure.

In order to improve the performance of lithium–sulfur batteries, researchers have conducted a large amount of research on the above problems. People first found that adding carbon materials to the cathode as a sulfur host can effectively improve the conductivity of the cathode. The high surface area and porosity of porous carbon materials can meet good electronic and ionic conductivity and anchor polysulfides in the composite material. Carbon-based materials can also buffer the huge volume change of active materials. These carbon-based materials mainly include graphene, carbon nanotubes, carbon nanospheres, carbon nanocages, porous carbon, etc. After Ji and others [2] made a significant breakthrough by confining sulfur in CMK-3 (a type of mesoporous carbon), porous carbon with a high specific surface area (SSA), adjustable pore structure, and controllable morphology has been widely studied and used as a sulfur host and functional coating to improve the performance of lithium–sulfur batteries. The performance of lithium–sulfur batteries is closely related to the pore structure of the carbon matrix. According to the pore size distribution, hierarchical porous carbon (HPC) materials can be divided into three types, namely micro/mesoporous carbon, micro/mesoporous/macroporous carbon, and mesoporous/macroporous carbon. Micro/mesoporous carbon is the most used in published papers, accounting for 63.5%, while the proportions of micro/mesoporous/microporous carbon and mesoporous/microporous carbon are 28.6% and 7.9%, respectively.

The pore volume of porous carbon materials is one of the important parameters affecting the sulfur content and battery performance in sulfur-carbon composite materials. The pore volume mainly depends on the size and proportion of mesopores and macropores. Through calculation [3], it can be obtained that when the sulfur content is 70% and 80%, if the volume expansion is not considered, sulfur is fully filled in the pores, the minimum pore volume of the sulfur-carbon composite material is 1.13 and 1.93 $cm^3\ g^{-1}$ respectively. When considering the volume expansion of sulfur, the expected pore volume of HPCs filled with the same sulfur content should belarger than 2.02 and 3.46 $cm^3\ g^{-1}$ respectively. Guo pointed out [4] that for porous carbon materials with similar specific surface areas, a higher pore volume can yield better battery performance. The authors synthesized a series of porous flower-like carbon nanospheres (FCNSs) using sea urchin-like silicon spheres as templates, as shown in Fig. 5.2a. The specific surface areas of FCNS2-750, FCNS2-800, FCNS2-900, and FCNS2-1000 (FCNS2-X, X represents the carbonization temperature) are 1103, 1148, 1151, and 1080 $m^2\ g^{-1}$, respectively, but the total pore volumes increase from 1.33 to 1.86, 1.94, and 1.74 $cm^3\ g^{-1}$, respectively. FCNS2-900/S with a high sulfur content (81wt%) has the highest discharge capacity and the best cycle stability. The reversible capacity of FCNS2-900/S at 1 C is 1104 mAh g^{-1}, and after two cycles of activation at 0.2 and 0.5 C, a capacity of 923 mAh g^{-1} and a high capacity retention rate of 85% can be obtained after 200 cycles at 1 C (Fig. 5.2b). In contrast, cathodes with smaller pore volumes show poorer cycling performance. The order of performance increase is FCNS2-750/S < FCNS2-1000/S < FCNS2-800/S81% < FCNS2-900/S, which is consistent with the increase in pore volume. To further explain the differences in electrochemical reversibility, the samples were immersed in a Li_2S_6 solution, and the adsorption of polysulfides by the synthesized porous carbon was detected by UV-visible spectroscopy. From Fig. 5.2c, d, the order of adsorption of polysulfides by porous carbon is: FCNS2-750 < FCNS2-1000 < FCNS2-800 < FCNS2-900, which matches the battery cycling performance. However, the sulfur cathode undergoes significant volume changes during charging and discharging, so in the above studies, the sulfur content has exceeded the optimal sulfur content in porous carbon, and high pore volume and porosity will affect the contact between sulfur and the carbon framework, resulting in poor conductivity. Therefore, when the pore volume of porous carbon is lower than the ideal sulfur content, the higher the pore volume, the better the performance of the sulfur-carbon composite cathode. When the pore volume is far higher than the optimal volume of sulfur content, further increasing the pore volume will not improve the performance of lithium–sulfur batteries. Specific surface area is important in many chemical reaction systems because a higher specific surface area can provide more active sites, and this rule also applies to the conversion reaction system of lithium–sulfur batteries. However, Sahore et al. [5] pointed out that to achieve a high sulfur utilization rate at high sulfur content, when the sulfur content is higher than the ideal sulfur content in porous carbon materials, the pore volume in the material becomes more important than the specific surface area. In their results, in carbon–sulfur

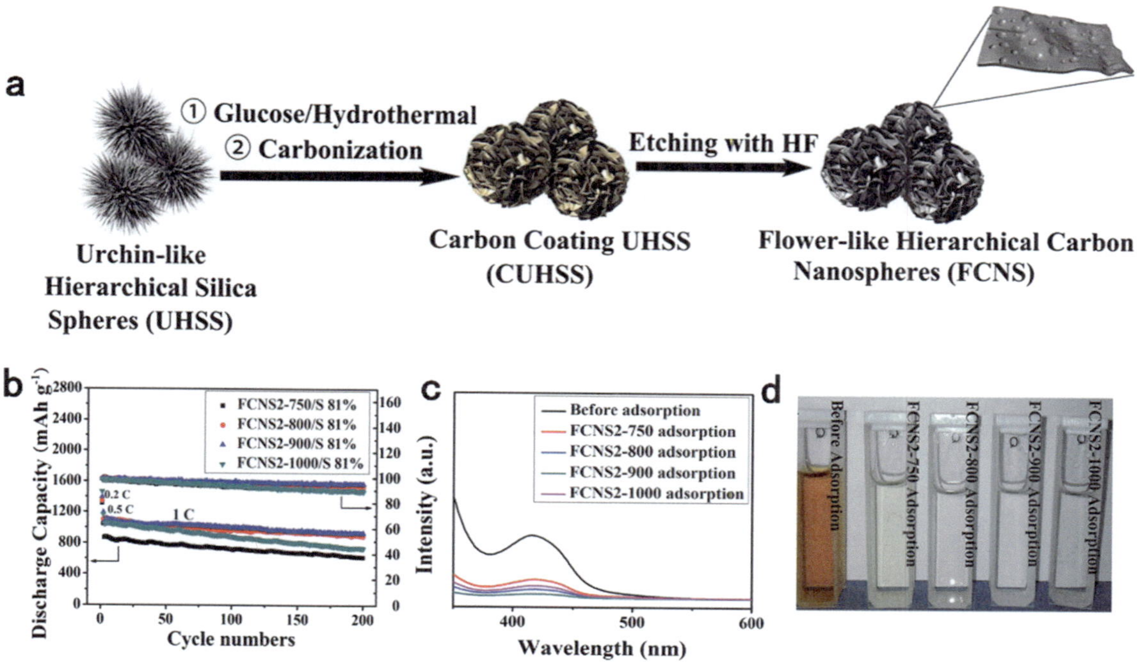

Fig. 5.2 **a** Schematic diagram of the FCNS preparation process [4]. **b** Long cycle performance of four types of composite cathodes at 1 C rate [4]. **c** UV-vis spectra of lithium polysulfide solution added with FCNS2-750, FCNS2-800, FCNS2-900, and FCNS2-1000 [4]. **d** Digital photos of the corresponding lithium polysulfide solutions [4]

composite materials with a sulfur content as high as 80%, the cathode capacity of those with low specific surface area and pore volume is particularly low, while porous carbon–sulfur composites with high specific surface area and high pore volume have high capacity. However, it can be found that porous carbon–sulfur composites with medium specific surface area and high pore volume also have high capacity, while carbon–sulfur composites with medium pore volume and high specific surface area have lower capacity than the former. According to their findings, if the specific surface area of porous carbon materials is less than 1500 $m^2\ g^{-1}$, the discharge capacity increases with the increase of specific surface area. However, when the specific surface area exceeds 1500 $m^2\ g^{-1}$, there is no very obvious correlation between the discharge capacity and the specific surface area (Table 5.1).

5.3.2 Modification Study of Two-Dimensional Materials in the Cathode of Lithium–Sulfur Batteries

Two-dimensional materials have many unique advantages compared to other nanomaterials. They typically have high specific surface areas, small sizes, and stable structures. Their lateral dimensions are largerthan 100 nm, even reaching several micrometers or larger sheet structures, but their thickness is only one or a few atoms thick (usually less than 5 nm). Due to the electron restrictions in two dimensions, they also have distinct physical, chemical properties and electronic characteristics. For example, the specific surface area of graphite is less than 50 $m^2\ g^{-1}$, while that of graphene is larger than 500 $m^2\ g^{-1}$. The conductivity of graphene is also several to tens of times that of graphite. Two-dimensional materials have adjustable and uniformly exposed crystal faces, which not only provide a sufficiently large contact reaction area for polysulfides, but also have their unique morphological advantages, such as the microporous structure of carbon materials, which can provide a huge free space for high sulfur loading and volume changes. The electrical properties of two-dimensional materials change significantly during the preparation process with thickness. On the one hand, some types of two-dimensional materials, such as MXenes, show excellent conductivity. On the other hand, modifying the surface chemistry or structure of 2D nanomaterials, such as doping metal atoms or constructing heterojunctions, can greatly improve electronic conductivity. Since Novoselov and others at the University of Manchester successfully isolated single-layer carbon atom graphene in 2004 [30], a series of two-dimensional materials such as graphene and their derivatives hexagonal boron nitride, graphite carbon nitride,

Table 5.1 The influence of specific surface area and pore volume on the performance of the battery in the sulfur@porous carbon composite cathode

Composite cathode	Specific surface area/m^2 g^{-1}	Pore volume / m^3 g^{-1}	Sulfur content/%	Cycle performance/mAh g^{-1}	Rate performance/mAh g^{-1}
S@HPC-graphene [6]	1558	1.1	74	1 C, 100 cycles 860	10 C 510
S@HPC [7]	2449	4.97	80	0.1 C, 100 cycles 860	0.5 C 766
S@HP graphene [8]	2700	2.5	72.3	0.5 C, 200 cycles 854	3 C 720
S@HPC [9]	957	1.46	65.3	0.2 C, 200 cycles 831	4 C 753.2
S@HPC [10]	4073	2.26	–	0.2 C, 200 cycles 980	2 C 600
S@N-HPC [11]	1290	3.04	76.2	1 C, 300 cycles 600	4 C 671
S@HPCNFs [12]	2480	1.84	66	0.5 C, 100 cycles 965	2 C 626.5
S@HPCs [13]	2600	2.46	66.7	0.1 C, 100 cycles 950	2 C 600
S@N,P-HPCs [14]	788	0.768	50	0.1 C, 100 cycles 729	2 C 213
S@N,S HPCs [15]	2488	0.99	85	0.5 C, 300 cycles 952	1 C 1201
S@HPC [16]	642	0.51	77.2	0.12 C, 150 cycles 762	1 C 720
S@O-HPC [17]	3068	1.37	80	0.1 C, 100 cycles 925	2 C 462
S@HPC [18]	2164	1.29	50	0.5 C, 200 cycles 943	2 C 758.7
S@N,O-HPC-rGO [19]	481	2.14	75	1 C, 300 cycles 848	5 C 685
S@HPC [20]	2441	1.69	58.8	1 C, 70 cycles 1023	3 C 686
S@N-HPC nanoflower [21]	1151	1.94	81	1 C, 200 cycles 923	5 C 829
S@N-HPCNTs [22]	292	0.37	66.7	0.2 C, 200 cycles 979	5 C 817
S@HPC [23]	915	4.15	78	3 C, 200 cycles 914	3 C 1134
S@HPC nanobelts [24]	3445	3.18	60	0.1 C, 100 cycles 752	5 C 800
S@HP-microsphere [25]	756	2.08	60	1 C, 1000 cycles 720	2 C 691
S@N-HPC [26]	2546	13.4	76	0.5 C, 1000 cycles 510	2 C 976
S@HPC [27]	897	0.68	59	0.1 C, 50 cycles 874	1 C 884
S@N,S-HPC [28]	1420	4.94	78	1 C, 100 cycles 829	8 C 581
S@Fe_{3-x} C-HPC [29]	197	–	75	0.2 C, 100 cycles 1089	5 C 609

transition metal dichalcogenides, MXenes, metal oxides, metal halides, etc., have been deeply studied.

Graphene is a typical two-dimensional carbon nanomaterial with a crystal structure formed by a single layer of carbon atoms. The combination of graphene and sulfur positive electrodes can greatly improve its conductivity, alleviate the volume change of sulfur during the discharge process, effectively anchor polysulfides, inhibit the shuttle effect, and thus improve the electrochemical performance of lithium–sulfur batteries. Graphene has outstanding conductivity advantages, but this conductivity is limited to a single-layer structure, and the conductivity between different layers is not ideal. Graphene layers are prone to stacking during the processing process, so the sulfur load is low, which does not meet the requirements of commercialization. Researchers have designed several different graphene/sulfur structures, such as planar, sandwich, core–shell, and three-dimensional (3D) structures, to maximize the use of active materials and promote electron transfer. In the planar structure, sulfur is directly deposited on the graphene sheet to be used as the cathode of the lithium–sulfur battery. However, under this configuration, graphene sheets are prone to stacking, cannot effectively wrap sulfur, and cannot limit the dissolution and shuttle of intermediate product lithium polysulfides. To solve these problems, researchers have successively developed structures such as nanotubes and nanopaper based on graphene. Because the planar structure has weak physical encapsulation ability for sulfur, leading to rapid capacity decay. Therefore, some researchers try to use the sandwich structure to improve graphene's capacity for sulfur. By restricting sulfur between two layers of graphene to form a G/S/G sandwich structure [31] (Fig. 5.3a), it can improve the physical encapsulation of the cathode for sulfur, alleviate the dissolution of lithium polysulfides, and also accommodate the volume change of sulfur during charging and discharging. However, after a long period of cycling, the dissolved polysulfides may still leak from the edge of the graphene, leading to a decrease in the content of active substances, thereby reducing the battery life. To overcome the sulfur leakage in the sandwich structure, a core-shell structure is used to encapsulate sulfur. The core-shell structure can maximize the advantage of graphene's flexibility. The core-shell structure can effectively alleviate the volume expansion/contraction of sulfur in the redox reaction process, compared with the planar and sandwich structures, it firmly encapsulates polysulfides in the shell, effectively preventing the shuttle effect from a spatial separation perspective. However, the synthesis method of this structure is lengthy and complicated, which hinders the development of sulfur cathodes to a certain extent.

The above methods are to improve the performance of the cathode by adjusting the structure, but the polarity of graphene is very low, and the anchoring ability for lithium polysulfides is not strong, and it is still difficult to solve the shuttle problem of lithium polysulfides by adjusting the structure alone. Therefore, some functional modifications can be made on the surface of graphene to enhance its anchoring ability for lithium polysulfides. The functionalization of graphene is mainly to convert sp^2 hybridized carbon into sp^3 hybridized carbon by attaching some functional groups to the surface. These functional groups mainly include hydroxyl, carboxyl, epoxy, amino, and sulfonic acid groups. Yu et al. [32] first synthesized phenyl sulfonated graphene/sulfur ($PhSO_3$-RG/S) composite material through in-situ redox reaction in aqueous solution in 2014 (Fig. 5.3b), compared to untreated graphene, functionalized graphene materials have better hydrophilicity and flexibility, making them easier to handle in aqueous solutions. At the same time, compared to non-functionalized graphene,

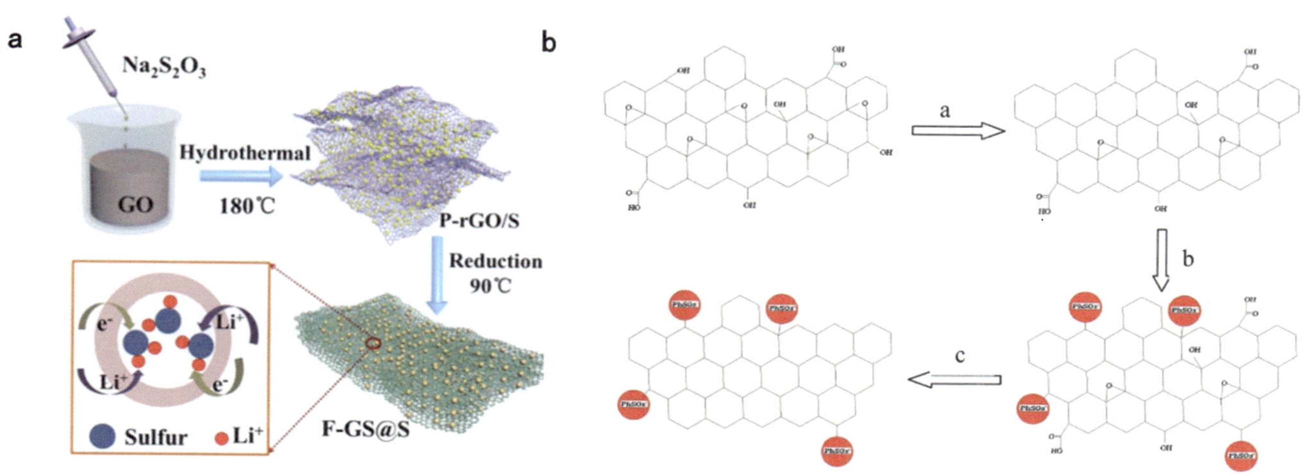

Fig. 5.3 **a** Schematic diagram of the preparation of sandwich-like F-GS@S composite material [31]. **b** Preparation diagram of $PhSO_3$-RG/S (**a**–**c**) [32]

functionalized graphene can chemically bond with polysulfides, to some extent inhibiting the loss of polysulfides. Due to these unique characteristics, functionalized graphene has been widely used in lithium–sulfur batteries. However, the role of functional groups on sulfur is not yet fully understood, and the preparation of functional group graphene is very complex, which to some extent hinders its further development.

Single atom doping of graphene is also one of the effective methods to regulate its conductivity and electroactivity, with nitrogen doping and boron doping commonly used. Graphene can be classified into pyridine-type, pyrrole-type, and graphite-type nitrogen-doped forms according to the bonding patterns of the surrounding carbon atoms. In Fig. 5.4a, Guo et al. [33] prepared a cathode composed of S and PVDF binder wrapped by N-doped graphene, without adding extra carbon black additives. They synthesized graphene oxide sheets using an improved Hummers method, which consists of carbon atoms modified with epoxy and hydroxyl groups and edge carbon atoms connected with carboxyl groups. Compared with the original graphene material, nitrogen-doped graphene material shows higher polarity, stronger chemical adsorption ability, can better alleviate the shuttle effect of lithium polysulfides, and improves the cycle stability of lithium–sulfur batteries. In addition to nitrogen-doped graphene, boron-doped graphene is also used as a sulfur host to improve battery performance. The electronegativity of boron is 2.04, less than that of carbon at 2.55. Electrons in the carbon-boron bond lean towards the carbon atom, so the boron atom carries a slight positive charge, while the surrounding carbon atoms carry a negative charge. Boron, with its slight positive charge, creates a more attractive position for lithium polysulfides, enhancing the adsorption of polysulfides on its surface. However, the proportion of doped atoms in sulfur host materials is usually less than 5%, so the effect of single atom doping is limited and not enough to meet the huge challenges of lithium–sulfur batteries. To better inhibit the shuttle of polysulfides, researchers propose doping multiple atoms (such as N/B, N/S) in graphene. Compared with single atom doped graphene, multi-atom doped graphene has stronger binding force and better adsorption ability for polysulfides. Qian et al. [34] synthesized BNG composite material with high proportions of B and N (B, 13.47%; N, 9.17%), forming abundant pyridine-N and N–B/N=B bonds, providing more electroactive sites. In Fig. 5.4b, it can also be seen that N–B dual doping has the strongest lithium polysulfide anchoring ability (Table 5.2).

Two-dimensional transition metal carbides/nitrides, known as MXenes, are a new family of two-dimensional

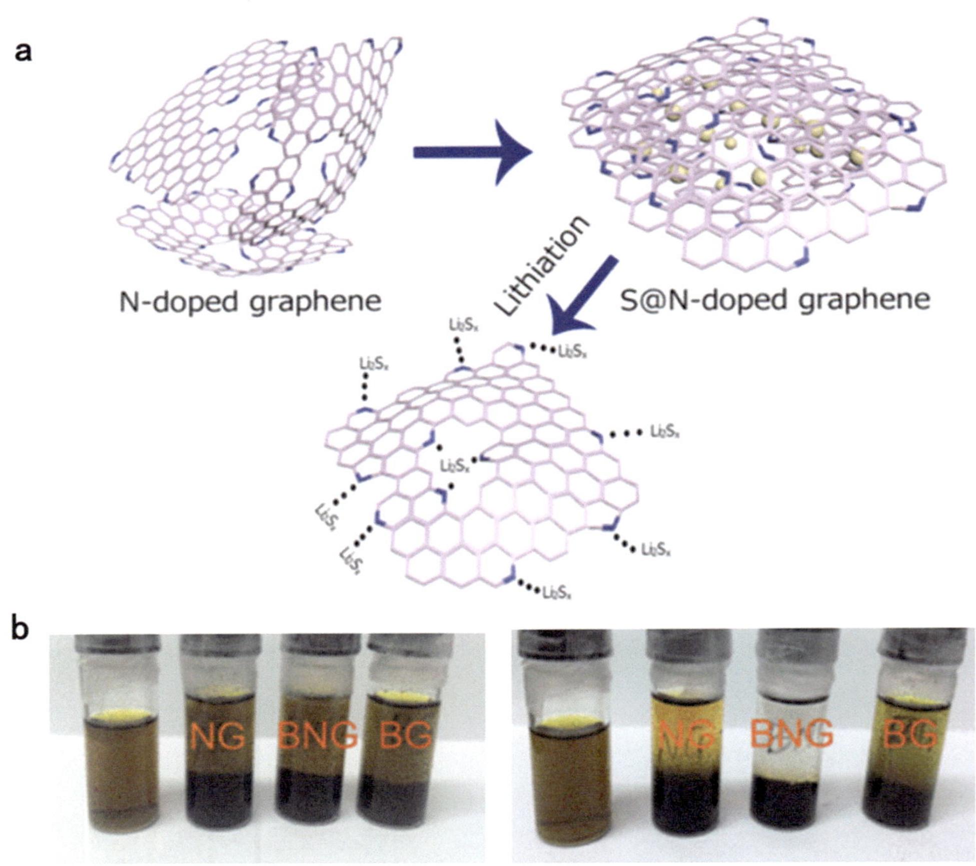

Fig. 5.4 **a** Synthesis route of S@NG nanocomposite material and schematic diagram of N doping anchoring polysulfides [33]. **b** Digital photos of lithium polysulfide solution added with NG, BNG and BG [34]

Table 5.2 Performance of various modified graphene@sulfur composite cathode batteries

Material	Sulfur content/%	Rate/C	Initial capacity/mAh g^{-1}	Cycle number	Retained capacity /mAh g^{-1}
F-GS@S [31]	63	2	601.7	500	424.2
$PhSO_3$-RG/S [32]	63	0.2	1023	400	460
S@NG [33]	60	2	789	2000	347
S@BNG [34]	75	1	882.7	500	765
AGNs10/S [35]	67	1	927	100	685
G-S [36]	73	1	900	100	615
R-rGO/nS [37]	78.2	0.1	1269.1	500	726.2
PSSN/S [38]	60	1	1013	500	735
EFG-S [39]	60–69	0.5	812	350	650
BGA-S [40]	59	0.2	1290	100	994
S@BEEG [41]	72.2	1	1018	130	838
NBCGN/S [42]	65	0.2	860	300	446
A-NSG@S [43]	72.4	0.2	1178	600	780

materials. MXene materials have key characteristics such as two-dimensional structure, metallic conductivity, high active surface, and dispersibility in water. MXenes show great potential in lithium-ion batteries, sodium-ion batteries, lithium–sulfur batteries, zinc-ion batteries, and supercapacitors. The metallic conductivity of MXenes can promote electron transfer and improve the utilization of sulfur. The two-dimensional structure of MXenes provides sufficient reaction sites for sulfur transformation, and can also be constructed into a 3D porous framework to achieve high sulfur loading. They have a strong binding energy with lithium polysulfides by forming metal-sulfur bonds. MXenes can also accelerate the redox reaction kinetics between lithium polysulfides and lithium sulfides. Since Gogotsi and Barsoum reported the first $Ti_3C_2T_x$ MXene in 2011, more than 30 stoichiometric MXenes have been successfully synthesized. Most MXenes have been proven to have excellent conductivity. The 2D structure of MXene can make full contact with sulfur. However, due to the action of van der Waals forces and hydrogen bonds, 2D MXene nanosheets tend to restack together, which limits the diffusion of Li^+ and hinders the full utilization of MXene. By cutting 2D nanosheets into 0D nanoparticles or 1D nanoribbons, and turning 2D constructing MXene nanosheets into a 3D porous framework can effectively expose the surface of the MXene nanosheets. In lithium–sulfur batteries, MXenes have been proven to effectively accelerate the mutual conversion between polysulfides and lithium sulfides, accelerating redox kinetics, which is beneficial for improving the rate performance of lithium–sulfur batteries. The catalytic ability of MXene is significantly affected by the terminations and vacancies on the MXene surface, and appropriate vacancies on the MXene surface can further enhance the catalytic effect. Therefore, MXenes are considered an ideal host material that can be combined with sulfur for high-performance lithium–sulfur batteries. By simply loading sulfur on MXene nanosheets, the electrochemical performance of lithium–sulfur batteries can be significantly improved. However, these electrodes are still plagued by the re-stacking tendency of 2D MXene nanosheets, which limits Li^+ diffusion and hinders the full utilization of MXene. To solve this problem and maximize the advantages of MXene in lithium–sulfur batteries, researchers have studied strategies including chemical modification and structural control of MXene, and the construction of MXene-based composite materials.

Chemical modification is a common strategy to improve the performance of MXenes. It has been reported that doping heteroatoms, such as nitrogen, sulfur, and metal atoms in conductive hosts can significantly enhance their catalytic effects. As can be seen from Fig. 5.5a [44], MXenes doped with N atoms significantly enhance the chemical adsorption effect on polysulfides. Carbon materials and transition metal compounds are traditional efficient sulfur hosts. Carbon materials with developed porous structures can physically limit the dissolution shuttle of polysulfides and provide sufficient space for high sulfur loading and volume expansion, while transition metal compounds can chemically adsorb polysulfides to suppress the shuttle effect. MXenes, due to their unique two-dimensional structure and high conductivity, become favorable candidates for composite materials with carbon materials and transition metal compounds. In the composite materials, MXenes provide a continuous conductive network for rapid charge transfer and adsorption and catalytic action on polysulfides, while other components, in addition to their inherent functions, can also act as "spacers" to alleviate the re-stacking of MXene and promote Li^+ diffusion. As can be seen from

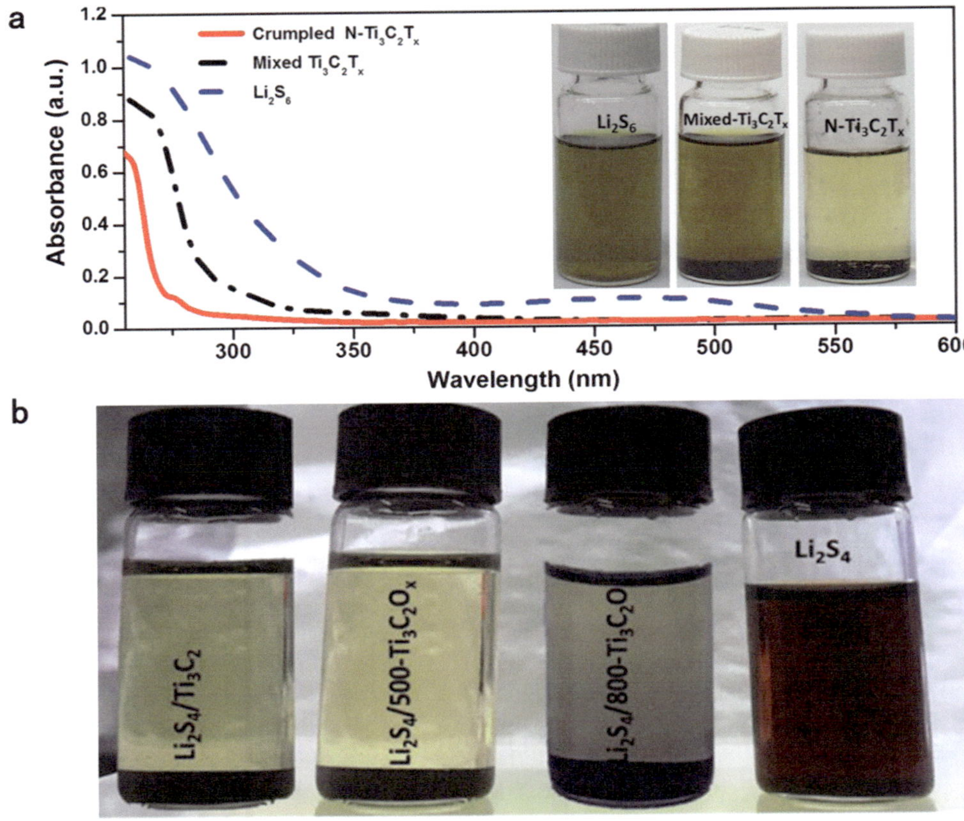

Fig. 5.5 **a** UV/visible absorption spectra of Li_2S_6 solution after adding N–$Ti_3C_2T_x$ powder and $Ti_3C_2T_x$ powder. The inset image shows the addition of N–$Ti_3C_2T_x$ powder and $Ti_3C_2T_x$ photos of Li_2S_6 solution before and 4 h after the addition of the powder [44]. **b** Photos of Li_2S_4 solution after the addition of S@Ti_3C_2, S@500-$Ti_3C_2O_x$, and S@800-$Ti_3C_2O_x$ powders [45]

Fig. 5.5b [45], titanium oxide-Ti_3C_2 has a strong anchoring effect on polysulfides, and MXene-based composite materials as sulfur hosts show good electrochemical performance, which is a very promising direction for the application of MXene in lithium–sulfur batteries (Table 5.3).

Table 5.3 Performance of various modified MXene@Sulfur composite cathode batteries

Material	Sulfur content/%	Rate/C	Initial capacity/ mAh g^{-1}	Cycle number	Retained capacity/ mAh g^{-1}
N–$Ti_3C_2T_x$/S [44]	73.8	0.2	1140	1000	765
S@500-$Ti_3C_2O_x$ [45]	80	0.5	1567	1000	803
Ti_2CT_x/S [46]	70	0.5	1090	650	735.75
S/$Ti_3C_2T_x$@PDA [47]	60	0.2	1430.1	150	1044
S@$Ti_3C_2T_x$ [48]	70	0.2	1184	800	724
S@V_2C–Li/C [49]	70	0.5	744	500	600
CNTs/Ti_2C [50]	83	0.5	929.7	1200	450
Mo_2C-CNTs/S [51]	87.1	0.1	1235	250	925
$Ti_3C_2T_x$/rGO/S [52]	70.4	0.5	1144.2	300	878.4
$Ti_3C_2T_x$@Meso-C/S [53]	72.88	0.5	1225.8	300	704.6
S@N-PC/Ti_3C_2 [54]	80	1	759	800	527

5.3.3 Improving Lithium–Sulfur Battery Performance Through Cooperative Catalysis

By calculating the Gibbs free energy of the intermediate transformations in lithium–sulfur batteries, it can be found that each stage of the intermediate transformation requires an appropriate catalyst to reduce the reaction barrier. The conversion of polysulfides to the final insoluble product, lithium sulfide, is the most difficult, and it is generally believed that the catalytic oxidation of lithium sulfide is the rate-determining step in lithium–sulfur batteries. The complex chemical properties of polysulfides, especially the unique electronegativity between the terminal Li and S, require more complex construction designs to achieve the ideal synergistic effect through various chemical interactions. Because the active material is an electrical insulator, leading to slow reaction kinetics and low sulfur utilization, the battery cannot operate normally at high current densities. Therefore, it is very important to anchor the soluble polysulfides on the conductive/polar synergistic surface and accelerate their reversible conversion. Both long-chain and short-chain polysulfides exhibit the same polarity, with terminal sulfur carrying a negative charge. Therefore, the polar surface can provide a strong affinity with polysulfides, effectively alleviating the shuttle effect. In addition to chemical adsorption, many studies report that heteroatom doping can not only enhance the electronic conductivity to a certain extent, promoting rapid charge transfer, but also catalyze the formation and transformation of Li_2S on the solid–liquid conductive interface. However, due to the limited polar sites of heteroatom doping, multifunctional metal compound nanomaterials have gradually been developed by researchers. For example, metal oxides, such as MnO_2 and VO_2, can not only firmly anchor polysulfides, but also regulate the transformation of polysulfides through the formed thiosulfate intermediates during the electrochemical reaction process. Transition metal nitrides, transition metal sulfides, transition metal phosphides, etc., compared with metal oxides with semiconductor properties, have higher conductivity and chemical stability, and are widely used as electronic transmission ties between polysulfides and conductive interfaces, making the transformation of polysulfides faster. In order to ensure the rapid anchoring and rapid redox transformation of polysulfides, the synergistic catalytic interface combining two functional metal compounds has also been widely explored and received more and more attention. Wu et al. [55] designed a Nb_4N_5–Nb_2O_5 heterostructure as a dual-functional main material for the sulfur positive electrode (Nb_4N_5–Nb_2O_5/S) and Li anode (Nb_4N_5–Nb_2O_5/Li). From the Fig. 5.6, it can be seen that the adsorption energies of Li_2S_4 on Nb_4N_5 (211) and Nb_2O_5 (001) surfaces are −4.8 and −6.2 eV, respectively, indicating that Nb_2O_5 has a higher adsorption affinity for lithium polysulfide. The superior conductivity of Nb_4N_5 ensures the rapid diffusion of lithium polysulfide from the Nb_2O_5 surface to the Nb_4N_5 surface through the interface. The results show that Nb_4N_5–Nb_2O_5 can effectively improve the electrochemical reaction, accelerate the oxidation of dissolved lithium polysulfide into solid Li_2S during the discharge process, ensuring good reversible charge and discharge. An effective synergistic mechanism is to create a coupled interface between two different catalyst components to achieve functional alignment or work coordination. Heterogeneous binary hosts can not only synergize the advantages of each single catalyst, but also enhance the catalytic effect, which may be due to the d-band center of the metal atoms in the catalyst being closer to the Fermi level, thereby generating more charge numbers to promote the conversion of lithium polysulfide and weaken the shuttle effect of lithium polysulfide. In recent years, people have deliberately designed different types of synergistic interfaces on conductive carbon to achieve anchoring and catalytic effects synchronously, such as adsorption-conductive type, ion–electron coupled type, electron coupled type, ion coupled diffusion type, etc. Another alternative strategy is to closely coordinate two catalysts without heterogeneous structures to selectively or stepwise catalyze the entire polysulfide conversion. Compared with doping engineering, the cooperation between two carefully selected catalysts enriches the adsorption and conductive sites, helps to promote redox kinetics, thereby reducing the electrolyte and active material ratio (E/S) to improve the practicality of lithium–sulfur batteries (Table 5.4).

5.4 Lithium Sulfide Cathode

5.4.1 Research Background of Lithium Sulfide Cathode

Although lithium–sulfur batteries have decent capacity and cycle performance after various modifications, the sulfur cathode still faces many challenges. Li_2S is a common alternative cathode material for lithium–sulfur batteries. The Li_2S cathode has a high theoretical specific capacity (1166 mAh g^{-1}), which is much higher than the cathode materials used in current lithium-ion batteries. This fully lithiated sulfur as a cathode material can avoid the problem of volume expansion during the cycle. More importantly, the Li_2S cathode can be combined with different types of anode materials other than lithium metal, such as silicon,

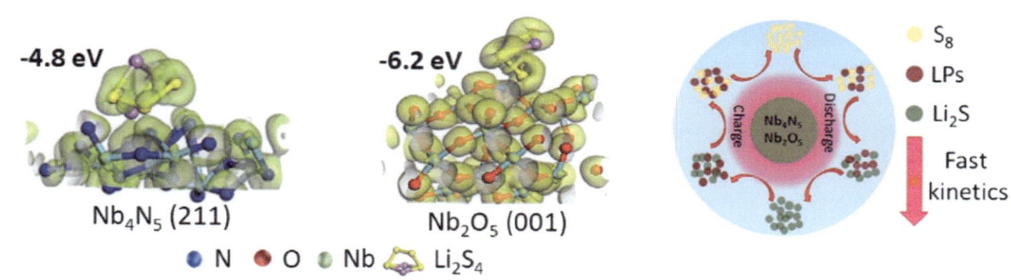

Fig. 5.6 Optimized geometric shapes of Li_2S_4 on Nb_4N_5 (211) and Nb_2O_5 (001) surfaces and their corresponding binding energies and schematic diagram of the reaction pathway of Nb_4N_5–Nb_2O_5 heterogeneous structure catalyzing sulfur cathode [55]

Table 5.4 Summary of reducing E/S ratio and high energy density lithium–sulfur batteries through synergistic catalysis

Material	Sulfur loading/mg cm^{-2}	E/S ratio/μL mg^{-1}	Cycle rate/C	Number of cycles	Reversible capacity mAh cm^{-2}
Nb_4N_5–Nb_2O_5 Heterostructure [55]	6.9	5.0	0.3	200	4.4
N,S-doped Porous Graphene [56]	4.0	5	0.1	–	10
Co,N doped graded conductive fabric [57]	13.5	4.5	0.05	55	8.6
TiO_2-MXene heterostructure [58]	10.5	5.0	0.05	100	5
$CoSe_2$–Co_3O_4@NC-CNT [59]	10.1	3.6	0.5	100	5.63
ZnS–SnS@NC heterostructure [60]	10.3	4.0	0.1	100	8.27
G-SnO_2–$SnSe_2$ hybrid [61]	10.8	3.8	0.1	100	9.06
ZnS and Co–N–C DEB [62]	6.25	4.0	0.06	80	5.82
V_8C_7–VO_2 heterostructure [63]	9.2	5.0	0.1	100	5.11

germanium, tin, and carbonaceous materials. Lithium–sulfur batteries based on Li_2S cathodes have more reliable safety features, avoiding serious safety issues in practical applications. The high melting point and large volume of Li_2S give it higher operating temperatures and more stable surface modification performance. Therefore, using Li_2S to replace sulfur as the cathode material to improve the performance of lithium–sulfur batteries is a promising strategy. However, Li_2S also faces some challenges. Li_2S itself has poor electrical conductivity and its ionic conductivity is not high. Therefore, 30–50 wt.% of carbon material (conductive additive) is usually added to the Li_2S cathode to increase the electronic conductivity, but adding conductive additives will reduce the weight and volumetric energy density of the lithium–sulfur battery. Due to poor conductivity, the transfer of electrons to Li_2S will be slow. During the charging and discharging process, the rate performance and utilization of Li_2S will decrease, and low conductivity will also cause poor cycling performance. In addition, polysulfides will inevitably be produced in the Li_2S cathode, these polysulfides will dissolve in the electrolyte and shuttle to the anode, leading to the loss of active material and shortening the battery life. The poor ionic conductivity of Li_2S and the high charge transfer resistance at the interface with the electrolyte mean that Li_2S will have a high barrier in the initial charging process, resulting in poor battery cycling performance. Li_2S is very sensitive to moisture in the air and must be handled in an inert gas, making the handling of Li_2S more expensive and complex than that of sulfur.

5.4.2 Lithium Sulfide Pre-melting Strategy

In order to overcome the obstacles of the practical application of the Li_2S cathode, researchers use many different strategies and methods to solve these obstacles. The electrochemical reactions of Li_2S materials are mostly inert and need to be activated at high charging potentials. In 2012, Yang's team used a high charging potential of 4.0 V to overcome the voltage barrier of the Li_2S cathode [64], achieving a capacity of 800 mAh g^{-1}. The normal cut-off charging potential of lithium–sulfur batteries (3.0 V) cannot make the Li_2S battery work, but the side reactions of electrolyte decomposition at 4.0 V charging will lead to poor cycle stability and low energy efficiency. Chiang and others proposed that morphology and particle size also affect the activation process of Li_2S particles [65]. They proposed that by reducing the particle size of Li_2S, the activation barrier can be reduced, thereby better utilizing Li_2S and improving battery performance. Some research groups have reduced the initial activation potential of Li_2S by ultra-high-speed ball milling, which increases the contact area of the electrode material, shortens the diffusion path of Li^+, and enhances the reactivity of Li_2S. However, the ultra-high-speed ball milling method brings safety issues, and the yield is low, the cost is high, and the electrode reaction interface is still solid–solid contact. Researchers have found that by pre-dissolving Li_2S, the solid–solid contact interface can be simply transformed into a better solid–liquid contact interface. This convenient liquid Li_2S loading method can provide new opportunities for industrial battery assembly. After pre-solvent treatment, the bulk Li_2S particles are transformed into liquid Li_2S, thereby transforming solid–solid contact into solid–liquid contact. The reaction kinetics and electron/ion transfer are significantly enhanced, the redox overpotential is reduced, and this also reduces the barriers to Li_2S activation and polysulfide transformation processes. Liang's team reported [66], P_2S_5-TEGDME system can accelerate the dissolution of large Li_2S (Fig. 5.7a), promote the reaction kinetics of Li_2S, optimize the surface of metallic lithium, and inhibit the shuttle effect of polysulfides. Based on P_2S_5, the Li_2S battery shows better reversible capacity (about 1300 mAh g^{-1}), after 40 cycles

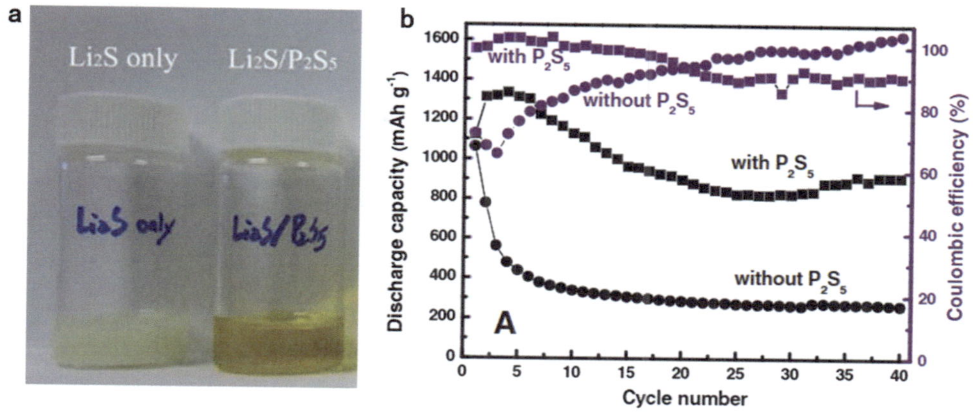

Fig. 5.7 **a** Insoluble Li_2S and dissolved Li_2S/P_2S_5. The weight ratio of solid to liquid in each vial is 1:10 [66]. **b** Cycling performance of Li_2S batteries with and without P_2S_5 [66]

at 0.1 C (about 98% coulombic efficiency), maintaining the stable performance of the Li–S battery (900 mAh g^{-1}) (Fig. 5.7b).

5.4.3 Lithium Sulfide–Carbon Composite Cathode

Liquid Li_2S cathode can reduce the initial activation barrier, avoid electrolyte decomposition, thus making the battery performance more stable and the cycle life longer. However, the intrinsically non-conductive nature of Li_2S and the shuttle behavior of polysulfides cannot be ignored. Therefore, researchers also use functional materials to enhance conductivity and prevent the "shuttle" effect. Among them, carbon materials, due to their strong conductivity, flexible manufacturing process, and abundant natural resources, have become an efficient and cheap choice. Carbon nanotubes (CNT) and reduced graphene oxide (rGO) layered electrodes can limit the migration of polysulfides. The combination of the two can form a three-dimensional (3D) structure of the cathode, with enough internal space, to allow the electrochemical reaction of polysulfides during the cycling process of lithium–sulfur batteries. Researchers have developed various methods to prepare liquid Li_2S–carbon composite cathodes. The simplest method is to directly drop liquid Li_2S onto the carbon material, another method is to combine Li_2S solution with a precursor for further carbonization to encapsulate Li_2S nanoparticles, thus obtaining a more stable structure at a lower cost. In addition, the capacity of lithium–sulfur batteries can be improved by chemically doping to enhance the active sites of carbon. The design of the carbon-based pre-dissolved Li_2S cathode is of great significance for improving the performance of lithium–sulfur batteries. Pure carbon is popular in lithium–sulfur battery research, but it hardly shows any electrocatalytic activity and is not an ideal positive electrode catalyst in high-performance lithium–sulfur batteries. Although the carbon doping strategy can improve the performance of lithium–sulfur batteries, it also increases the cost of material synthesis.

5.4.4 Lithium Sulfide-Functional Material Composite Cathode

Compared with the doping of carbon materials, functional materials (metal sulfides, metal oxides, etc.) directly coupled with the cathode can further improve the performance of Li_2S batteries. Nanostructured functional materials can undertake more polysulfides to physically/chemically limit their dissolution shuttle and shorten the ion/electron diffusion length, thereby promoting redox electrochemistry and accelerating electrochemical reaction kinetics. The Li atoms of polysulfide lithium have a strong anchoring characteristic with the S atoms of metal sulfides, metal sulfides are usually metals or semimetals, their interface active sites can promote the conversion of polysulfides through the adsorption effect on the substrate, thereby inhibiting the migration of polysulfides. Metal oxides also have a strong anchoring effect on polysulfide lithium, but metal oxides generally have poor conductivity, so they are often combined with carbon materials to form composite materials to enhance the performance of the battery. Functional organic materials are also applied to lithium–sulfur batteries due to their high theoretical capacity, sustainability, and low cost. In order to further improve the multi-electron reduction reaction of Li_2S, Fu's team used dimethyl trisulfide as the cathode material [67], by forming $LiSCH_3$ (two kinds of thiomethanol lithium) and Li_2S to enhance its redox reaction in rechargeable lithium batteries. The liquid dimethyl trisulfide (DMTS) material forms a cathode electrolyte with the electrolyte, showing a 4e$^-$ reduction process (Fig. 5.8a) during the redox reaction process, with a theoretical capacity of 849 mAh g^{-1}, and the current collector is composed of binder-free multi-walled carbon nanotube (MWCNT) paper (Fig. 5.8a).

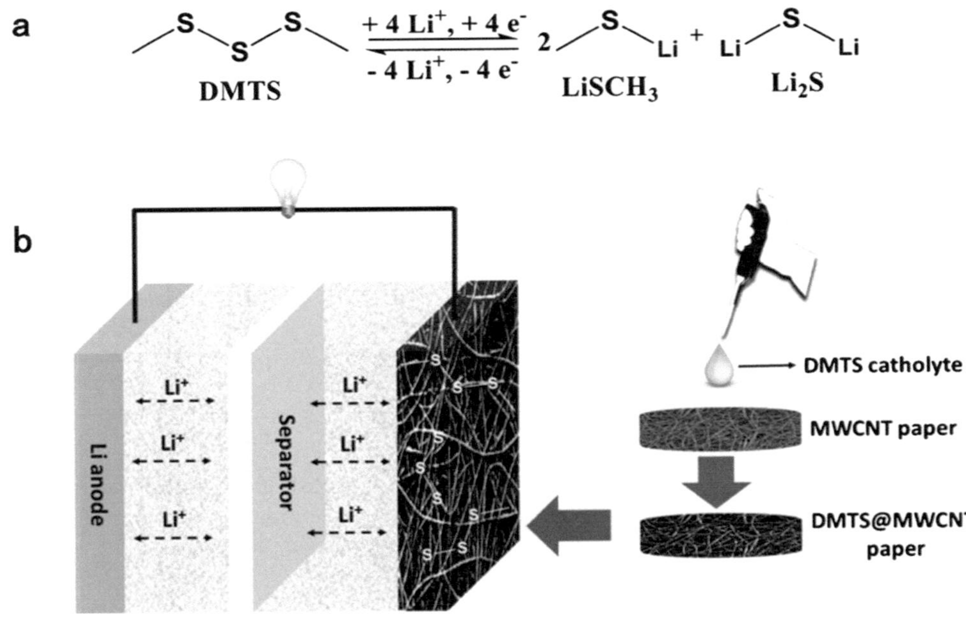

Fig. 5.8 **a** Redox reaction of DMTS in lithium batteries [67]. **b** Schematic diagram of the battery structure and electrolyte with added DMTS on multi-walled carbon nanotube (MWCNT) paper [67]

5.5 Conclusion

With the rapid development of lithium–sulfur battery technology, considerable progress has been made in the development of new sulfur cathode materials. Carbon nanotubes, graphene, porous carbon materials, metal compounds, and polymers have played a prominent role due to their unique physical and chemical structures. The excellent design of cathode material structure ensures the superior performance of these sulfur host materials. Under high sulfur load, these sulfur host materials have their own advantages and disadvantages. Although lithium–sulfur batteries have made some progress, there is still a long way to go to realize them in real life. Therefore, to achieve the practical application of lithium–sulfur batteries, many problems and challenges need to be solved. We need to make more efforts to reduce costs, improve battery capacity, sulfur utilization, cycle stability, and rate performance. Therefore, the manufacture of batteries with high sulfur utilization, high sulfur load, high energy density, and high cycle performance is still an urgent problem.

Li_2S cathode batteries have also made some good progress, but there is still a long way to go to use them in real life. We should focus more efforts on the research of high energy density full batteries, not just the research of Li_2S cathode materials. Therefore, to achieve the practical application of Li_2S batteries, some challenges need to be solved. Firstly, Li_2S active materials are highly sensitive to moisture and need to be synthesized in an inert atmosphere, resulting in high costs, so we need to explore more cost-effective Li_2S synthesis methods. Secondly, Li_2S needs to be paired with a non-lithium metal anode, choices include graphite, silicon, and tin. The low weight capacity of graphite will undoubtedly reduce the energy density of Li_2S batteries. Silicon, due to its high capacity, is considered an ideal anode to match with Li_2S batteries. However, the poor cycle performance of silicon will damage the life of Li_2S batteries, and improving the cycle performance of silicon is closely related to the practical application of Li_2S batteries. Finally, the mass load of Li_2S batteries should be increased to 6.0 mg cm^{-2}, which puts higher requirements on conductive additives and current collectors. Therefore, further research is needed on Li_2S batteries.

References

1. Manthiram A, Chung SH, Zu CX (2015) Lithium–sulfur batteries: progress and prospects. Adv Mater 27(12):1980–2006
2. Ji XL, Lee KT, Nazar LF (2009) A highly ordered nanostructured carbon-sulphur cathode for lithium–sulphur batteries. Nat Mater 8(6):500–506
3. Xiang YY, Lu LQ, Kottapalli AGP et al (2022) Status and perspectives of hierarchical porous carbon materials in terms of high-performance lithium–sulfur batteries. Carbon Energy 4(3):346–398
4. Guo DY, Chen XA, Wei HF et al (2017) Controllable synthesis of highly uniform flower-like hierarchical carbon nanospheres and their application in high performance lithium–sulfur batteries. J Mater Chem A 5(13):6245–6256
5. Sahore R, Levin BDA, Pan M et al (2016) Design principles for optimum performance of porous carbons in lithium–sulfur batteries. Adv Energy Mater 6(14):1600134
6. Zhao CY, Liu LJ, Zhao HL et al (2014) Sulfur-infiltrated porous carbon microspheres with controllable multi-modal pore size distribution for high energy lithium–sulfur batteries. Nanoscale 6(2):882–888

7. Hoffmann C, Thieme S, Bruckner J et al (2014) Nanocasting hierarchical carbide-derived carbons in nanostructured opal assemblies for high-performance cathodes in lithium–sulfur batteries. ACS Nano 8(12):12130–12140
8. Zhu L, Zhu W, Cheng X-B et al (2014) Cathode materials based on carbon nanotubes for high-energy-density lithium–sulfur batteries. Carbon 75:161–168
9. Xu GY, Ding B, Nie P et al (2014) Hierarchically porous carbon encapsulating sulfur as a superior cathode material for high performance lithium–sulfur batteries. ACS Appl Mater Interfaces 6(1):194–199
10. To JWF, Chen Z, Yao HB et al (2015) Ultrahigh surface area three-dimensional porous graphitic carbon from conjugated polymeric molecular framework. ACS Cent Sci 1(2):68–76
11. Lyu Z, Xu D, Yang L et al (2015) Hierarchical carbon nanocages confining high-loading sulfur for high-rate lithium–sulfur batteries. Nano Energy 12:657–665
12. Gong Z, Wu Q, Wang F et al (2016) A hierarchical micro/mesoporous carbon fiber/sulfur composite for high-performance lithium–sulfur batteries. RSC Adv 6(44):37443–37451
13. Strubel P, Althues H, Kaskel S (2016) Zinc-salt templating of hierarchical porous carbons for low electrolyte high energy lithium–sulfur batteries (LE-LiS). Carbon 107:705–710
14. Wu H, Mou J, Zhou L et al (2016) Cloud cap-like, hierarchically porous carbon derived from mushroom as an excellent host cathode for high performance lithium–sulfur batteries. Electrochim Acta 212:1021–1030
15. Chen M, Jiang S, Cai S et al (2017) Hierarchical porous carbon modified with ionic surfactants as efficient sulfur hosts for the high-performance lithium–sulfur batteries. Chem Eng J 313:404–414
16. Liu X, Huang W, Wang D et al (2017) A nitrogen-doped 3D hierarchical carbon/sulfur composite for advanced lithium sulfur batteries. J Power Sour 355:211–218
17. Zhang L, Zhao W, Yuan S et al (2021) Engineering the morphology/porosity of oxygen-doped carbon for sulfur host as lithium–sulfur batteries. J Energy Chem 60:531–545
18. Wu P, Chen L-H, Xiao S-S et al (2018) Insight into the positive effect of porous hierarchy in S/C cathodes on the electrochemical performance of Li–S batteries. Nanoscale 10(25):11861–11868
19. Zhang H, Gao Q, Qian W et al (2018) Binary hierarchical porous graphene/pyrolytic carbon nanocomposite matrix loaded with sulfur as a high-performance Li–S battery cathode. ACS Appl Mater Interfaces 10(22):18726–18733
20. Zhao S, Li C, Wang W et al (2013) A novel porous nanocomposite of sulfur/carbon obtained from fish scales for lithium–sulfur batteries. J Mater Chem A 1(10):3334–3339
21. Guo D, Chen XA, Wei H et al (2017) Controllable synthesis of highly uniform flower-like hierarchical carbon nanospheres and their application in high performance lithium–sulfur batteries. J Mater Chem A 5(13):6245–6256
22. Deng W, Hu A, Chen X et al (2016) Sulfur-impregnated 3D hierarchical porous nitrogen-doped aligned carbon nanotubes as high-performance cathode for lithium–sulfur batteries. J Power Sour 322:138–146
23. Wang M, Zhang H, Zhou W et al (2016) Rational design of a nested pore structure sulfur host for fast Li/S batteries with a long cycle life. J Mater Chem A 4(5):1653–1662
24. Zeng S-Z, Zeng X, Tu W et al (2017) Facile and tailored synthesis of ultrahigh-surface-area clews of carbon nanobelts for high-rate lithium–sulfur batteries. J Mater Chem A 5(44):23209–23220
25. Qin X, Wu J, Xu Z-L et al (2019) Electrosprayed multiscale porous carbon microspheres as sulfur hosts for long-life lithium–sulfur batteries. Carbon 141:16–24
26. Yang M, Hu X, Fang Z et al (2017) Bifunctional MOF-derived carbon photonic crystal architectures for advanced Zn–Air and Li–S batteries: highly exposed graphitic nitrogen matters. Adv Func Mater 27(36):1701971
27. Yan Y, Wei Y, Zhao C et al (2019) A modified template-removal process to improve the specific surface area and hierarchical porosity of carbon materials. J Solid State Chem 269:24–29
28. Zhang J, You C, Wang J et al (2019) Confinement of sulfur species into heteroatom-doped, porous carbon container for high areal capacity cathode. Chem Eng J 368:340–349
29. Zhang Y, Li G, Wang J et al (2020) Hierarchical defective Fe_3-xC@C hollow microsphere enables fast and long-lasting lithium–sulfur batteries. Adv Func Mater 30(22):2001165
30. Novoselov KS, Geim AK, Morozov SV et al (2004) Electric field effect in atomically thin carbon films. Science 306(5696):666–669
31. Li Y, Guan Q, Cheng J et al (2020) Ultrafine nanosulfur particles sandwiched in little oxygen-functionalized graphene layers as cathodes for high rate and long-life lithium–sulfur batteries. Nanotechnology 31(24):245404
32. Zhou L, Lin X, Huang T et al (2014) Binder-free phenyl sulfonated graphene/sulfur electrodes with excellent cyclability for lithium sulfur batteries. J Mater Chem A 2(14):5117–5123
33. Qiu Y, Li W, Zhao W et al (2014) High-rate, Ultralong cycle-life lithium/sulfur batteries enabled by nitrogen-doped graphene. Nano Lett 14(8):4821–4827
34. Cai W, Zhou J, Li G et al (2016) B, N–Co-doped graphene supported sulfur for superior stable Li–S half cell and Ge–S full battery. ACS Appl Mater Interfaces 8(41):27679–27687
35. Ding B, Yuan C, Shen L et al (2013) Chemically tailoring the nanostructure of graphene nanosheets to confine sulfur for high-performance lithium–sulfur batteries. J Mater Chem A 1(4):1096–1101
36. Lin T, Tang Y, Wang Y et al (2013) Scotch-tape-like exfoliation of graphite assisted with elemental sulfur and graphene–sulfur composites for high-performance lithium–sulfur batteries. Energy Environ Sci 6(4):1283–1290
37. Yeon JS, Yun S, Park JM et al (2019) Surface-modified sulfur nanorods immobilized on radially assembled open-porous graphene microspheres for lithium–sulfur batteries. ACS Nano 13(5):5163–5171
38. Tang H, Yang J, Zhang G et al (2018) Self-assembled N-graphene nanohollows enabling ultrahigh energy density cathode for Li–S batteries. Nanoscale 10(1):386–395
39. Wang Z, Dong Y, Li H et al (2014) Enhancing lithium–sulphur battery performance by strongly binding the discharge products on amino-functionalized reduced graphene oxide. Nat Commun 5(1):5002
40. Xie Y, Meng Z, Cai T et al (2015) Effect of boron-doping on the graphene aerogel used as cathode for the lithium–sulfur battery. ACS Appl Mater Interfaces 7(45):25202–25210
41. Shi P, Wang Y, Liang X et al (2018) Simultaneously exfoliated boron-doped graphene sheets to encapsulate sulfur for applications in lithium–sulfur batteries. ACS Sustain Chem Eng 6(8):9661–9670
42. Chen L, Feng J, Zhou H et al (2017) Hydrothermal preparation of nitrogen, boron co-doped curved graphene nanoribbons with high dopant amounts for high-performance lithium sulfur battery cathodes. J Mater Chem A 5(16):7403–7415
43. Xu J, Su D, Zhang W et al (2016) A nitrogen–sulfur co-doped porous graphene matrix as a sulfur immobilizer for high performance lithium–sulfur batteries. J Mater Chem A 4(44):17381–17393
44. Liang X, Garsuch A, Nazar LF (2015) Sulfur cathodes based on conductive MXene Nanosheets for high-performance lithium–sulfur batteries. Angew Chem Int Ed 54(13):3907–3911

45. Yao Y, Feng W, Chen M et al (2018) Boosting the electrochemical performance of Li–S batteries with a dual polysulfides confinement strategy. Small 14(42):1802516
46. Tang H, Li W, Pan L et al (2018) In situ formed protective barrier enabled by Sulfur@Titanium Carbide (MXene) ink for achieving high-capacity, long lifetime Li–S batteries. Adv Sci 5(9):1800502
47. Bao W, Liu L, Wang C et al (2018) Facile synthesis of crumpled nitrogen-doped MXene Nanosheets as a new sulfur host for lithium–sulfur batteries. Adv Energy Mater 8(13):1702485
48. Chen Z, Yang X, Qiao X et al (2020) Lithium-ion-engineered interlayers of V_2C MXene as advanced host for flexible sulfur cathode with enhanced rate performance. J Phys Chem Lett 11(3):885–890
49. Liang X, Rangom Y, Kwok CY et al (2017) Interwoven MXene nanosheet/carbon-nanotube composites as Li–S cathode hosts. Adv Mater 29(3):1603040
50. Lv L-P, Guo C-F, Sun W et al (2019) Strong surface-bound sulfur in carbon nanotube bridged hierarchical Mo2C-based MXene nanosheets for lithium–sulfur batteries. Small 15(3):1804338
51. Bao W, Xie X, Xu J et al (2017) Confined sulfur in 3D MXene/reduced graphene oxide hybrid nanosheets for lithium–sulfur battery. Chem-A Eur J 23(51):12613–12619
52. Bao W, Su D, Zhang W et al (2016) 3D Metal Carbide@ Mesoporous carbon hybrid architecture as a new polysulfide reservoir for lithium–sulfur batteries. Adv Func Mater 26(47):8746–8756
53. Zhang Y, Tang W, Zhan R et al (2019) An N-doped porous carbon/MXene composite as a sulfur host for lithium–sulfur batteries. Inorg Chem Front 6(10):2894–2899
54. Pan H, Huang X, Zhang R et al (2019) Titanium oxide-Ti_3C_2 hybrids as sulfur hosts in lithium–sulfur battery: fast oxidation treatment and enhanced polysulfide adsorption ability. Chem Eng J 358:1253–1261
55. Shi H, Qin J, Lu P et al (2021) Interfacial engineering of bifunctional niobium (V)-based heterostructure nanosheet toward high efficiency lean-electrolyte lithium–sulfur full batteries. Adv Func Mater 31(28):2102314
56. Peng L, Wei Z, Wan C et al (2020) A fundamental look at electrocatalytic sulfur reduction reaction. Nat Catal 3(9):762–770
57. Cai W, Li G, Luo D et al (2018) The dual-play of 3D conductive scaffold embedded with Co, N Codoped hollow polyhedra toward high-performance Li–S full cell. Adv Energy Mater 8(34):1802561
58. Hou R, Zhang S, Zhang P et al (2020) Ti_3C_2 MXene as an "energy band bridge" to regulate the heterointerface mass transfer and electron reversible exchange process for Li–S batteries. J Mater Chem A 8(47):25255–25267
59. Chu R, Nguyen TT, Bai Y et al (2022) Uniformly controlled treble boundary using enriched adsorption sites and accelerated catalyst cathode for robust lithium–sulfur batteries. Adv Energy Mater 12(9):2102805
60. Yao W, Zheng W, Xu J et al (2021) ZnS–SnS@NC heterostructure as robust Lithiophilicity and Sulfiphilicity mediator toward high-rate and long-life lithium-sulfur batteries. ACS Nano 15(4):7114–7130
61. Yao W, Xu J, Cao Y et al (2022) Dynamic intercalation-conversion site supported ultrathin 2D mesoporous $SnO_2/SnSe_2$ hybrid as bifunctional polysulfide immobilizer and lithium regulator for lithium–sulfur chemistry. ACS Nano 16(7):10783–10797
62. Zhao C, Xu G-L, Yu Z et al (2021) A high-energy and long-cycling lithium–sulfur pouch cell via a macroporous catalytic cathode with double-end binding sites. Nat Nanotechnol 16(2):166–173
63. Cai J, Jin J, Fan Z et al (2020) 3D printing of a V_8C_7–VO_2 bifunctional scaffold as an effective polysulfide immobilizer and lithium stabilizer for Li–S batteries. Adv Mater 32(50):2005967
64. Yang Y, Zheng G, Misra S et al (2012) High-capacity micrometer-sized Li_2S particles as cathode materials for advanced rechargeable lithium-ion batteries. J Am Chem Soc 134(37):15387–15394
65. Fan FY, Carter WC, Chiang Y-M (2015) Mechanism and kinetics of Li_2S precipitation in lithium–sulfur batteries. Adv Mater 27(35):5203–5209
66. Lin Z, Liu Z, Fu W et al (2013) Phosphorous Pentasulfide as a novel additive for high-performance lithium–sulfur batteries. Adv Func Mater 23(8):1064–1069
67. Wu M, Cui Y, Bhargav A et al (2016) Organotrisulfide: a high capacity cathode material for rechargeable lithium batteries. Angew Chem Int Ed 55(34):10027–10031

6 Lithium Manganese Oxide Cathode

Cobalt-based and nickel-based cathode materials, such as lithium cobaltate (LCO) and nickel cobalt manganese ternary cathode (NCM), can achieve high specific capacity. However, the toxicity, high cost, and complex supply chain of cobalt and nickel hinder their application as the main cathode materials for lithium-ion batteries. For the next-generation lithium-ion batteries (LIBs), it is important to use lower-cost and safer cathode materials. Therefore, manganese-based cathode materials have emerged. Spinel lithium manganese oxide (LMO) is an economical and efficient LIB cathode material with fast Li^+ diffusion kinetics. However, manganese-based cathode materials are often affected by the Jahn-Teller effect and are easily corroded by hydrofluoric acid produced by electrolyte hydrolysis. These existing problems hinder the practical application of lithium manganese oxide cathode materials. It is important to analyze and summarize the research on lithium manganese oxide cathodes. Thus, this section mainly reviews the structure, preparation, existing problems, material modification, and electrolyte additives of spinel lithium manganese oxide cathodes, and proposes insights for future research on lithium manganese oxide cathodes.

6.1 Structure of Lithium Manganese Oxide Cathode

Lithium manganese oxide is divided into spinel structure lithium manganese oxide ($LiMn_2O_4$) and layered structure lithium manganese oxide ($LiMnO_2$). Spinel-type $LiMn_2O_4$ was first synthesized by Hunter in 1981, with an Fd-3 m space group, and its theoretical specific capacity is 148 mAh/g. As shown in Fig. 6.1, in the crystal structure of spinel $LiMn_2O_4$, Li ions fill one-eighth of the tetrahedral sites (8a), Mn ions fill half of the octahedral sites (16d), and O ions fill 32e sites, forming a cubic close-packed array. In $LiMn_2O_4$, the interstitial octahedra 16c sites provide a three-dimensional pathway for Li^+ diffusion, allowing Li^+ to jump from 8a to 16c and then to 8a [1]. The spinel lithium manganese oxide cathode is a promising lithium-ion cathode material because it has good structure and cost-effectiveness, and can achieve fast lithium diffusion kinetics. However, $LiMn_2O_4$ suffers from severe Mn dissolution due to the Jahn-Teller distortion and degeneracy of Mn^{3+}, resulting in poor cycling performance of lithium manganese oxide. The poor cycle stability and low capacity limit the application of lithium manganese oxide cathodes. Therefore, it has not been widely commercialized to date, and its market share is much lower than that of NCM and $LiFePO_4$ cathode materials. The layered structure of lithium manganese oxide ($LiMnO_2$) is a non-thermodynamically stable phase, which is difficult to synthesize under equilibrium conditions, and its cyclic stability is extremely poor, making it difficult to apply in practice. Therefore, this chapter mainly introduces the spinel structure lithium manganese oxide cathode material.

6.2 Preparation of Lithium Manganese Oxide Cathode

The natural Mn resources are abundant, and the spinel $LiMn_2O_4$ is easy to synthesize. The preparation methods mainly include high-temperature solid-phase method, microwave synthesis method, molten salt method, sol-gel method, hydrothermal method, co-precipitation method, etc. The synthesis method and process directly affect the micro-morphology and crystallinity of $LiMn_2O_4$ materials, so it is often necessary to optimize the synthesis process and simplify the synthesis process in industry to further obtain electrode materials with superior electrochemical performance.

(1) The high-temperature solid-phase method is a common method for preparing spinel-type $LiMn_2O_4$, which is obtained by mixing lithium salt and manganese oxide

X. Li et al., *Electrode Development for Lithium-ion Batteries*, https://doi.org/10.1007/978-981-96-3021-9_6

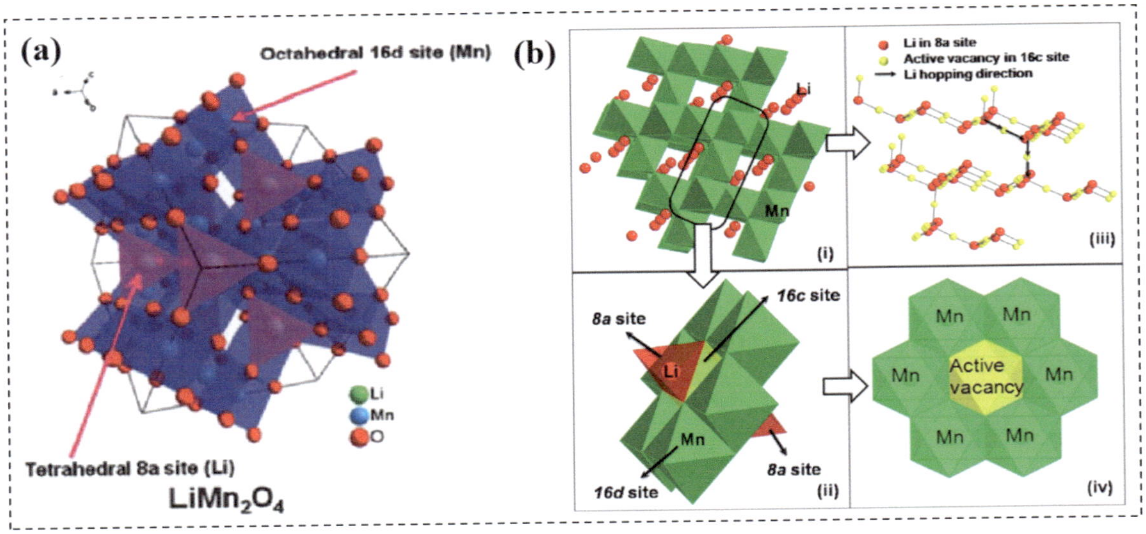

Fig. 6.1 **a** Crystal structure of $LiMn_2O_4$ (LMO); **b** Crystal structure of $LiMn_2O_4$ (LMO) [2, 3]

in a certain ratio and sintering at high temperature for a certain time. During the preparation process of the high-temperature solid-phase method, the sintering temperature, cooling rate, and raw material characteristics have a great impact on the electrochemical performance of the final product. The most critical steps of the solid-phase method are uniform material mixing and heat treatment. The high-temperature solid-phase method has simple preparation steps and can meet industrial production standards, so it is widely used in industrial production.

(2) The microwave synthesis method is to cause the internal movement of the material through microwaves, causing molecules and atoms to polarize and collide with each other to produce violent reactions and rapidly increase the temperature of the material. The microwave heating method has the characteristics of fast heating rate and high energy utilization rate. The microwave-assisted synthesis method is fast and convenient, but compared with other methods, it has defects such as lower capacity, cycle stability and unstable electrochemical performance. Usually, the microwave synthesis method is combined with other methods to prepare lithium manganese oxide cathode materials with finer and more uniform grains, which is the most promising aspect.

(3) The molten salt method is similar to the high-temperature solid-phase method. It heats the low-melting-point lithium salt and flux to the melting point. The molten lithium salt penetrates into the pores of manganese oxide for full contact, and the spinel $LiMn_2O_4$ is obtained by sintering at high temperature, and the excess salt is removed by water washing. This method increases the contact area between the precursors, enhances the heterogeneity of the solid lithium salt and manganese oxide, and accelerates the progress of the lithiation reaction. However, this method is relatively complex, and the experimental parameters are not easy to control, making it difficult to achieve industrial controllable preparation.

(4) The sol-gel method involves dissolving a certain proportion of manganese source and lithium salt using a complexing agent, then solidifying through sol-gel, and finally obtaining lithium manganese oxide cathode powder through sufficient drying and sintering at relatively low temperatures. The sol-gel method is the most commonly used method to obtain ultrafine powder. The sol-gel technique has the advantages of simple equipment, easy process control, low sintering temperature, and the ability to produce materials with high chemical uniformity and high chemical purity. However, it has disadvantages such as a long preparation cycle, relatively complex synthesis process, high cost, and difficulty in industrial production.

(5) The hydrothermal method involves mixing a manganese salt and lithium salt solution and synthesizing $LiMn_2O_4$ under high temperature and high pressure conditions. Therefore, the hydrothermal method requires a chemical reaction to be completed in a sealed hydrothermal reaction kettle. The hydrothermal method has the advantages of low synthesis temperature and more controllable morphology, but the reaction conditions are harsh and not conducive to industrial production.

(6) The co-precipitation method involves adding a precipitant to a solution of various soluble lithium salts and manganese salts, co-precipitating at a certain pH value, and then filtering, washing, drying, and calcining to obtain spinel-type $LiMn_2O_4$ cathode. The particles of the

Fig. 6.2 **a** FE-SEM image of 3-D LMO single-crystalline samples; **b** SEM image of $LiMn_2O_4$ prepared by microwave heating method; **c** SEM image of $LiMn_2O_4$ compounds prepared at 800 °C (The scale bars are 1 mm for all the compounds), **d** SEM image of LMO; **e** SEM image of $LiMn_2O_4$ at 120 °C hydrothermal reaction temperatures; **f** FESEM image of $Li_{1.1}Mn_2O_4$ powders calcined at 870 °C for 15 h [4–9]

$LiMn_2O_4$ cathode material prepared by the co-precipitation method are uniform and nanoscale, with short synthesis time, low energy consumption, and good yield.

In addition to the above methods, there are unconventional synthesis methods such as electrospinning, carbon combustion, and ultrasonic spray pyrolysis (Fig. 6.2).

6.3 Problems with Lithium Manganese Oxide Cathodes

Spinel $LiMn_2O_4$ is one of the ideal cathode materials for lithium-ion batteries, with broad development prospects, but its current market share is relatively small. This is largely related to the unstable cycling performance and poor high-temperature performance of the current $LiMn_2O_4$ cathode material. The $LiMn_2O_4$ cathode has problems such as the Jahn-Teller effect, oxygen defects, Mn dissolution, and electrolyte decomposition during the charge and discharge process, which cause a significant decrease in its capacity and reduce its cycle life [10].

6.3.1 Jahn-Teller Effect

During the electrochemical process, the structural decay from cubic phase $LiMn_2O_4$ to tetragonal phase $Li_2Mn_2O_4$ occurs along with the Jahn-Teller effect (Fig. 6.3a). At the same time, the disproportionation reaction of manganese caused by the Jahn-Teller effect causes some manganese to dissolve in the electrolyte in the form of Mn^{2+}. $LiMn_2O_4$ contains two types of molar ratios of Mn, Mn^{3+} and Mn^{4+}. Mn^{3+}, due to its high spin state and large magnetic moment, causes the octahedral structure formed by the surrounding oxygen atoms to elongate, transforming from a cubic phase to a tetragonal phase (as shown in Fig. 6.3b). The distorted tetragonal crystal system is incompatible with the cubic crystal system inside the particles, destroying the integrity of the structure, thereby hindering the transmission of reversible lithium ions and thus deteriorating the cycling performance of the material. At the beginning of discharge, the manganese ions in lithium manganate exist as Mn^{4+}. During the discharge process, Mn^{4+} is reduced to Mn^{3+}. As the discharge process continues, the number of Mn^{3+} increases, especially when the discharge is excessive, more tetravalent manganese ions Mn^{4+} in the material are reduced to Mn^{3+}. At this time, the average valence state of the manganese ions in the lithium manganate cathode material is less than +3.5, which exacerbates the impact of the Jahn-Teller effect, causing serious effects on the specific capacity and cycle stability of the lithium manganate cathode material.

6.3.2 Dissolution of Manganese and Erosion of Electrolyte

The small amount of water present in the electrolyte reacts with $LiPF_6$ in the electrolyte to produce HF, causing

Fig. 6.3 .

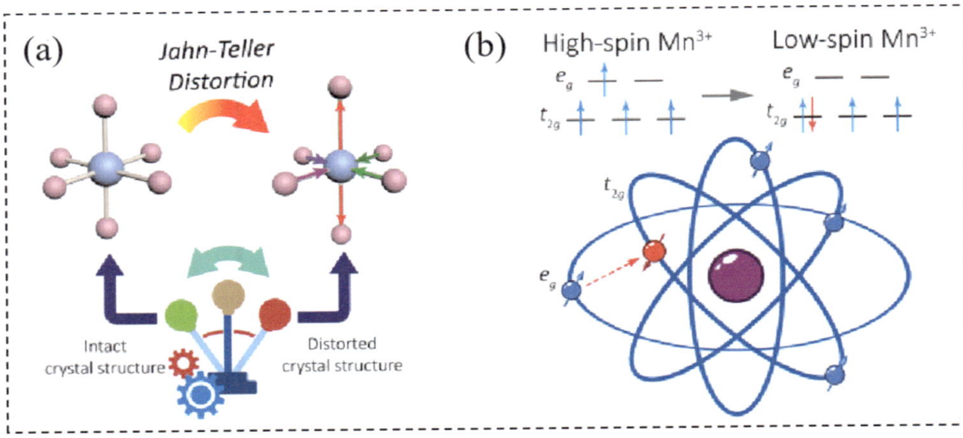

$LiMn_2O_4$ to undergo a disproportionation reaction to produce Mn^{2+} dissolved in the electrolyte, and causing damage to the spinel structure. The Mn^{2+} dissolved in the electrolyte is reduced to metallic Mn on the surface of the graphite anode, catalyzing the decomposition of the SEI film and damaging the anode interface. Some Mn blocks the lithium intercalation channels in graphite, and some Mn even deposits on the interface between the Cu foil and the anode coating, causing the anode to peel off. These factors all lead to the capacity decay of the $LiMn_2O_4$ battery, and even rapid capacity decay. The dissolution and loss of manganese is one of the direct causes of the capacity decay of the $LiMn_2O_4$ cathode material. Especially at high temperatures (above 50 °C), the dissolution rate of manganese will accelerate, and the higher the temperature, the more serious the dissolution loss.

The electrolyte decomposes at high potentials and forms a Li_2CO_3 film on the surface of $LiMn_2O_4$, increasing the polarization of the battery and thus causing the capacity decreases of the spinel $LiMn_2$ O_4 during the cycling process. At the same time, the residual lithium on the surface and the generated Li_2CO_3 will also undergo side reactions with the electrolyte, oxidizing and releasing gas during the charging and discharging process. In addition, the moisture in the electrode will hydrolyze with the electrolyte component $LiPF_6$ to produce highly corrosive HF, which ultimately leads to surface corrosion of the cathode material and dissolution of manganese ions, and even causes battery safety issues (as shown in Fig. 6.4). On the other hand, at high temperatures, the $LiMn_2O_4$ material has a certain catalytic effect on the electrolyte, causing the spinel to lose oxygen, which is also one of the reasons for the poor high-temperature electrochemical performance of the $LiMn_2O_4$ battery [11].

6.4 Research on the Modification of Lithium Manganese Oxide Cathode

Despite the many advantages of spinel $LiMn_2O_4$, issues such as the Jahn-Teller effect and the dissolution of Mn lead to structural collapse of $LiMn_2O_4$ during long-term cycling, causing continuous capacity decay, which severely restricts its lifespan as a type battery material. To address these issues, common modification methods such as bulk doping, surface coating, morphology control, and electrolyte optimization are used to improve the performance of $LiMn_2O_4$.

6.4.1 Bulk Doping

Bulk doping is a method of introducing a certain amount of other elements into the lattice of the material to improve its electrochemical performance. It is the most commonly used modification method for $LiMn_2O_4$ cathodes. Doping modification has the advantages of being simple and effective, usually not requiring additional operation steps and equipment. For the doping modification of $LiMn_2O_4$ material, it can be divided into cation doping, anion doping, and multi-ion composite co-doping, among which cation doping is the most used. Cation doping can replace Mn ions in the lattice, inhibit the generation of oxygen defects during the synthesis process, and inhibit the Jahn-Teller effect and reduce Mn dissolution during the charging and discharging process, which is beneficial for the cathode material to maintain structural stability. The main cation doping modifications include Ni^{2+}, Cr^{2+}, Mg^{2+}, Co^{3+}, Al^{3+}, B^{3+}, Sm^{3+}, La^{3+}, Gd^{3+}, etc. [13–18]. The most studied anion doping is F^- and S^{2-}, mainly by doping oxygen sites to replace

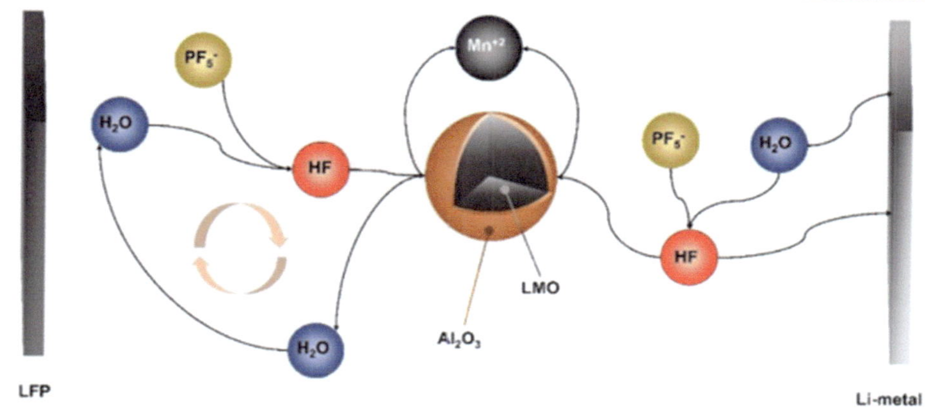

$$2\ Mn^{3+} \rightarrow Mn^{2+}(l) + Mn^{4+}(s) \quad (1)$$

$$LiPF_6\ (s) \rightleftharpoons LiF\ (s) + PF_5\ (g) \quad (2)$$

$$H_2O\ (l) + PF_5(g) \rightarrow 2\ HF\ (l) + POF_3\ (g) \quad (3)$$

$$2\ LiMn_2O_4\ (s) + 4\ H^+\ (l) \rightarrow 3\ MnO_2\ (s) + Mn^{2+}\ (l) + 2\ Li^+\ (l) + 2\ H_2O\ (l) \quad (4)$$

$$Al_2O_3\ (s) + 6\ HF\ (l) \rightarrow 2\ AlF_3\ (s) + H_2O\ (l) \quad (5)$$

Fig. 6.4 A schematic illustration of HF as a root cause for Mn-ion dissolution from a pristine LMO as well as Al_2O_3 coated LMO. Al_2O_3 coating in the presence of HF can form H_2O, and consequently H_2O leads to the hydrolysis of $LiPF_6$ to produce more HF. Li-metal counter electrode consumes impurities such as HF and H_2O that lead to Mn dissolution in the electrolyte [12]

some oxygen, reducing the impact of oxygen defects. Composite doping can more comprehensively solve various problems of $LiMn_2O_4$, so it is also a current research hotspot. Doping a large amount of Ni elements can stabilize the structure, and Ni^{2+} also has electrochemical activity. For example, the spinel structure $LiNi_{0.5}Mn_{1.5}O_4$ as a promising cathode material has become one of the research hotspots [19]. As shown in Fig. 6.5a–c, Guo et al. [20] prepared a spinel $LiMn_{1.9-x}Ni_{0.1}B_xO_4$ cathode material with a truncated octahedral morphology by the solid-state combustion method. Due to the Ni–B double doping replacing Mn, the MnO_6 octahedron in $LiMn_2O_4$ shrinks, leading to an increase in the average oxidation state of Mn and the suppression of the Jahn-Teller effect. In addition, with the truncated octahedral morphology, Ni–B co-doping shows preferential growth of the (111) and (100) planes, and the growth of the (110) crystal face is suppressed by Ni–B co-doping, thereby alleviating the dissolution of Mn. As shown in Fig. 6.5d–g, Li et al. [7] prepared a Cu–Al co-doped $Li_{0.94}Al_{0.02}Mn_{1.75}Cu_{0.25}O_4$ cathode using the sol-gel method, with Cu and Al elements doped into the Mn and Li sites of $LiMn_2O_4$ respectively. After doping, the structure stability, ion/electron conductivity of $LiMn_2O_4$ are enhanced, the Jahn-Teller effect is reduced, and it shows excellent electrochemical performance.

6.4.2 Surface Coating

Surface coating modification is achieved by applying a coating to $LiMn_2O_4$. A stable protective layer is formed on the surface of the material through physical or chemical means, which can isolate the body from direct contact with the outside world (water, electrolyte, etc.). Surface coatings generally require the surface protective layer to have a relatively stable chemical structure and good electrical conductivity and electrochemical activity of lithium ion diffusion, which is conducive to the conduction of electrons in the electrode and the diffusion of lithium ions. Surface modification usually includes surface coating, surface body material reconstruction, and washing, etc. Surface coating is the most direct and common surface modification method. The coating materials include metal oxide coatings (MgO, Al_2O_3, ZrO_2, TiO_2, etc.), non-oxide coatings (phosphates, fluorides, carbon materials, etc.) [21, 22], and fast ion conductor coatings, etc. [23], among which oxide coatings are

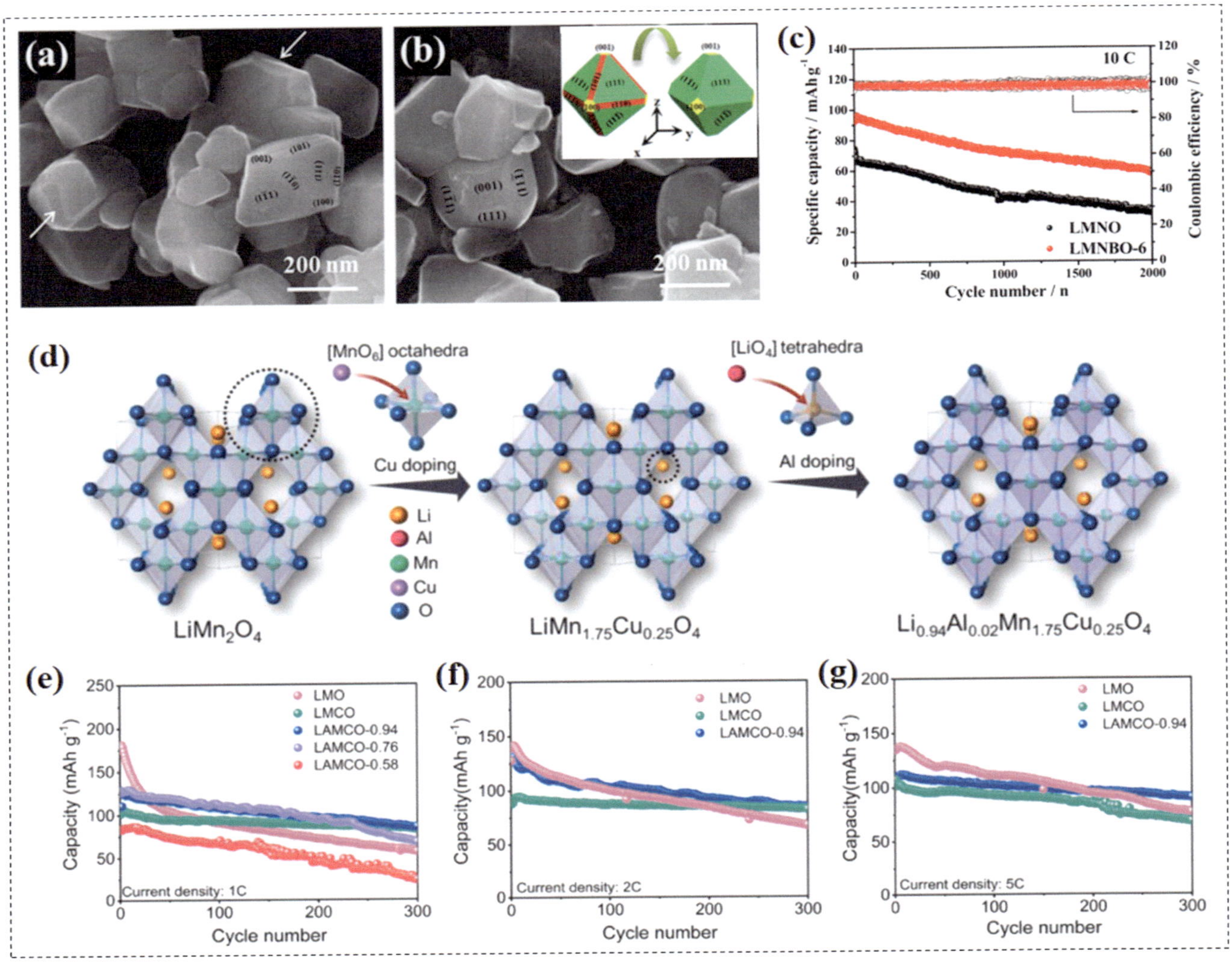

Fig. 6.5 SEM images of **a** LMNO and **b** LMNBO-6, **c** schematic illustrations of crystal planes transformation; **d** schematic illustration of crystal structure evolution from LMO to LAMCO-0.94; **e** cyclic performance of LMO, LMCO and LAMCO at a current density of 1 C. Cyclic performance of LMO, LMCO and LAMCO-0.94 at current densities of **f** 2 C and **g** 5 C

the most common. Mn dissolution is one of the main reasons for the poor high-temperature performance of LMO. Coating a layer on the surface of $LiMn_2O_4$ that can conduct Li^+ interface layer and mechanically isolate the electrolyte from $LiMn_2O_4$ contact can improve the high-temperature storage and cycling performance of $LiMn_2O_4$. As shown in Fig. 6.6a–c, Wang et al. [24] added $WO_{2.72}$ as an oxygen vacancy inducer to $LiMn_2O_4$ and formed a Li_2WO_4 coating on its surface. The results show that $WO_{2.72}$ can induce the formation of oxygen vacancies on the material surface, which is conducive to the adsorption and storage of active oxygen, and reduces the energy barrier of lithium ion migration. In addition, the formation of the Li_2WO_4 coating helps to eliminate residual lithium and improve the interface between the electrode and the electrolyte. Therefore, $WO_{2.72}$ modified $LiMn_2O_4$ has a more stable bulk structure and a better electrode–electrolyte contact interface, thus showing superior electrochemical performance. As shown in Fig. 6.6d, e, Bai et al. [22] modified commercial LMO with saccharin (SAC) by hydrothermal method in this paper. SAC modified LMO shows significantly improved rate performance and cycle stability, improves the stability of the spinel structure during the cycle, and inhibits the dissolution of Mn ions.

6.4.3 Morphology Control

For $LiMn_2O_4$, the morphology control of materials is achieved by changing the structure to make $LiMn_2O_4$ more rapid and effective in electrochemical reactions, thereby improving performance. For $LiMn_2O_4$, the size and shape of particles have a significant impact on the electronic conductivity of the material and the diffusion migration of

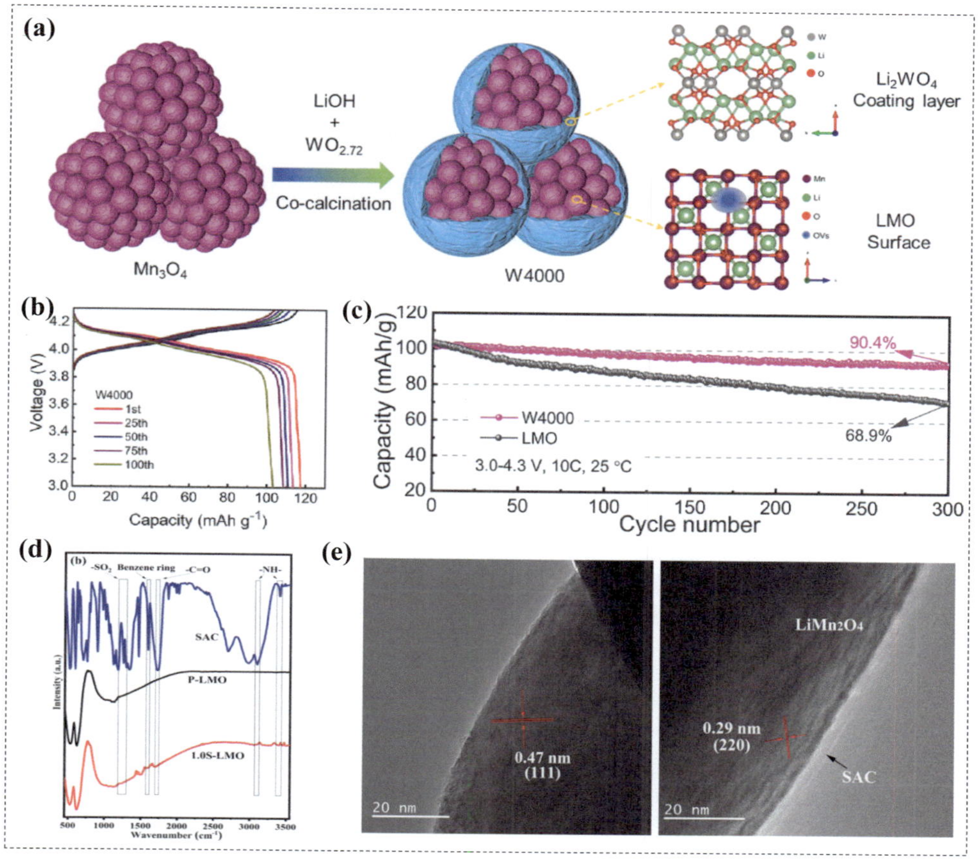

Fig. 6.6 **a** Schematic of oxygen-vacancy formation; **b** charge–discharge curves of W4000; **c** long-term cycling performance of the bare LMO and W4000 at 10 C; **d** FTIR spectra of SAC, P-LMO, and 1.0S-LMO; **e** TEM images of P-LMO and 1.0S-LMO

lithium ions. Different precursors and synthesis conditions can yield various polycrystalline LMOs, with common crystal morphologies being octahedrons and polyhedrons. Single-crystal morphology $LiMn_2O_4$ cathode materials can reduce the particle surface area to decrease the contact area between particles and electrolyte, thereby reducing the dissolution of Mn in the electrolyte and the occurrence of other side reactions. At the same time, single-crystal particles can achieve higher electrode compaction density, which is beneficial for improving battery energy density. Therefore, the currently best-performing high-end modified $LiMn_2O_4$ are single-crystal particles. In addition, nano-morphology structure $LiMn_2O_4$ cathode materials, such as nanowires, nanospheres, nanorods, nanotubes, etc., these special morphology structures shorten the diffusion path of lithium ions, providing higher current density [25]. As shown in Fig. 6.7, Sun et al. [4] synthesized single-crystal $LiMn_2O_4$ cathodes with three different morphology structures: one-dimensional nanorods, two-dimensional nanosheets, and three-dimensional octahedrons, and studied the impact of morphology structure on the electrochemical performance of $LiMn_2O_4$ cathodes. The results showed that the one-dimensional nanorod structure $LiMn_2O_4$ exhibited the best electrochemical performance. To address the issues of $LiMn_2O_4$ disproportionation and Jahn-Teller distortion leading to structural degradation and capacity decay, Xu et al. [26] designed single-crystal truncated octahedron $LiMn_2O_4$ (TO-LMO) with (111), (100), and (110) facets exposed. The synergistic effect of the single crystal and the exposed crystal orientation significantly reduced the disproportionation of Mn^{3+} ions, thereby improving its structural stability. The results showed that the cycle stability of single-crystal TO-LMO was significantly improved, with a capacity retention rate of 84.3% after 2000 cycles, which is significantly better than the 61.2% of octahedral $LiMn_2O_4$.

6.4.4 Electrolyte Optimization

HF in the electrolyte is the main culprit causing Mn dissolution, and the matching of the electrolyte to $LiMn_2O_4$ performance is crucial. The addition of cathode film-forming additives to the electrolyte, such as 1-(dimethylamino) pyrrole, tri(2,2,2-trifluoroethyl) phosphite, polythiophene, lithium styrene, etc., can form a more stable surface phase, inhibiting the dissolution of Mn [27–30]. It is worth noting that compared with the widely used doping and coating strategies, the development of advanced electrolytes and additives for $LiMn_2O_4$ is relatively less, so there is a lot of room for the development of new liquid electrolytes, which

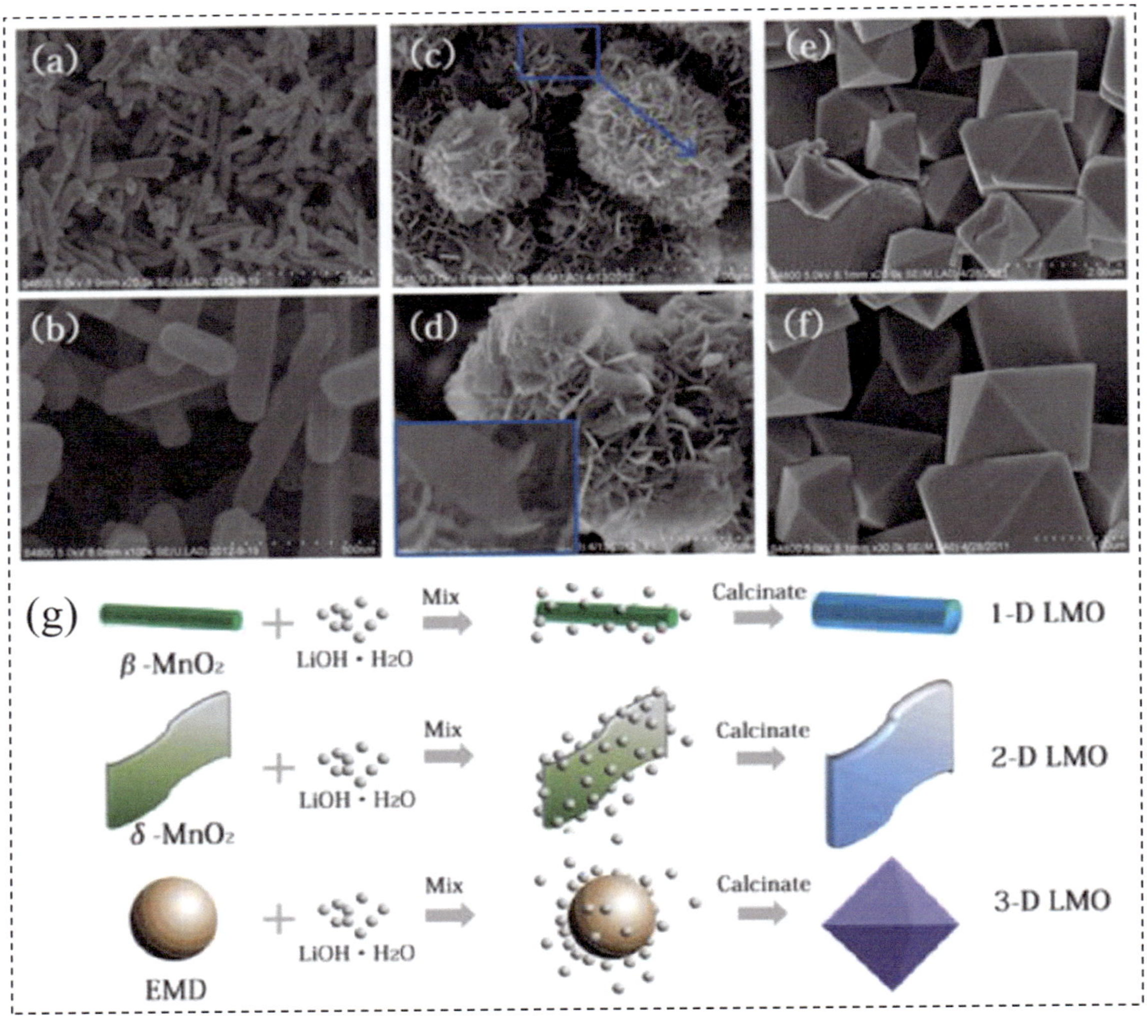

Fig. 6.7 FE-SEM images of samples at different magnifications: **a** and **b** 1-D LMO, **c** and **d** 2-D LMO, **e** and **f** 3-D LMO single-crystalline samples; **g** scheme of the formation mechanism of 1-D, 2-D and 3-D $LiMn_2O_4$ single-crystalline samples

may provide a good solution for the electrochemical stability of spinel $LiMn_2O_4$. The $LiMn_2O_4$ cathode has serious manganese dissolution in the electrolyte, and the dissolved manganese ions can migrate to the anode through the electrolyte and deposit, accelerating capacity decay. It was found that compared with traditional electrolytes, a large amount of manganese loss was found during cycling in traditional electrolytes, and ionic liquids have a significant advantage in suppression. Kar et al. [31] synthesized a new lithium salt, 1,1,1,3,3-(tetra)six-fluoroisopropoxyboronic acid lithium (LiBHfip), whose crystal structure is shown in Fig. 6.8a–d. The oxidative stability of this lithium salt electrolyte in a mixture of ethylene carbonate and dimethyl carbonate is 5.0 V relative to Li^+/Li, significantly improved compared to $LiPF_6$ electrolyte. The assembled $LiMn_2O_4$ cathode battery shows stable cycling performance over 200 cycles, with a capacity retention rate of 90%. Anouti et al. [29] found that Li or Gr anodes prepared with TTFP and FEC have a protective effect against the dissolution of manganese and erosion. The results show that for the surface phenomena of $LiMn_2O_4$, an appropriate proportion of fluorinated additives or a combination of solvents can both avoid the oxidative reactivity of the $LiMn_2O_4$ cathode and prevent its harmful effects on lithium metal or graphite anodes by controlling the solvation of lithium ions.

6.5 Summary and Outlook

With the dramatic increase in the consumption of cobalt and nickel resources and the increasingly expensive trend, spinel lithium manganese oxide, with its low cost and easy

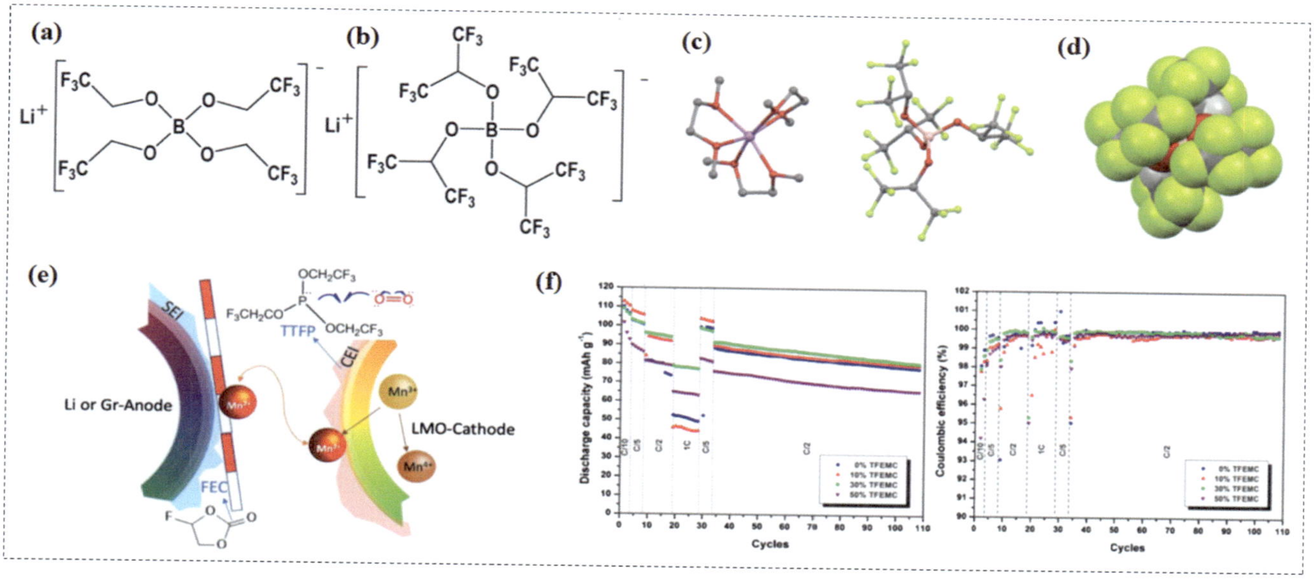

Fig. 6.8 Structure of lithium salts: **a** lithium tetrakis(2,2,2-trifluoroethoxy)borate (LiBTfe) and **b** lithium tetrakis(1,1,1,3,3,3- hexafluoroisopropoxy)borate (LiBHFip). **c** X-ray crystal structure of DME chelated LiBHFip · 3DME salt (gray, red, purple, pale green and pink spheres represent C, O, Li, F and B atoms, respectively; H atoms are omitted for clarity), **d** a space filled representation of the [BHFip]-anion from the crystal structure; **e** Schematic presentation for the protective action of delithiated LMO cathode by TTFP and Li or Gr anode by FEC toward manganese dissolution and attack; **f** Discharge capacities of Gr//LMO cells cycled between 3.0 and 4.7 V at room temperature with $LiPF_6$ 1.5 M in EC/EMC/TFEMC (3/7−x/x) with x = 0, 1, 3, 5. Corresponding coulombic efficiencies

preparation, is expected to make up for the shortcomings of other cathode materials in the market. Spinel lithium manganese oxide has a good structure and can achieve rapid lithium diffusion kinetics. However, $LiMn_2O_4$ is due to the Jahn-Teller distortion and Mn^{3+} degeneracy lead to severe Mn dissolution, resulting in poor cycle life and high-temperature performance of lithium manganate. Current research mainly modifies lithium manganate through doping, surface coating, and morphology control to enhance its structural stability. However, the Mn dissolution and poor high-temperature performance of lithium manganate are both caused by side reactions with the electrolyte, so the optimization of the matching electrolyte is also very important. Therefore, the following aspects are crucial in the research related to spinel lithium manganate: (1) More fundamental research is needed to understand the degradation mechanism of spinel cathodes, especially the main issues such as TM dissolution and surface/interface instability. (2) The further development of lithium manganate should pay special attention to the optimization of the electrolyte and the development of new electrolytes. (3) The combination of new technologies such as machine learning and intelligent synthesis may help optimize the combination of components and various production process parameters.

References

1. Huang Y, Dong Y, Li S et al (2020) Lithium manganese spinel cathodes for lithium-ion batteries. Adv Energy Mater 11(2):2000997
2. Park OK, Cho Y, Lee S et al (2011) Who will drive electric vehicles, olivine or spinel? Energy Environ Sci 4(5):1621
3. Xu B, Meng S (2010) Factors affecting Li mobility in spinel $LiMn_2O_4$—A first-principles study by GGA and GGA+U methods. J Power Sources 195(15):4971–4976
4. Sun Y, Xu C, Li B et al (2014) Synthesis of single-crystalline $LiMn_2O_4$ with different dimensional nanostructures for Li-ion batteries. Int J Electrochem Sci 9(11):6387–6401
5. Liu X, Liu P, Wang H (2019) Preparation of spinel $LiMn_2O_4$ with porous microscopic morphology by simple coprecipitation-microwave synthesis method. Ionics 25(11):5213–5220
6. Zhao X, Reddy MV, Liu H et al (2012) Nano $LiMn_2O_4$ with spherical morphology synthesized by a molten salt method as cathodes for lithium ion batteries. RSC Adv 2(19):7462
7. Li B, Wang M, Zhang Y et al (2023) Strengthening reversibility at high rate of spinel $LiMn_2O_4$ by aluminum and copper co-doping for lithium ion battery. Electrochim Acta 464:142898
8. Jiang Q, Wang X, Zhang H (2016) One-pot hydrothermal synthesis of $LiMn_2O_4$ cathode material with excellent high-rate and cycling properties. J Electron Mater 45(8):4350–4356
9. Chan HW, Duh JG, Sheen SR (2003) $LiMn_2O_4$ cathode doped with excess lithium and synthesized by co-precipitation for Li-ion batteries. J Power Sour 115(1):110–118
10. Iskandar Radzi Z, Helmy Arifin K, Zieauddin Kufian M et al (2022) Review of spinel $LiMn_2O_4$ cathode materials under high cut-off voltage in lithium-ion batteries: challenges and strategies. J Electroanal Chem 920:116623

11. Manthiram A (2020) A reflection on lithium-ion battery cathode chemistry. Nat Commun 11(1):1550
12. Tesfamhret Y, Liu H, Chai Z et al (2021) On the manganese dissolution process from $LiMn_2O_4$ cathode materials. ChemElectroChem 8(8):1516–1523
13. Li P, Luo S, Wang Y et al (2022) Effect of Co^{3+} doping on crystal structure and electrochemical properties of Spinel $LiMn_2O_4$ cathode material. Energy Fuels 36(16):9329–9336
14. Saini B, Krishnapriya R, Laishram D et al (2023) Impact of gadolinium doping into the frustrated antiferromagnetic lithium manganese oxide spinel. iScience 26(1):105869
15. Zhang B, Zhang Y, Zhang Q et al (2022) Spinel $LiMn_2O_4$ with remarkable electrochemical performances by synergistic enhancement of double-cation (Sm^{3+}, Mo^{6+}) doping for Li-ion batteries. JOM 74(12):4672–4681
16. Iqbal MJ, Ahmad Z (2008) Electrical and dielectric properties of lithium manganate nanomaterials doped with rare-earth elements. J Power Sour 179(2):763–769
17. Tao Y, Liu Q, Guo Y et al (2022) Regulation of morphology evolution and Mn dissolution for ultra-long cycled spinel $LiMn_2O_4$ cathode materials by B-doping. J Power Sour 524:231073
18. Xu H, Chen Y, Li Y et al (2018) Synthesis of single-crystal magnesium-doped spinel lithium manganate and its applications for lithium-ion batteries. J Solid State Electrochem 22(12):3735–3742
19. Liang G, Peterson VK, See KW et al (2020) Developing high-voltage spinel $LiNi_{0.5}Mn_{1.5}O_4$ cathodes for high-energy-density lithium-ion batteries: current achievements and future prospects. J Mater Chem A 8(31):15373–15398
20. Luo Y, Liu Q, Guo Y et al (2022) Improved long cycle performance of $LiMn_2O_4$ cathode material by Ni–B co-doping combined with truncated octahedron morphology. J Alloy Compd 924:166537
21. Tomon C, Sarawutanukul S, Phattharasupakun N et al (2022) Core-shell structure of $LiMn_2O_4$ cathode material reduces phase transition and Mn dissolution in Li-ion batteries. Commun Chem 5(1):54
22. Mu C-L, Qi Y-X, Liu W et al (2023) Stabilizing commercial $LiMn_2O_4$ cathode by constructing protective saccharin coating. ACS Appl Electron Mater 5(3):1793–1803
23. Kebede MA (2023) An investigation of the lattice parameter and micro-strain behaviour of $LiMn_2O_4$ coated with $LiMn_{1.5}Ni_{0.5}O_4$ to achieve high-rate capability and cycling stability. J Energy Storage 72:108602
24. Wang S, Jiang S, Li Y et al (2023) Interface dual-engineering strategy enhancing the excellent high-temperature electrochemical properties of $LiMn_2O_4$. J Power Sour 579:233292
25. Ahuja A, Kumar A, Sengupta A et al (2022) Single-crystal spinel $Li_{1.08}Mn_{1.92}O_4$ octahedra cathode covered with Li-ion permeable robust NMC thin-layer protection for high voltage lithium-ion batteries. Energy Storage Mater 52:169–179
26. Hou P, Lin Z, Li F et al (2023) General synthesis of single-crystal spinel cathodes with the tailored orientation of exposed crystal planes for advanced lithium-ion batteries. Small 2304482
27. Kang B-G, Jeong Y-W, Ryu S-W (2022) Effect of lithium 4-styrene sulfonate-based self-doped polymer electrolyte on $LiMn_2O_4$ electrodes in lithium-ion secondary batteries. Electrocatalysis 14(2):213–220
28. Yang H, Shao M, Zhang W et al (2021) Scalable fabrication of $LiMn_2O_4$/polythiophene with improved electrochemical performance for lithium-ion batteries. Ind Eng Chem Res 60(39):14185–14192
29. Flamme B, Yeadon DJ, Phadke S et al (2021) Effect of fluorinated additives or co-solvent on performances of graphite//$LiMn_2O_4$ cells cycled at high potential. J Energy Chem 52:332–342
30. Yang J, Rodrigues M-TF, Yu Z et al (2022) Design of a scavenging pyrrole additive for high voltage lithium-ion batteries. J Electrochem Soc 169(4):020556
31. Roy B, Cherepanov P, Nguyen C et al (2021) Lithium borate ester salts for electrolyte application in next-generation high voltage lithium batteries. Adv Energy Mater 11(36):2101422

Ternary Layered Oxide Cathode

7

7.1 Introduction to Ternary Cathode Materials

Ternary cathode materials are developed from layered structure cathode materials such as $LiCoO_2$, $LiNiO_2$, $LiMnO_2$, through the synergistic effect of Ni–Co–Mn [1], combining the advantages of three layered oxide cathode materials: the excellent cycling performance of $LiCoO_2$, the high capacity of $LiNiO_2$, and the safety and low cost of $LiMnO_2$. Therefore, NCM materials have high capacity, good thermal stability, high energy density, good structural stability, and the cost is also controlled [2]. In addition, the discharge voltage platform of ternary lithium batteries is as high as 3.7 V, much higher than the 3.4 V of lithium iron phosphate and 2.3 V of lithium titanate. From the perspective of energy density, ternary lithium batteries have advantages over cathode materials such as lithium iron phosphate, lithium manganate, or lithium titanate [3–5], as shown in Table 7.1. At present, it has been widely commercialized in fields such as electric vehicles, drones, and energy storage, and is a lithium battery cathode material with great development potential.

Ternary materials show different properties according to the ratio changes of the three elements Ni–Co–Mn, and have derived a variety of cathode materials. Mainly can be divided into two categories:

(1) Nickel-manganese equivalent type: such as $LiNi_{1/3}Co_{1/3}Mn_{1/3}O_2$(NCM111), $LiNi_{0.4}Co_{0.2}Mn_{0.4}O_2$ (NCM424), $LiNi_{0.416}Co_{0.166}Mn_{0.416}O_2$ (NCM525). In these materials, Co is +3 valence, Ni is +2 valence, Mn is +4 valence, and the valence state of Mn remains +4 during the charging and discharging process, where Mn mainly stabilizes the structure of the material. During the charging process, N changes from +2 valence to +4 valence to maintain the high capacity of the battery;

(2) Nickel-rich type: mainly includes $LiNi_{0.5}Co_{0.2}Mn_{0.3}O_2$(NCM523), $LiNi_{0.6}Co_{0.2}Mn_{0.2}O_2$(NCM622), $LiNi_{0.8}Co_{0.1}Mn_{0.1}O_2$(NCM811) etc., in these materials, Co is in +3 valence, Ni is in +2/+3 valence, Mn is in +4 valence [6]. During the charging and discharging process, $Ni^{2+/3+}$, Co^{3+} undergo oxidation, Mn^{4+} does not change, playing a role in stabilizing the structure in the material. The three elements play different roles in the material. When the charging voltage is below 4.4 V (relative to Li/Li^+), it is generally believed that mainly Ni^{2+} participates in the electrochemical reaction, forming Ni^{4+}; continue charging, at a higher voltage, Co^{3+} participates in the reaction, and Co^{4+} appears in the material. Therefore, when charging and discharging below 4.4 V, the higher the Ni content, the larger the reversible specific capacity of the material. The Co content significantly affects the ionic conductivity of the material, the higher the Co content, the better the ionic conductivity of the material, and the better the charge and discharge rate [7].

In addition to NCM, there is also a high-nickel ternary material NCA. However, in NCA materials, Al^{3+} replaces Mn^{4+}, in NCA, Co, Ni, Al are all +3 valence, during the charging and discharging process, Al remains unchanged in valence, also playing a role in stabilizing the structure. The nickel content of NCA materials is generally above 80%, and the change in charging and discharging voltage at 4.35 V mainly depends on Ni^{2+}/Ni^{3+}, so the reversible capacity is as high as 200 mAh g^{-1}, Al^{3+} does not affect the valence state of nickel, and 5% Al content is enough to stabilize the layered structure of the material, so NCA has advantages over NCM [8]. However, the technical barriers of NCA are higher, the preparation is more difficult, and the consistency of subsequent sintering of the sample is difficult to guarantee, currently only a few power battery manufacturers use it [9].

The different components of ternary materials lead to differences in their theoretical specific capacity, which is roughly 280 mAh g^{-1}, and different components of ternary materials have a certain pattern when charging and

X. Li et al., *Electrode Development for Lithium-ion Batteries*, https://doi.org/10.1007/978-981-96-3021-9_7

Table 7.1 Basic characteristics of commonly used lithium–ion battery cathode materials

Material	$LiCoO_2$	$LiNiO_2$	$LiMnO_2$	$LiFePO_4$	NCM/NCA
Theoretical/actual capacity (mAh g^{-1})	274/~150	275/≥150	148/~120	170/~160	280/~220
Voltage platform (V vs. Li/Li^+)	3.8	3.8	4.1	3.4	3.7
Cycle performance (time)	≥500	–	≥500	≥2000	≥800
Tap density (g cm^{-3})	2.8–3.0	2.4–2.6	2.2–2.4	1.0–1.7	2.0–3.0
Cost	High	Medium	Low	Low	Medium
Safety	Low	Low	Medium	High	Medium
Toxicity	High	Medium	Low	Low	Medium
Synthesis difficulty	Simple	Difficult	Simple	Medium	Medium

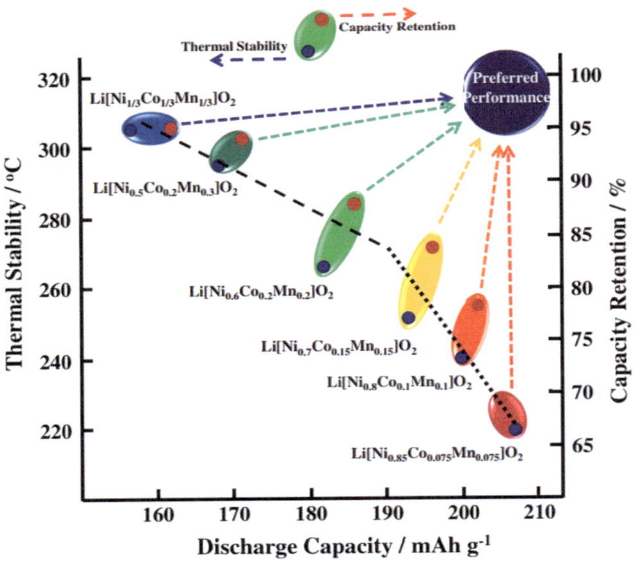

Fig. 7.1 Relationship between discharge capacity, thermal stability, and capacity retention of $Li/LiNi_xCo_yMn_zO_2$ (x = 1/3, 0.5, 0.6, 0.7, 0.8 and 0.85)

discharging in the voltage range of 2.7 ~ 4.2 V (relative to Li/Li^+), that is, the higher the Ni content, the higher the actual discharge specific capacity, but the thermal stability and capacity retention rate decrease with it [10], as shown in Fig. 7.1.

7.2 Structure and Electrochemical Performance of Ternary Materials

7.2.1 Structure of Ternary Materials

The $LiNi_xMn_yCo_{1-x-y}O_2$ crystal belongs to the hexagonal crystal system, and its structure is similar to that of the $LiCoO_2$ material, which is a layered structure, R $\overline{3}$ m space group. In the structure of ternary materials, Li occupies the 3a position, the transition metal M (Ni, Co, Mn) is at the 3b position, O is at the 6c position, the MO_6 octahedral structure is composed of O and transition metal ions, and Li and O form the LiO_6 structure, with the position of Li^+ located in the middle of the octahedral layer [11], as shown in Fig. 7.2.

7.2.2 Several Important Features of NCM Material Electrochemical Performance

7.2.2.1 Capacity Cycling Performance

The capacity cycling performance is one of the basic factors to measure the performance of lithium-ion batteries, and it can also be used to judge the quality of the cathode material. The $LiNi_xCo_yMn_zO_2$ series cathode materials have different capacities for different components, and the specific capacity increases with the increase of Ni content within the range of 2.7 ~ 4.2 V (relative to Li/Li^+). Because the oxidation and reduction voltages of Ni and Co are different, there is no very flat voltage platform. The Sun research group studied the $LiNi_xCo_yMn_zO_2$ with different Ni, Co, and Mn contents in the range of 3 ~ 4.3 V (25 °C, current is 20 mA g^{-1}, 0.1 C rate) (x = 1/3, 0.5, 0.6, 0.7, 0.8, 0.85) specific capacity, tested using a half-cell (with a lithium sheet as the negative electrode). The results show that as the Ni content increases, the specific capacity increases, and the first discharge specific capacities of the materials when x = 0.85, 0.8, 0.7, 0.60, 0.5, 1/3 are 206 mAh g^{-1}, 203 mAh g^{-1}, 194 mAh g^{-1}, 187 mAh g^{-1}, 175 mAh g^{-1}, 163 mAh g^{-1} respectively. However, as the Ni content increases, the cycle performance decreases [10]. As shown in Fig. 7.3.

For the common NCM523 material, the cycle performance of the material can also be improved by fine-tuning the Co and Mn content. The research results of Kim et al. show that on the basis of NCM523 material, the Ni content remains unchanged, Co decreases by 0.04, Mn increases by 0.04, which can effectively improve the cycle performance and thermal stability of the material under high temperature and high voltage conditions [11], as shown in Fig. 7.4. This is mainly due to the increase in the content of inactive Mn^{4+}, which stabilizes the layered structure of the material.

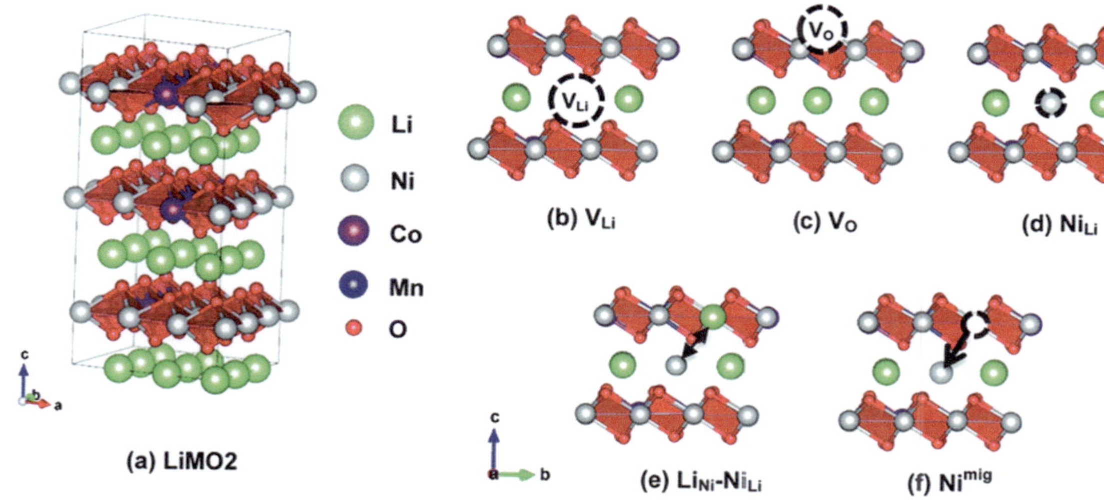

Fig. 7.2 Schematic diagram of the crystal structure of $LiNi_xMn_yCo_zO_2$ [11]

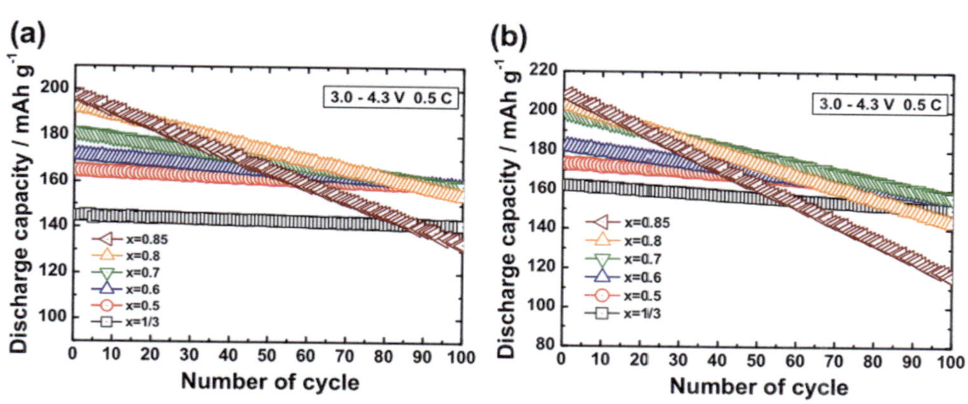

Fig. 7.3 The relationship between the discharge capacity and the number of cycles of Li/$LiNi_xCo_yMn_zO_2$ (x = 1/3, 0.5, 0.6, 0.7, 0.8 and 0.85) batteries at **a** 25 °C and **b** 55 °C. The current density is 100 mA g^{-1} (0.5 C), and the voltage range is 3.0 ~ 4.3 V [10]

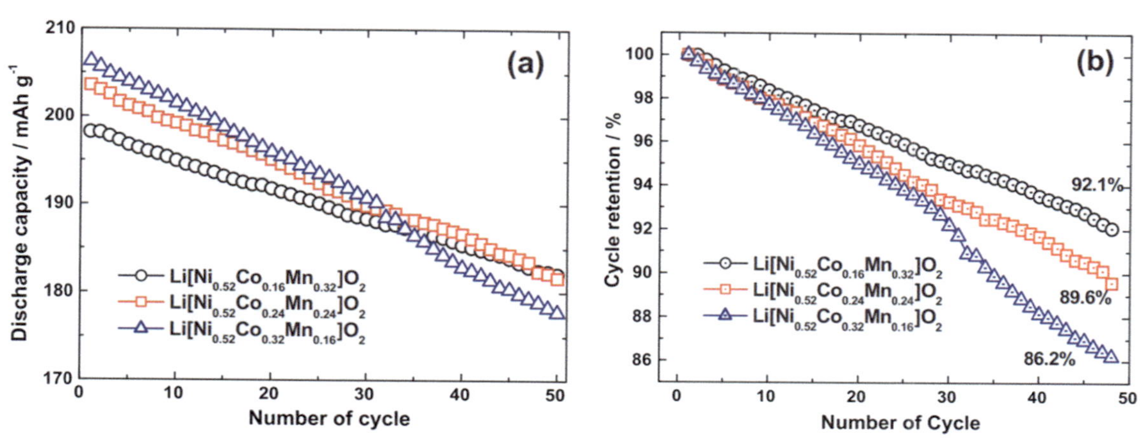

Fig. 7.4 **a** Recyclability and **b** cycle retention over 50 cycles; applied current density is 93 mA g^{-1} (0.5C), 2.7 ~ 4.5 V, temperature is 55 °C [11]

7.2.2.2 Rate Performance

The rate performance is mainly affected by the charge transfer and the diffusion rate of lithium ions. Different proportions of NCM in ternary materials can also lead to differences in rate performance. Generally speaking, the higher the Co content, the better the rate performance. Although the ternary material has the same layered structure as $LiCoO_2$, due to the larger radius of Ni^{2+}, the polarization force of Mn^{4+} and Ni^{2+} is less than that of Mn^{3+} and Ni^{3+}, which makes the covalent bond of the O–M–O layer weaker than $LiCoO_2$, leading to the weakening of M–O bonds and the strengthening of Li–O bonds, thereby increasing the activation energy of ion diffusion [12]. In addition, the electronic configuration of transition metal ions in layered $LiMnO_2$ materials affects the charge transfer rate. For example, in $LiNiO_2$, the electron grouping state of Ni^{3+}'s d orbitals is $t_{2g}{}^6e_{g^1}$, because the e.g., orbitals point to the σ bond formed by the coordinating oxygen and the 2p orbitals of oxygen, while the t_{2g} orbitals of Co^{3+} form π bonds with the 2p orbitals of oxygen. Therefore, the full band overlap and band gap of the d orbitals between Ni–Ni layers are not as good as those between Co–Co layers. Hence, the mixed valence state charge transfer of $Ni^{4+/3+}$ is more than that of $LiMnO_2$ materials. The $Co^{4+/3+}$ in $LiCoO_2$ is much more difficult, so its high-rate discharge capacity is relatively worse than $LiCoO_2$. Therefore, the electrochemical reaction rate of these materials is slower. For ternary materials, the rate performance will deteriorate with the increase of nickel content. Further research has found that the charge transfer process is the controlling step of the electrochemical reaction. As the temperature rises, the charge transfer activity increases, the electrochemical insertion and removal reactions speed up, thereby improving the electrochemical performance. The rate performance of the cathode material directly affects the application of secondary lithium-ion batteries in high-power appliances.

Venkatraman and others analyzed the factors affecting the delamination rate of layered $LiNi_{1-y-z}Co_yMn_zO_2$ materials, and the time required for complete lithium ion stripping increases with the increase of nickel content. In $Li_{1-x}Ni_{0.5-0.5y}Co_yMn_{0.5-0.5y}O_2$, the time required for complete lithium ion removal increases with the increase of cobalt content, and there are two reasons for this phenomenon [7]. First, the "cation disorder" phenomenon causes some nickel ions to be distributed in the lithium layer, hindering the diffusion channel of Li^+, thereby slowing down the release of lithium; second, the thickness of the Li–O layer (d_{LiO2}) also affects the ion removal rate, the larger d_{LiO2}, the faster the lithium insertion removal rate.

7.2.2.3 Low-Temperature Performance

Although significant progress has been made in lithium-ion batteries in recent years, the performance degradation at low temperatures is still prominent, which has become one of the main obstacles limiting their application in high-altitude or high-latitude areas, as well as certain defense and space applications [13, 14]. A common 18650 commercial lithium-ion battery, at −40 °C, the battery's discharge capacity is only about 5% of that at 20 °C, not to mention the power density at this time [15]. Therefore, if ternary cathode materials are to be used on a large scale, the most important thing is to achieve the goal of normal operation at low temperatures.

In general, the low-temperature performance of solid solution cathodes (such as NCM) is better than that of phase transformation cathodes (such as LFP), although both types of cathodes follow the intercalation/deintercalation mechanism [16]. Sun et al. grafted NCM622 with diazonium tetrafluoroborate salt ($C_6H_5N_2+BF^{4-}$) to form polystyrene-coated NCM622, improving its low-temperature performance (Fig. 7.5a). At −20 °C, the capacities of polystyrene-coated NCM622 at 0.1 C and 1 C rates are 148 mAh g^{-1} and 120 mAh g^{-1} [17] respectively. The full cell of NCM622 anode and exfoliated graphene (Gr) anode shows excellent low-temperature cycle stability after 300 cycles at −30 °C, with no capacity loss [18] (Fig. 7.5b, c). It is not difficult to see that the low-temperature charge and discharge process shows obvious polarization phenomena, the most significant polarization comes from the process of lithium ions passing through the active material/electrolyte interface and the charge transfer process, rather than the diffusion process of lithium ions inside the electrode material. Therefore, improving the low-temperature performance of NCM and NCA materials should start from reducing surface film impedance and electrochemical reaction impedance.

7.2.2.4 Other Performance

Compared with $LiCoO_2$, NCM materials also have higher specific capacity and good thermal stability. Ohzuku's team compared $Li_{1-x}CoO_2$ (charge capacity 140 mAh g^{-1}) and $Li_{1-x}Ni_{1/3}Co_{1/3}Mn_{1/3}O_2$ through mass spectrometry analysis. (With a charging capacity of 185 mAh g^{-1}) Lithium thermal decomposition gas in the material. The results show that the oxygen release initiation temperature of NCM material is higher than that of $Li_{1-x}CoO_2$ (Fig. 7.6), so it is believed that NCM material has better thermal stability [12]. The product of $LiNi_{1/3}Co_{1/3}Mn_{1/3}O_2$ charged at 4.45 V was determined by high-temperature X-ray diffraction

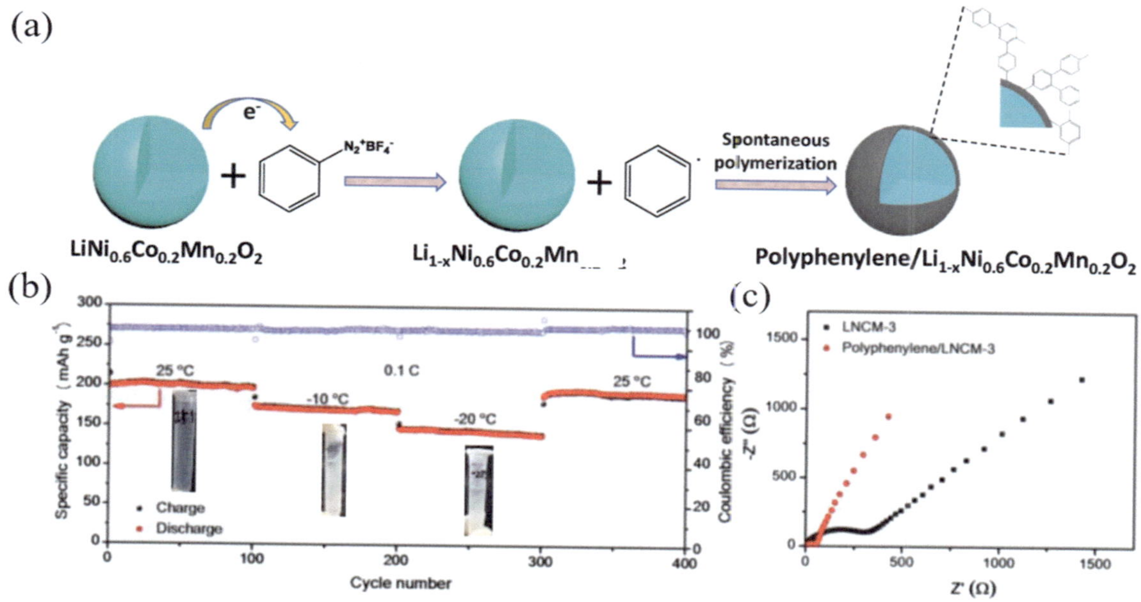

Fig. 7.5 **a** Schematic diagram of the preparation of polystyrene/LNCM-3 cathode and the corresponding **b** 25, −10 and −20 °C low-temperature performance; **c** comparison of EIS of original LNCM-3 and polystyrene/LNCM-3 [17]

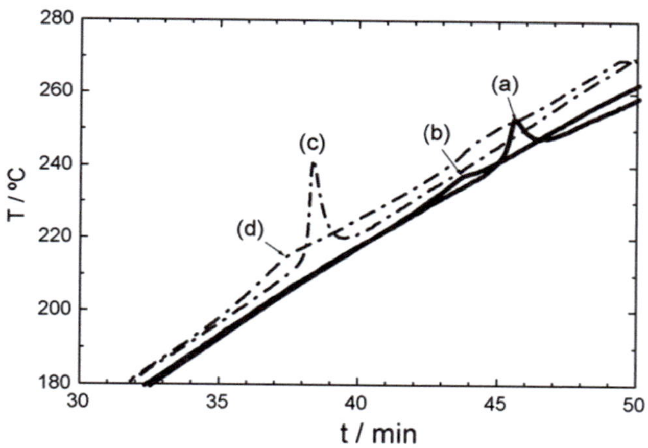

Fig. 7.6 Hot-pot test results of graphite anode/$LiNi_{1/3}Co_{1/3}Mn_{1/3}O_2$ and $LiCoO_2$ cathode at different charging end voltages. **a** $Li_{1-x}Ni_{1/3}Co_{1/3}Mn_{1/3}O_2$ charged to 4.4 V, **b** $Li_{1-x}Ni_{1/3}Co_{1/3}Mn_{1/3}O_2$ charged to 4.7 V, **c** $Li_{1-x}CoO_2$ charged to 4.2 V, **d** $Li_{1-x}CoO_2$ charged to 4.7 V [12]

(XRD) to be a cubic spinel phase containing Ni, Co, and Mn. The transition from layered structure to spinel structure effectively inhibits the oxygen precipitation in the material matrix structure. Therefore, it shows better high-rate performance, cycle performance, and safety performance. Moreover, after a series of research and development, it was found that for ternary materials with different Ni contents, due to the different effects of each element on the thermal stability of the material (mainly Co and Mn), their thermal stability shows a great difference.

In order to verify the high-temperature storage performance of high-nickel ternary NCA batteries, Watanaber and others assembled 1865 batteries with NCA cathode (NCR18650) and $LiCoO_2$ cathode (CGR18650E), and took them out after storing at 45 °C for 2 years, comparing the capacity degradation of the two batteries in long-term storage experiments [19, 20]. The experiment studied the capacity degradation of two types of batteries and made a comparison. The results of this study show that the storage performance of the NCR18650 battery in the storage experiment is far superior to that of the CGR18650E battery. The increase in impedance and capacity decay during storage is mainly due to the degradation of the positive electrode, and the degradation of the LCO positive electrode is greater than that of the NCA positive electrode. They used XPS to analyze the SEI film on the surface of the NCA and LCO positive electrodes. The elements of these two positive electrodes are almost the same, and the thickness is also the same, indicating that SEI is not the main factor for the difference in degradation behavior between NCA and LCO. As can be seen from Fig. 7.7a, after 2 years of storage, in NCA, the deposition of Co at the positive electrode of the LCO battery is greater than the deposition of Ni at the negative electrode, which indicates that during the storage

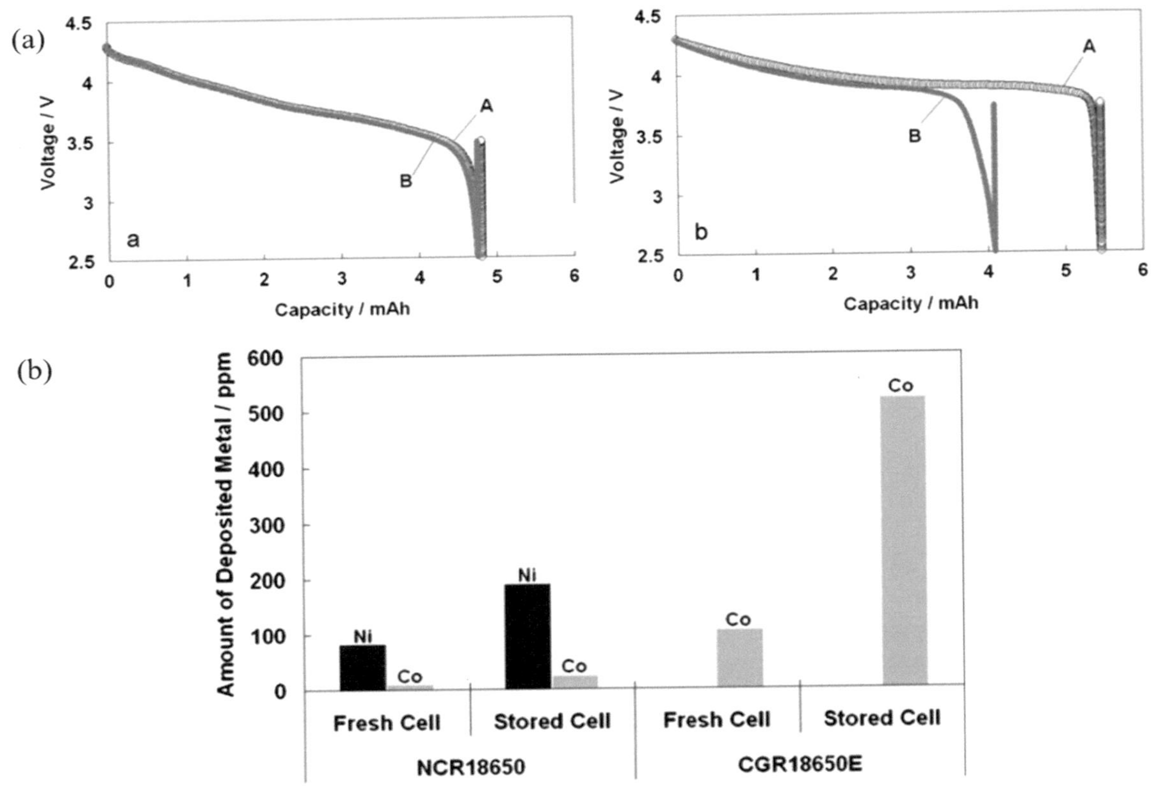

Fig. 7.7 NCA positive electrode and LCO positive electrode. **a** Discharge curves obtained from new batteries and batteries stored for 2 years; **b** 45 °C stored for 2 years, metal deposition on the negative electrode surface [19]

process, the dissolution of Co in LCO is greater, as shown in Fig. 7.7b, which destroys the structure of the material and leads to the deterioration of battery performance.

7.3 Synthesis Method of High Nickel Ternary Positive Electrode Material

The current commercial method for synthesizing ternary positive electrode materials is to first synthesize the ternary precursor by co-precipitation, then mix it evenly with lithium salt, and finally synthesize the final product by high-temperature sintering. Other synthesis methods have also been reported in the literature, such as direct high-temperature solid-phase method, sol-gel method, hydrothermal method, etc.

7.3.1 Sol-Gel Method

The sol-gel method is a soft chemical method developed to solve the problems of uniformity of components between reactants and slow reaction speed in the high-temperature solid-phase method. Compared with the traditional solid-phase high-temperature sintering method, the materials prepared by the sol-gel method have uniform particle size and even element distribution. The sol is a colloidal solution, with solid or macromolecules dispersed in the system. The gel belongs to the solid, composed of flowing components, solid components in the internal network structure, and high polymer three states. The chemical process of the sol-gel method first disperses the raw materials in the liquid phase, then generates active monomers through hydrolysis or polymerization reactions, the active monomers undergo polymerization reactions to form a gel, and a gel with a certain spatial structure is formed under appropriate conditions, and the target product is obtained after heat treatment. Although the sol-gel method has the characteristics of uniform distribution and small powder particles, it is very difficult to produce in industrialization due to the long formation time of the sol and the high price of raw materials [21].

7.3.2 Hydrothermal Method and Solvent Thermal Synthesis Method

Hydrothermal and solvent thermal synthesis technologies belong to a type of wet chemical method, with a synthesis temperature range of 100 ~ 1000 °C, and a pressure range of 1 MPa ~ 1 Gpa conditions for the synthesis carried out by the chemical substances in the solution. Hydrothermal and solvent thermal reactions are carried out in a closed system, so the relationship between temperature and pressure is very important for hydrothermal synthesis. The synthesis of powder by the hydrothermal method mainly depends on the different solubilities of oxides in various states [22].

7.3.3 High-Temperature Solid-Phase Method

The high-temperature solid-phase method refers to the process of preparing the target material by thoroughly grinding and mixing the raw materials, and then sintering at high temperatures. The general reaction temperature is above 600 °C, suitable for preparing thermodynamically stable compounds. This method is simple and easy to implement in industrial production, but it requires a high sintering temperature and a long sintering time. These conditions can cause lithium salts to form impurities and volatilize severely during the sintering process. Also, because the sintering temperature is too high, the final product has uneven particle sizes, which leads to poor chemical properties of the material. In the current synthesis process of high-nickel ternary materials, few manufacturers use the high-temperature solid-phase method [23].

7.3.4 Co-precipitation Method

The co-precipitation method is the mainstream method for commercial production of ternary cathode materials. The co-precipitation method refers to adding an appropriate precipitant to the source solution, causing the existing components in the solution to precipitate together in a stoichiometric ratio, or obtaining an intermediate product in the solution first, and then sintering to obtain the final target sample. This process can accurately control the particle size and morphology of the product, and the effective components in the product can achieve atomic and molecular level mixing. However, when synthesizing ternary NCM materials by the co-precipitation method, due to the rapid formation of manganese hydroxide, it is difficult to directly synthesize precursors with spherical morphology, so it is necessary to control the supersaturation, PH value, ammonia concentration, temperature, stirring ratio, etc. in the system reaction [24].

7.4 Problems with High-Nickel Ternary Materials and Their Modifications

7.4.1 Problems with High-Nickel Ternary Materials

High-nickel ternary materials have a significant advantage in energy density compared to other cathode materials, but they still have drawbacks, such as the transformation of the material's layered structure into a spinel structure during the cycle, the easy generation of element segregation during the synthesis process, cation mixing, etc. These shortcomings have hindered the large-scale application of high-nickel ternary cathode materials in power batteries [25].

(1) As the Ni content increases, the material's cycle performance deteriorates

As the proportion of nickel in ternary materials increases, the battery's cycle performance deteriorates. This phenomenon is mainly due to the multiple phase changes that occur in the ternary materials during the charge and discharge process as the nickel content increases. Noh H-J conducted a systematic analysis of the capacity differential curve of the first charge–discharge curve of ternary materials [10]. When the nickel content is 1/3, the oxidation and reduction peaks of the material's first cycle are 3.76 V and 3.72 V, respectively. As the Ni content increases, a new oxidation peak appears at 3.64 V in the curve; when the nickel content is higher than 0.8, the material exhibits four pairs of oxidation peaks, which are 3.62 V, 3.78 V, 4.04 V, and 4.23 V, respectively. They correspond to the transformation of the hexagonal phase to the monoclinic phase ($H1 \rightarrow M$), the transformation of the monoclinic phase to the hexagonal phase ($M \rightarrow H_2$), and the transformation of the hexagonal phase to the hexagonal phase ($H_2 \rightarrow H_3$). After 100 cycles, the $H_2 \rightarrow H_3$ structural transformation causes the material structure to collapse, which is the main reason for capacity decay [26].

(2) As the Ni content increases, the surface alkalinity increases

For high-nickel ternary materials, especially those with Ni content higher than 0.8, they easily combine with H_2O and CO_2 in the air, forming LiOH and Li_2CO_3 on the material surface. When assembled into a battery, the LiOH on the material surface reacts with the electrolyte in the battery, generating corrosive HF, and Li_2CO_3 easily reacts with the electrolyte to produce gas [27–29].

(3) As the Ni content increases, the thermal stability of the material deteriorates

As the nickel content in the ternary material increases, the stability of the battery worsens. This is because the lithium released by high-nickel materials at the same potential is higher than that of low-nickel ternary materials, resulting in a higher content of Ni^{4+}. Ni^{4+} has a strong reducibility and easily transforms into Ni^{3+}. Therefore, to maintain charge balance, the O^{2-} in the ternary material is reduced and releases oxygen, causing the material stability to deteriorate [30]. The thermal stability of the cathode material determines the safety performance of the battery. As the nickel content increases, the stability of the cathode material changes, becoming a key factor affecting the safety performance of the battery (Fig. 7.8) [10].

(4) Surface reaction non-uniformity

In pursuit of higher energy density, power batteries often use high-nickel ternary as the cathode material of the battery. Hwang and others found that high-nickel ternary materials show rapid capacity decay and increased surface impedance during long-term cycling. Based on kinetic analysis, they believe that the larger amount of Li released from the particle surface leads to material instability (Fig. 7.9) [31]. This instability causes an imbalance in the valence of transition metals, some transition metal ions are reduced, therefore to maintain the charge balance of the material, pores and new phases are generated on the material surface [32, 33]. In addition, the side reaction of the reduced

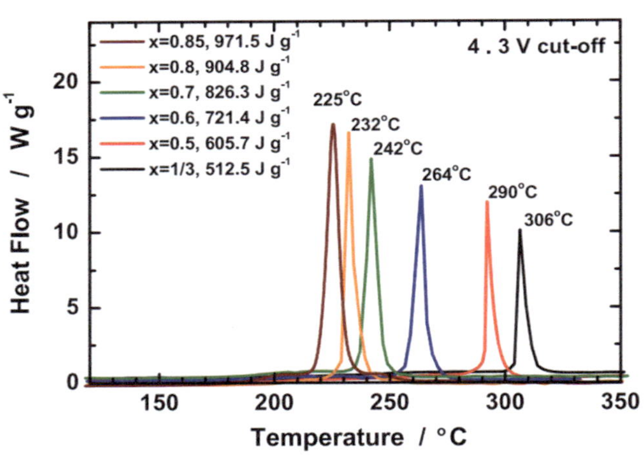

Fig. 7.8 DSC results of $Li_{1-\delta}Ni_xCo_yMn_zO_2$ material (x = 1/3, 0.5, 0.6, 0.7, 0.8 and 0.85) [10]

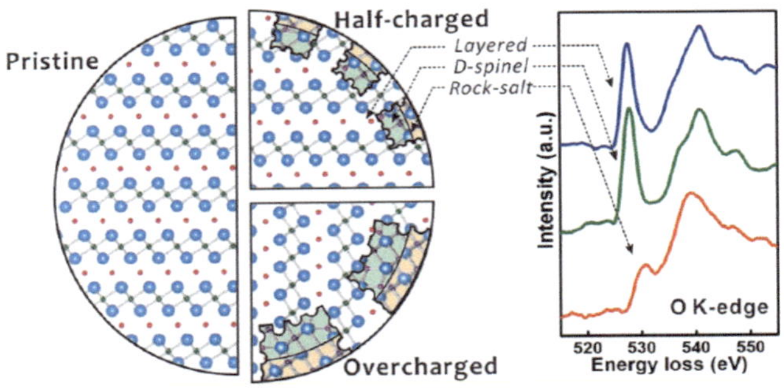

Fig. 7.9 Schematic diagram of the structural changes of NCA material during the charging process [31]

transition metal ions on the surface with the electrolyte will also increase the degree of battery polarization [34].

(5) Matching of the electrolyte

With the increase in research on lithium-ion batteries, researchers have found that electrolytes have a significant impact on ternary power batteries, but solid-state lithium-ion batteries are still in the research and development stage and cannot be mass-produced commercially. The mainstream electrolyte in commercial batteries is still electrolyte liquid. In the electrolyte liquid, the corrosion of ternary cathode materials and the decomposition of the electrolyte seriously affect the efficiency of ion and electron transmission in the electrode/electrolyte [35].

7.4.2 Modification of NCM or NCA Cathode Materials

High-nickel ternary materials have various inherent defects, which seriously hinder the commercial development of batteries. Therefore, researchers have proposed various modification methods to solve these problems, among which the most widely used methods are ion doping, surface coating, and material gradient design.

7.4.2.1 Doping

Doping some metal ions and non-metal ions into the lattice of ternary materials can not only improve the electronic conductivity and ionic conductivity, increase the output power density of the battery, but also improve the stability of the ternary material structure (especially thermal stability). There are usually three replacement positions: lithium positions in the lithium-ion layer, transition metal positions in the transition metal layer, and oxygen positions in the transition metal layer.

(1) Doping of lithium positions in the lithium-ion layer

Due to the irreversible structural transformation of NCM materials during the charge and discharge cycle, its capacity inevitably decays. At the same time, the mixing of cations will limit the effective diffusion of Li^+, which is not conducive to the stability of the layered structure and the electrochemical performance of the electrode material [36]. At present, many researchers have modified the surface state of nickel-rich layered oxides to improve their structural stability as a good strategy. Feng and others synthesized Na^+ -doped gradient-modified $LiNi_{0.6}Co_{0.2}Mn_{0.2}O_2$ cathode material by a two-step grinding-spray calcination method. The XRD, EDS, and XPS test results show that Na^+ doping increases the interlayer spacing between Li^+, reduces the degree of ion disorder, and is conducive to the formation of a regular interlayer structure. The Na^+ -doped sample has a higher capacity (179 mAh g^{-1} at 0.2 C), excellent rate performance (152.6 mAh g^{-1} at 1 C, 133.9 mAh g^{-1} at 5 C), and excellent cycle stability at high current density (82.8% at 1 C, 73.7% at 5 C) (Fig. 7.10). Due to the pillar support of large-radius Na^+, the modified sample still maintains a complete spherical shape after long-term cycling [37].

(2) Doping at the transition metal position in the transition metal layer

When uneven cation doping is used, it will cause the valence state of the transition metal ions in the ternary material to increase or decrease, generating vacancies or electrons, changing the band structure of the material, and thus increasing the intrinsic electron conductivity of the material. Liao and others studied the doping of Mg ions in ternary materials, forming $LiNi_{0.6-y}Co_{0.25}Mn_{0.15}Mg_yO_2$ and discussed the impact of Mg doping on the structural stability and local performance of O_2 [38]. They found that as the amount of Mg doping increased, the lattice parameters a and c both increased, the material formed a better layered structure, the oxidation state of Ni also increased, and the doping of Mg greatly improved the cycle performance of the battery (Fig. 7.11a). Xu and others reported using Cr^{3+} as a dopant to stabilize the structure of $LiNi_{0.6}Co_{0.2}Mn_{0.2}O_2$ and improve its cycle life. The appropriate amount of Cr^{3+} doping can not only expand the insertion and extraction pathways of Li^+, but also inhibit the mixing of cations (Fig. 7.11b). The electrochemical performance of the material is significantly improved, especially under harsh conditions. When the battery is charged and discharged in the voltage range of 2.7 ~ 4.3 V, after 200 cycles at 10 C with 1 mol% Cr^{3+} doped in $LiNi_{0.6}Co_{0.2}Mn_{0.2}O_2$ material, the discharge capacity is 100.6 mAh g^{-1}, equivalent to 93.1% of the first cycle capacity. When charged and discharged in the voltage range of 2.7 ~ 4.5 V, after 200 cycles (3 C and 50 °C), the discharge capacity of the positive electrode is 166.8 mAh g^{-1}, maintaining 75.0% of the initial capacity, while the original doped $LiNi_{0.6}Co_{0.2}Mn_{0.2}O_2$ the discharge capacity of the material quickly drops to 31.5 mAh g^{-1} [39]. In the doping process of NCM materials, transition metal ions are widely used by researchers due to their good properties.

In addition to NCM materials, NCA materials are also of great importance in the development of ternary materials, and it is a common strategy to modify and optimize their performance using doping methods. Nurpeissova and others studied the process of Ni being partially replaced by Ti in the transition metal layer (especially the non-main surface area of $LiNi_{0.8}Co_{0.15}Al_{0.05}O_2$) through transmission electron microscopy combined with X-ray energy spectrum, electron energy loss spectrum, and X-ray diffraction analysis, as well as electrochemical performance and interface phenomena [40]. The results show that Ti-doped $LiNi_{0.8}Co_{0.15}Al_{0.05}O_2$ inhibits the formation of NiO passivation phase, making the electrode have a higher capacity in cycle testing (Fig. 7.11c).

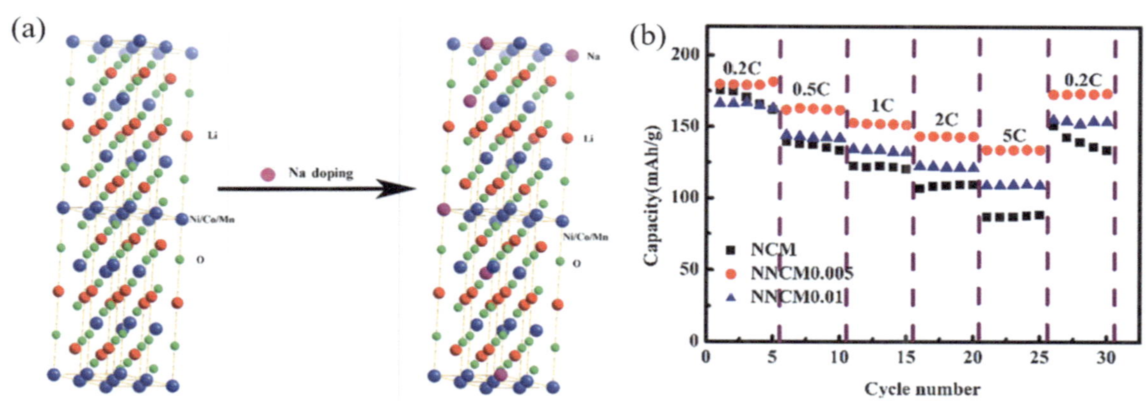

Fig. 7.10 **a** $LiNi_{0.6}Co_{0.2}Mn_{0.2}O_2$ type and $Li_{1.05-x}Na_xNi_{0.6}Co_{0.2}Mn_{0.2}O_2$ type structural features; **b** rate performance of bare NCM and NCM doped with different proportions of sodium [37]

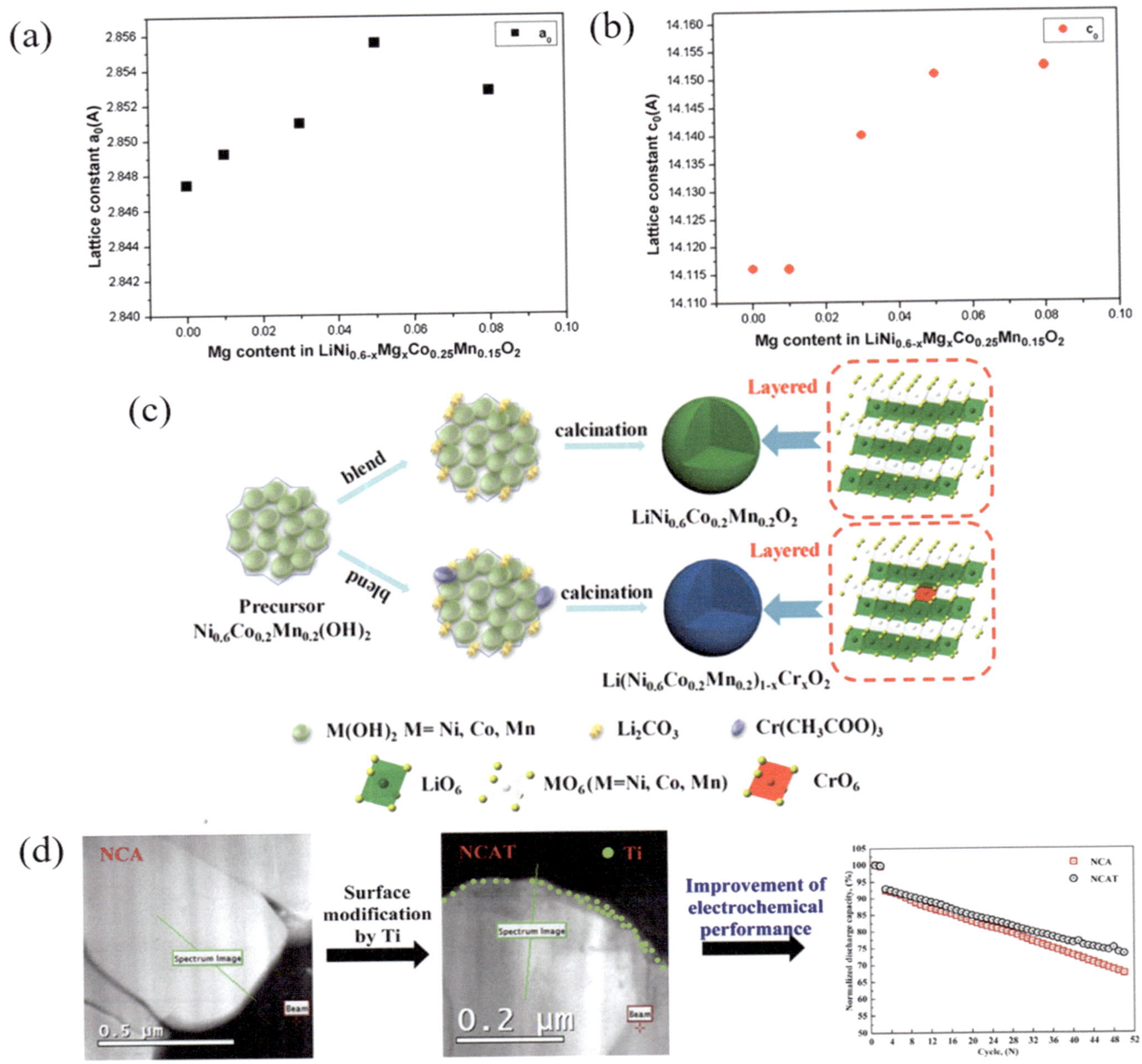

Fig. 7.11 **a** Lattice constants a of $LiNi_{0.6-y}Co_{0.25}Mn_{0.15}Mg_yO_2$ and **b** relationship between lattice constant c and Mg content [38]; **c** schematic diagram of the synthesis route of LNCM and Cr-doped materials [39]; **d** TEM images and electrochemical cycling performance of undoped NCA and doped NCA-Ti [40]

(3) Oxygen site doping exists in the transition metal layer

F replacing O can make the positive electrode material form a spinel structure or layered structure, thereby effectively improving its electrochemical performance [41]. Zhang and others prepared $LiNi_{0.8}Co_{0.1}Mn_{0.1}O_2$ material with a single crystal structure through multi-step calcination, and further controllably doped F to improve the storage performance of Li^+. F doping not only improved the crystal space, but also increased the lattice oxygen content of Co^{3+}, promoting the diffusion and stability of Li^+ crystal structure [42]. With appropriate F doping, the specific capacity of $LiNi_{0.8}Co_{0.1}Mn_{0.1}O_2$ single crystal material is 202.7 mAh g^{-1} (under 0.1 C), and after 100 cycles at a 1 C rate, the capacity retention rate is as high as 86.6%, which is much higher than the control sample, confirming that F doping has a good application prospect in lithium-ion batteries (Fig. 7.12).

In NCM or NCA materials, in addition to the three substitution sites mentioned above, there is also a situation of co-doping, that is, two elements simultaneously replace the transition metal site or replace different sites separately. This effect can make NCM or NCA materials obtain better electrochemical performance or better structural stability, which is worth our continued attention in subsequent research work. In addition, we have summarized some commonly used doping elements and their electrochemical properties, see Table 7.2.

In summary, doping is a simple and effective method of modifying chemical composition, which can be manipulated during the synthesis of NCM or NCA. The substitution of cations and anions can expand the distance between Li plates, facilitate the rapid diffusion of Li^+, enhance the bonding strength between metal ions and oxygen ions, hinder the migration of Ni to the Li plate, prevent unnecessary phase changes, eliminate the loss of metal and oxygen, or suppress the generation of side reactions.

7.4.3 Surface Modification

Considering the impact of the surface stability of nickel-rich layered oxides on battery performance, the lithium oxide remaining on the surface of synthesized NCM particles will rapidly react with H_2O, O_2 and CO_2 in the air. In addition, the continuous side reactions occurring on

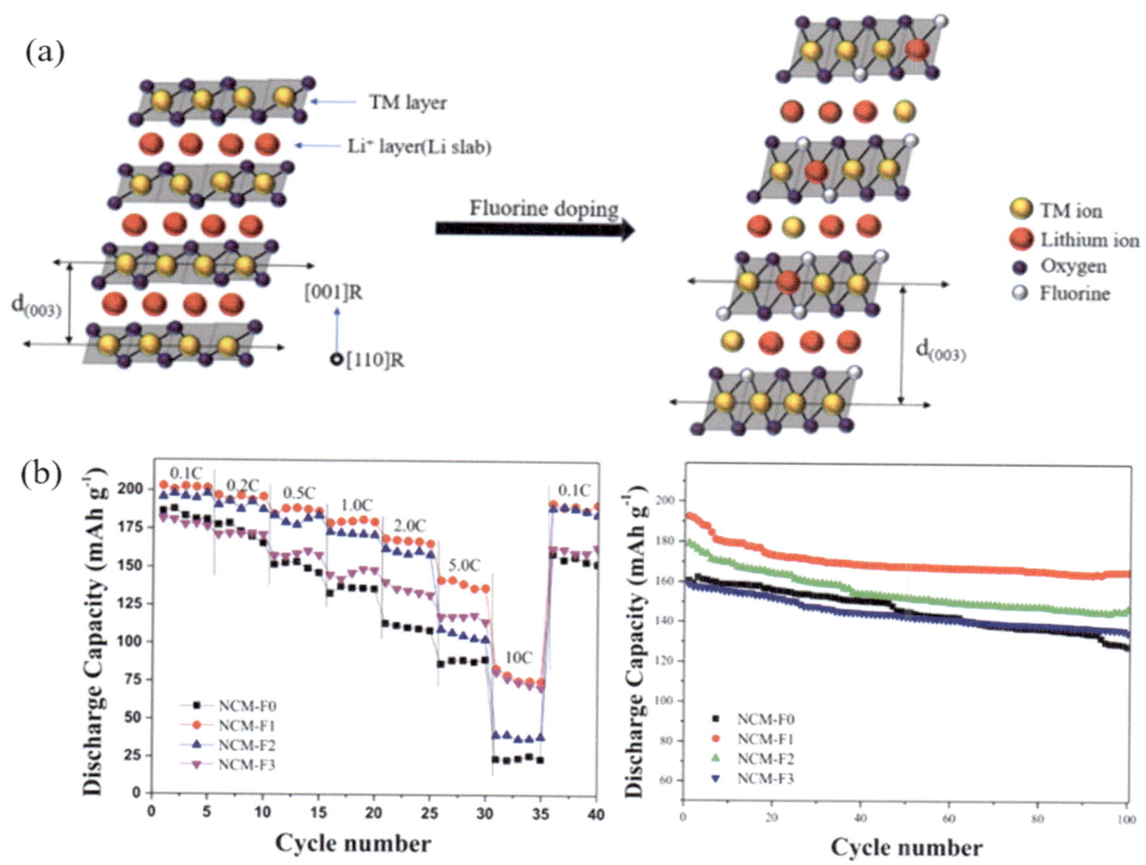

Fig. 7.12 **a** Schematic diagram of the effect of fluorine doping on crystal space and lithium insertion; **b** rate performance and cycle performance of NCM811 doped with F [42]

Table 7.2 Electrochemical characteristics of some common doping elements

Element	Cathode material	Voltage range (V)	First cycle/ capacity retention	Rate	References
Mo	NCM622	2.7–4.6	203.1 mAh g^{-1} 82.9% 100 cycle	1 C	[43]
Ti	NCM811	3.0–4.6	165.6 mAh g^{-1} 84% 100 cycle	1 C	[44]
Cu	NCM622	2.7–4.4	137.6 mAh g^{-1} 86% 200 cycle	1 C	[45]
W	NCM622	3.0–4.5	152.8 mAh g^{-1} 96.7% cycle	1 C	[46]
La	NCM811	2.75–4.5	160.9 mAh g^{-1} 91% 100 cycle	1 C	[47]
Nb	NCM622	3.0–4.5	179.8 mAh g^{-1} 90.1% 200 cycle	1 C	[48]
F	NCM333	3–4.6	174.2 mAh g^{-1} 97.3% 30 cycle	32 mA g^{-1}	[49]
Mg/F	NCM424	2.8–4.6	89 mAh g^{-1} 96% 50 cycle	20 mA g^{-1}	[50]
Ga	NCM811	2.8–4.5	161.1 mAh g^{-1} 89.5% 100 cycle	1 C	[51]

the NCM surface will gradually form a detrimental layer (including LiOH, Li_2CO_3 and LiF) can lead to the degradation of active substances during the decomposition of $LiPF_6$ salt, thereby hindering the diffusion of Li^+ and causing the release of gas, which may cause serious safety issues. Surface coating modification provides a solution to these problems [52] (Fig. 7.13).

(1) Metal Oxides

Al_2O_3 is considered the most common oxide in oxide coatings. Al_2O_3 coating is an insulator of ions and electrons. After heat treatment, a Li–Al–Co–O layer is formed. This layer will resist the corrosion of HF on active materials, and can reduce surface impedance and improve cycle stability. Al_2O_3 coating can be achieved by various preparation methods, with the feature of simple preparation. Chen et al. prepared $LiNi_{0.6}Co_{0.2}Mn_{0.2}O_2$ powder by ultrasonically coating nano Al_2O_3 particles, with a coating thickness of about 20–25 nm (Fig. 7.14a). A suitable amount of Al_2O_3 coating is conducive to forming a better layered structure, and the degree of cation disorder is small. Electrochemical test results show that $LiNi_{0.6}Co_{0.2}Mn_{0.2}O_2$ coated with 1.0 wt.% Al_2O_3 is 197.1 mAh g^{-1}, with an initial Coulombic efficiency of 91.2% at 0.2 C under a voltage of 2.8 ~ 4.5 V, and 82.9% after 30 cycles at 1C; especially, the discharge capacity of the sample coated with 1.0 wt.% Al_2O_3 significantly increased, reaching as high as 153.5 mAh g^{-1} after 5 cycles at 10 C, while the uncoated battery only had 126.3 mAh g^{-1} (Fig. 7.14b and c). However, too thick a coating can also have negative effects, such as higher overpotential and lower capacity [53].

In addition to Al_2O_3 coating, ZnO coating, which does not block the diffusion of lithium ions during the charge and discharge process and improves the discharge specific capacity of the material, has also become a focus of attention. Lai and others directly sputtered a uniform nano-zinc

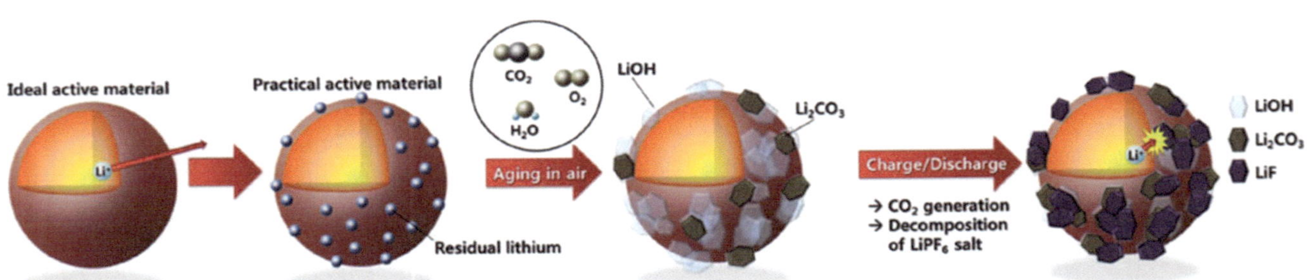

Fig. 7.13 Schematic diagram of particle damage caused by the reaction of lithium oxide remaining on the particle surface with H_2O, O_2, and CO_2 in the air [52]

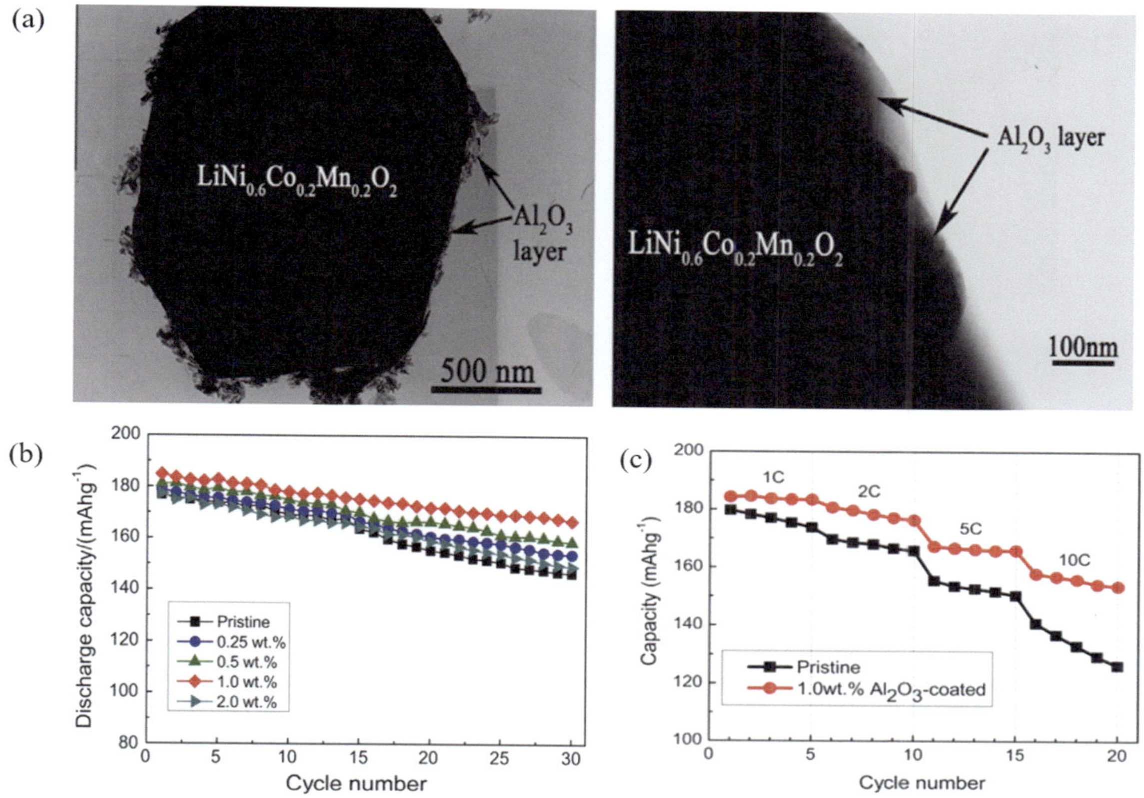

Fig. 7.14 **a** TEM image and local magnification image of 1.0wt.% Al_2O_3 coated $LiNi_{0.6}Co_{0.2}Mn_{0.2}O_2$ powder; **b** cycling performance of $LiNi_{0.6}Co_{0.2}Mn_{0.2}O_2$ materials with different Al_2O_3 coating amounts; **c** rate performance of $LiNi_{0.6}Co_{0.2}Mn_{0.2}O_2$ with less than 1.0 wt.% Al_2O_3 coating [53]

oxide film on the NCA electrode surface by magnetron sputtering (MS). The ZnO film was characterized by scanning electron microscopy (SEM), transmission electron microscopy (TEM), and X-ray photoelectron spectroscopy (XPS). The results showed that the ZnO film was completely and uniformly covered on the NCA electrode. After 90 cycles at 1 C, the discharge capacity of the optimized MS-2 min coated NCA electrode was 169 mAh g^{-1}, much higher than the 127 mAh g^{-1} of the original NCA electrode. In addition, the battery at 3 C, the discharge capacity also reached 166 mAh g^{-1}, while the original electrode discharge capacity was 125 mAh g^{-1} (Fig. 7.15). The excellent performance of the MS-ZnO film reduced the charge transfer resistance, prevented the dissolution of cations from the NCA electrode, and thus improved the electrochemical performance of the NCA electrode [54].

(2) Fluorides

Fluorine modification is also an effective method to improve the electrochemical performance of layered compounds. Due to the presence of Al_2O_3, the coating layer will gradually transform into stable AlF_3 through Al–O–F, thus protecting the active material from HF corrosion. Kim and others used AlF_3 as a new coating material for $LiNi_{0.8}Co_{0.15}Al_{0.05}O_2$ particles are modified. Although the initial discharge capacity of the coated $LiNi_{0.8}Co_{0.15}Al_{0.05}O_2$ material is the same as the original material, the capacity retention rate and thermal stability are significantly improved under high oxidation states. This effect is attributed to the thin AlF_3 coating protecting the oxidized positive electrode particles from HF erosion in the electrolyte [55] (Fig. 7.16a–c). Moreover, Myung et al. coated NCM material with AlF_3 material, and studied the thermal stability of the AlF_3 coating on $Li_{0.35}Ni_{1/3}Co_{1/3}Mn_{1/3}O_2$ material after chemical modification from room temperature to 600 °C. Thermogravimetric analysis results show that the weight loss rate is reduced due to the reduction of O_2 precipitation after coating [56]. The weight loss of uncoated AlF_3 powder is accompanied by an irreversible phase transition from $R\bar{3}m$ phase to cubic spinel phase $Fd\bar{3}m$. High-temperature XRD experiments show that the coating layer delays the phase transition, the main exothermic peak moves to high temperature, the heat release decreases, which is mainly due to the formation of Li–Al–O layer on the surface (Fig. 7.16d).

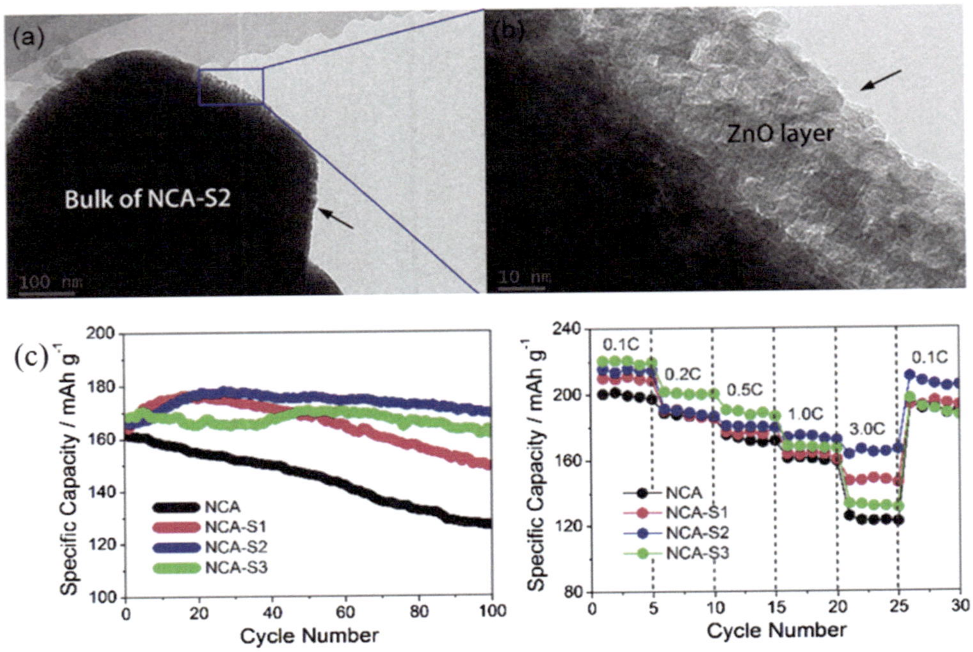

Fig. 7.15 **a** TEM image of NCA-S2; **b** local high magnification image; **c** cycling performance and rate performance of bare NCA and NCA coated with different proportions of ZnO [54]

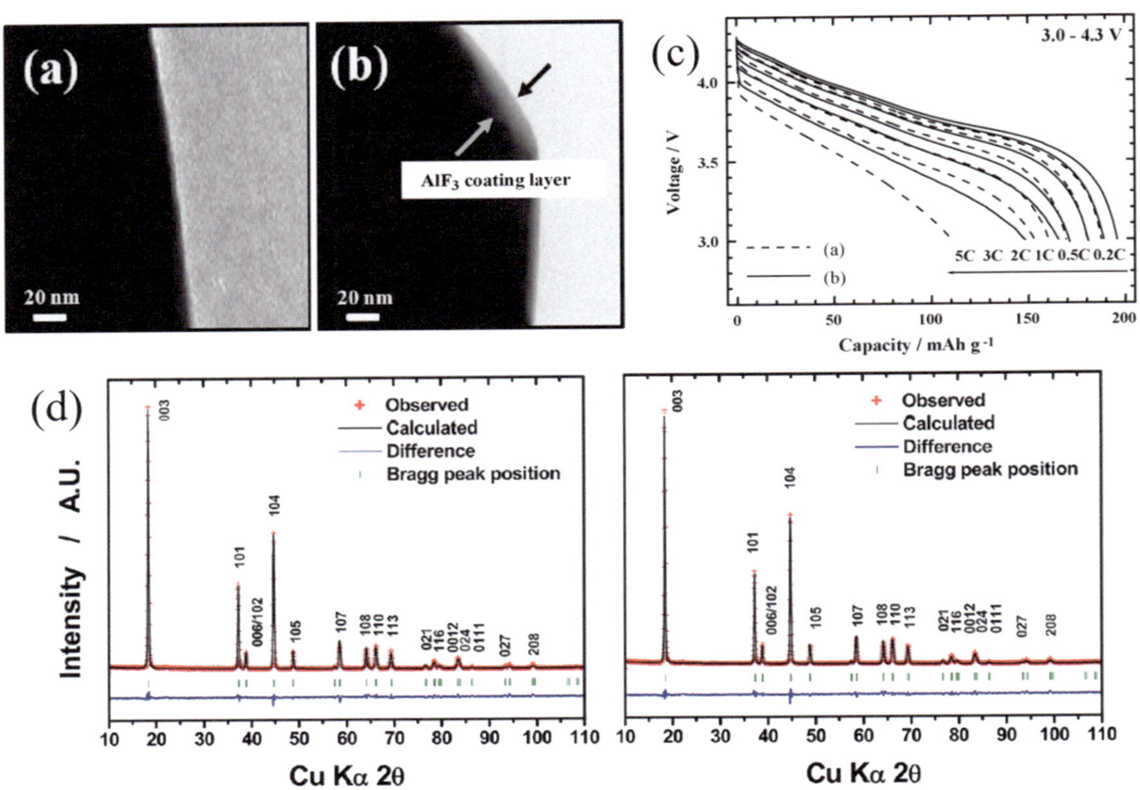

Fig. 7.16 **a** TEM images of bare and **b** AlF_3 coated $LiNi_{0.8}Co_{0.15}Al_{0.05}O_2$; **c** magnification performance of bare and AlF_3 coated $LiNi_{0.8}Co_{0.15}Al_{0.05}O_2$ [55]; **d** rietveld refinement images of XRD data of chemically delithiated uncoated and AlF_3 coated $Li_{0.35}Ni_{1/3}Co_{1/3}Mn_{1/3}O_2$ [56]

(3) Phosphates

Phosphates are a major category of composite anion coatings. In the coating, the P=O bond can enhance the chemical stability of the material, protect the electrode material from acid corrosion by the electrolyte, and the strong covalent bond of PO $_4$ with metal ions can improve thermal stability. $AlPO_4$, $Co_3(PO_4)_2$, $SnPO_4$ can enhance the cyclic and thermal stability of the electrode, but some researchers believe that such coatings have poor conductivity and will hinder the diffusion of lithium ions. However, since Li_3PO_4 is an ionic conductor, it can theoretically optimize the electrochemical performance of the coated material. Xia and others used a liquid phase method to prepare uniformly distributed, highly crystalline nano $FePO_4$, and coated it on the surface of $LiNi_{0.8}Co_{0.15}Al_{0.05}O_2$ (NCA) to adjust the electrode interface performance. Surface modification of $FePO_4$ can effectively improve the cycling performance and thermal stability of NCA material. The sample with a coating amount of 2 wt.% maintained a capacity retention rate of over 85.82% after 100 cycles. Even at 5 C, it showed good rate capacity performance (Fig. 7.17). Electrochemical impedance spectroscopy (EIS) analysis showed that the coating reduced the charge transfer resistance, improved the cycling performance, and thus achieved stronger electrochemical performance [57].

(4) Carbon coating and ionic conductor coating

Carbon materials, as relatively cheap and easy-to-prepare materials, have long been used by researchers for ternary material coatings and have achieved good results. Graphene (Gr) material, as a new type of carbon material, has received widespread attention in various fields of materials in recent years. Liao and others prepared layered $LiNi_{0.8}Co_{0.1}Mn_{0.1}O_2$ @Gr composite material by chemical method, the results showed that $LiNi_{0.8}Co_{0.1}Mn_{0.1}O_2$ was wrapped by graphene nanosheets, the structure was intact, and the crystal characteristics remained unchanged [58]. Compared with pure $LiNi_{0.8}Co_{0.1}Mn_{0.1}O_2$ Compared to, the $LiNi_{0.8}Co_{0.1}Mn_{0.1}O_2$ modified by Gr has better electrochemical performance. Its initial discharge capacity can reach 212.9 mAh g^{-1}, with good cycle performance and rate performance. The excellent electrochemical performance is inseparable from the highly conductive graphene substrate, which improves the conductivity and structural stability of the material (Fig. 7.18a). In recent years, researchers have proposed a clever phase method, which uses the lithium residue on the electrode surface as the Li source, and uses

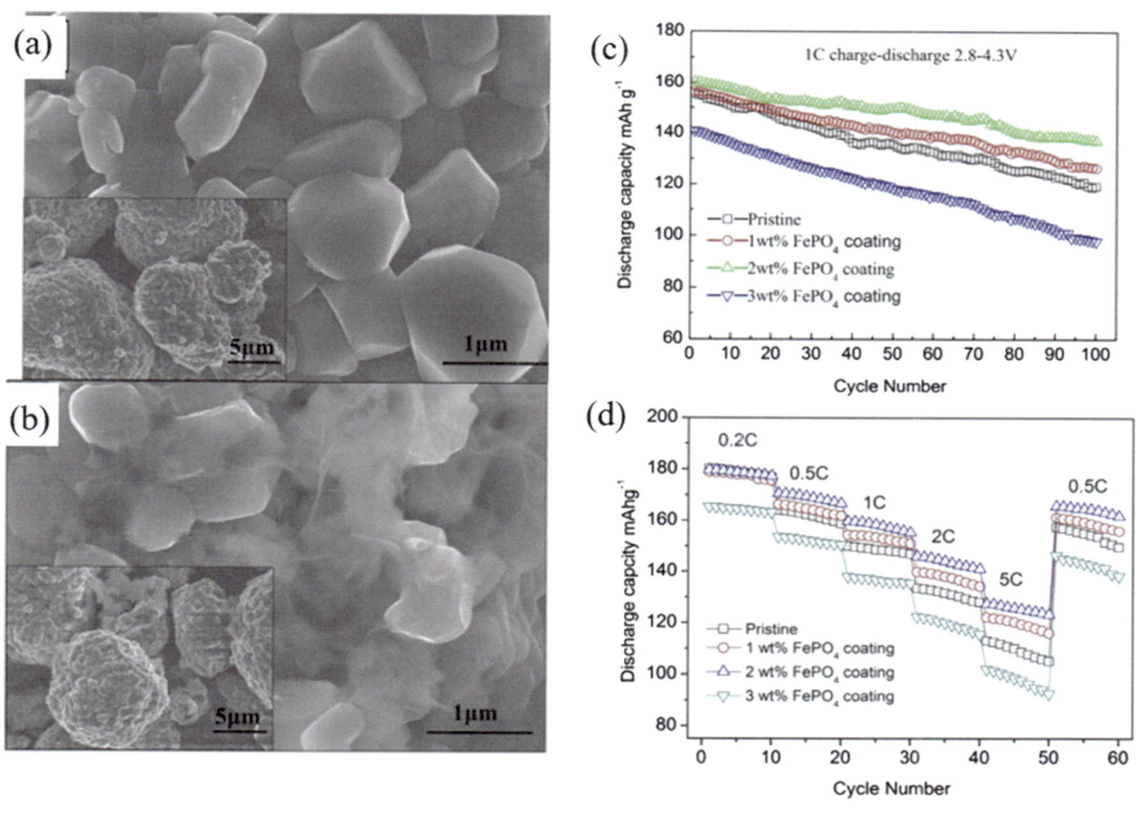

Fig. 7.17 SEM image of $FePO_4$ coated NCA sample **a** bare sample; **b** 2 wt.%; **c** cycling performance of NCA samples coated with different amounts of $FePO_4$; **d** rate performance of bare and 2 wt.% $FePO_4$ coated NCA samples [57]

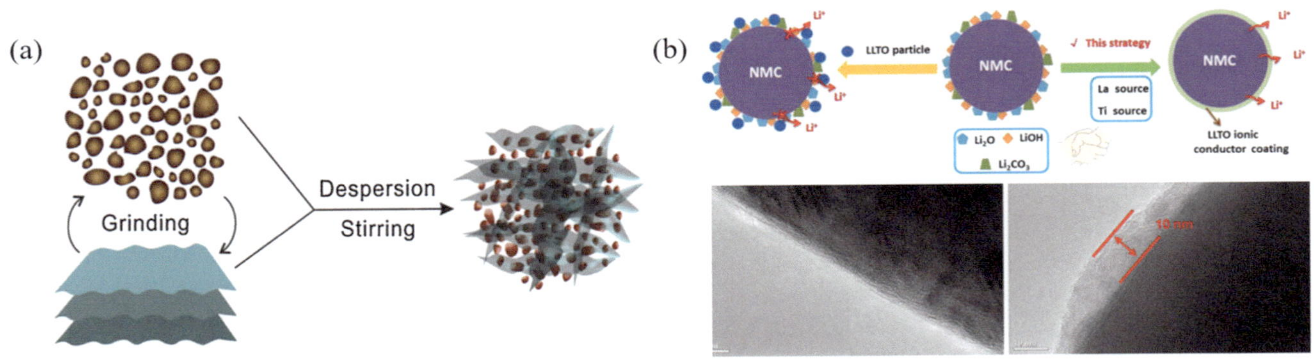

Fig. 7.18 **a** Schematic and TEM images of PSB-NCM synthesis [58]; **b** (Top) Schematic and (bottom) TEM images of bare sample and NCM modified with 1 mol% LLTO [59]

additional reagents to convert lithium impurities into Li^+ conductor coating, which is expected to improve electrochemical performance, simplify the process, and improve the Li^+ transfer rate between the electrode/electrolyte. Liu and others have studied a new surface modification strategy, forming a lithium lanthanum titanate (LLTO) coating with a thickness of less than 10 nm on the material surface (Fig. 7.18b). Among them, the $LiNi_{0.6}Co_{0.2}Mn_{0.2}O_2$ modified with 1 mol% LLTO has the best electrochemical performance, with a capacity retention rate of 87.2% after 200 cycles at 0.5 C, which is better than the original $LiNi_{0.6}Co_{0.2}Mn_{0.2}O_2$. Moreover, the cycle performance at higher charging voltage (4.5 V) and the storage performance against humidity and air are significantly improved after modification. The high ionic conductivity of LLTO and the uniformity of the coating are the main reasons for the enhancement of its electrochemical performance. It can inhibit the increase of interface resistance of the positive electrode material during the charge and discharge cycle and stabilize the crystal structure of the positive electrode material [59].

(5) Other materials

More and more materials are now being used for the development of electrode materials. These materials often have good characteristics or excellent structures, which can help solve various difficulties encountered in electrode materials. Wang and others used tetraethyl orthosilicate (TEOS) to polymerize on the surface of NCM (ENCM) under mild conditions without adding any additives, to form a polyepoxy functionalized polysiloxane (EPS) wrapped NCM (ENCM) positive electrode. This solved the $LiNi_{0.6}Co_{0.2}Mn_{0.2}O_2$ The issue of water molecules on the surface of the (NCM) cathode and HF in the electrolyte. The formed amorphous coating can provide a protective shell, improve the thermal stability of NCM, inhibit the thickening of the solid electrolyte interface (SEI), and remove HF in the electrolyte. After 100 cycles of charging and discharging in the voltage range of 2.8–4.3 V at a discharge rate of 1 C at 55 °C, the E-NCM composite material containing 2 mol% EPS can still maintain 84.9% of the capacity (Fig. 7.19a) [60]. In addition, the results of electrochemical impedance spectroscopy tests show that the polystyrene

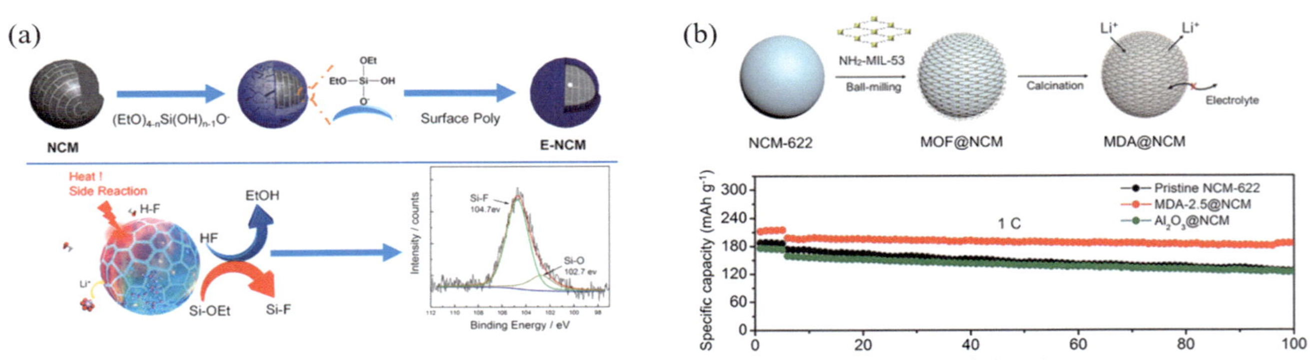

Fig. 7.19 **a** Schematic diagram and XPS spectrum of TEOS-coated NCM and TEOS inhibiting hydrolysis on the surface of NCM [60]; **b** schematic diagram of the synthesis of Al-based MOF-coated NCM and the cycle performance of bare NCM, Al_2O_3-coated NCM, and Al-based MOF-coated NCM at 1 C [61]

coating can alleviate the impedance increase during the cycle, especially at high temperatures. Therefore, the prepared E-NCM cathode material has good long-term cycle performance and thermal stability, and has good application prospects in high-energy lithium-ion batteries. Li and others used Al-based MOF material to coat the surface of $LiNi_{0.6}Co_{0.2}Mn_{0.2}O_2$. The electrochemical performance of the coated material at high voltage is improved [61]. In the voltage range of 3.0–4.5 V, the discharge specific capacities at 0.2 C and 1 C are 214.6 mAh g$-^1$ and 196.5 mAh g^{-1}, respectively. In addition, the initial discharge capacity, rate performance, and cycle performance have also been greatly improved (Fig. 7.19b). There are many types of surface-modified materials, which are widely used in NMC and NCA batteries. It is impossible to list them all in this article. However, we will briefly summarize some widely used surface modification materials in Table 7.3.

In summary, in order to improve the electrochemical performance of classical layered oxides, various surface modification methods have been adopted. The general idea is to reduce potential side reactions and their effects, including transformation into rock salt/spinel, dissolution of transition metal ions, and the formation of poor SEI-like films between the electrode and the electrolyte at high working voltages. Surface modification can be considered as an artificial SEI, serving as a Li^+ conductive passivation layer. A good surface coating should have the following key characteristics: 1. The coating should be a thin layer evenly distributed on the particle surface to minimize weight. 2. The coating material should be an excellent lithium ion conductor and electron conductor. 3. The coating material should maintain electrochemical stability within the working voltage window. 4. The coating should have sufficient mechanical strength to prevent fractures caused by electrode shape/volume changes during the (de)intercalation process.

7.4.4 Material Gradient Design

Nickel-rich cathode materials ($LiNi_xM_{1-x}O_2$; M = Mn, Co, Al; x > 0.5) usually undergo layered phase ($R\bar{3}m$) degradation into spinel phase ($Fd\bar{3}m$) and rock salt phase ($Fm\bar{3}m$) when cycled at high pressure (>4.3 V) and high temperature (>50 °C). Therefore, secondary particles formed by

Table 7.3 Electrochemical characteristics of some common coating materials in ternary batteries

Coating material	Cathode material	Voltage range (V)	First cycle/capacity retention	Rate	References
WO_3	NCM622	2.8–4.3	140 mAh g$-^1$ 84.8% 200 cycle	1 C	[62]
TiO_2	NCM622	2.8–4.5	135.8 mAh g^{-1} 88.7% 50 cycle	1 C	[63]
ZrO_2	NCM333	3.0–4.5	173.3 mAh g$-^1$ 99.1% 100 cycle	75 mA g^{-1}	[64]
Cr_2O_5	NCA	2.0–4.3	173.3 mAh g$-^1$ 69.3% 50 cycle	5 C	[65]
CuO	NCM532	3.0–4.6	164.6 mAh g^{-1} 89% 50 cycle	5 C	[66]
FeF_3	NCM333	3–4.8	154.7 mAh g$-^1$ 85% 50 cycle	200 mA g^{-1}	[67]
ZrFx	NCM333	3.0–4.6	152.3mAh g$-^1$ 89.9% 100 cycle	1 C	[68]
$AlPO_4$	NCM811	3.0–4.3	159.2 mAh g^{-1} 92% 200 cycle	1 C	[69]
Li_3PO_4	NCM433	3.0–4.8	145.4 mAh g$-^1$ 73.9% 50 cycle	1 C	[70]
C	NCM811	3.0–4.5	158.4 mAh g^{-1} 87.8% 80 cycle	0.5 C	[71]
Li_2ZrO_3	NCM811	2.7–4.5	163.8 mAh g$-^1$ 94.5% 200 cycle	1 C	[72]
Li_2TiO_3	NCM811	2.8–4.3	160.7 mAh g^{-1} 98% 170 cycle	1 C	[73]
H_2SiO_3	NCM622	2.75–4.5	101.5 mAh g$-^1$ 67% 200 cycle	5 C	[74]
PEDOT	NCM333	2.8–4.5	133.7 mAh g^{-1} 85.9% 80 cycle	150 mA g$-^1$	[75]

the aggregation of micron-sized primary particles continuously produce micro-cracks. These micro-cracks cut off the conductive paths between primary particles and cause thermal instability due to oxygen evolution during phase transition, leading to deterioration of electrochemical performance. The state of the material interface under electrochemical reaction is very important. Core-shell ternary materials mainly use nickel-rich materials as the core and relatively stable manganese-based materials as the shell. This structure is more conducive to the smooth transition of Li^+ in electrochemical reactions, making the material have higher specific capacity and better cycle performance [76, 77]. However, after long-term cycling, structural mismatch between the core and shell can lead to the formation of voids at the core/shell interface, thereby reducing cycle stability. In the concentration gradient structure, the concentration of transition metals changes continuously from the body to the surface, which can minimize lattice parameter mismatch at the core/shell interface [78, 79]. The structure and electrochemical performance of concentration gradient NCM cathode materials are shown in Fig. 7.20a and b [80]. The electrochemical performance of the full concentration gradient (FCG) core with different shell layers under harsh test conditions was compared. It was found that the low-nickel shell improved the capacity retention rate. Kim et al. proposed a nanoscale surface treatment of the primary particles of the $LiNi_{0.6}Co_{0.2}Mn_{0.2}O_2$ cathode material using a mixed ethanol solution of dissolved lithium and cobalt precursors [81]. The results showed that $LiNi_{0.6}Co_{0.2}Mn_{0.2}O_2$. The positive electrode surface has a high cobalt content and a nanometer-thick cation mixed phase ($Fm\bar{3}m$), which inhibits the structural change from layered to rock salt phase, thereby alleviating micro-cracks (Fig. 7.20c). In addition, the high oxidation state of Mn^{4+} on the surface reduces oxygen evolution at high temperatures. This method improves the structure and thermal stability, showing a rate performance comparable to the original sample under harsh cycling test conditions at 60 °C and high temperatures between 3.0 and 4.45 V.

7.5 Future Research and Development Directions

Due to the rapid development of electric vehicles in recent years, the demand for energy density of power batteries has also significantly increased. Advanced developed countries and our country have proposed the development goals of power batteries. By 2030, the specific energy of power batteries based on advanced systems should reach 500 Wh kg^{-1} or more, and the range of pure electric vehicles is comparable to that of gasoline vehicles. According to the above

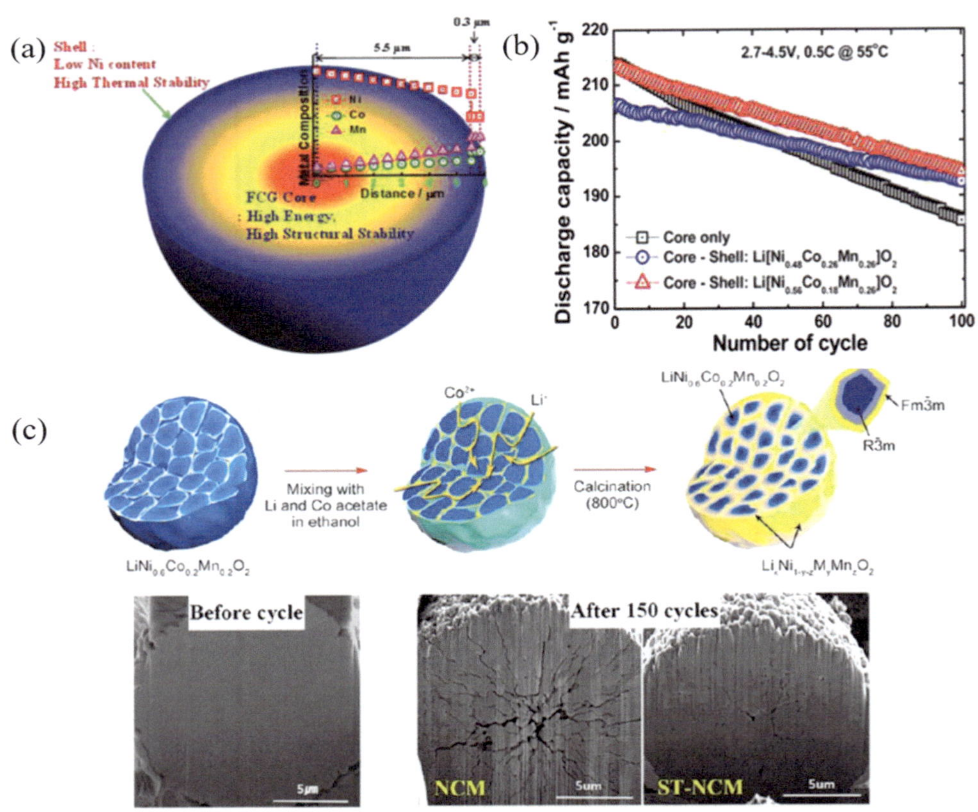

Fig. 7.20 **a** Schematic diagram of the mesostructure of core-shell particles with a full concentration gradient (FCG) core and a thin shell with low nickel content; **b** discharge capacity and cycle life of FCG core with different shells under harsh test conditions [80]; **c** concentration gradient NCM structure formation process and TEM images before and after cycling of bare and modified samples [81]

goals, it seems that transition metal oxide cathodes have more advantages in improving the specific energy of batteries compared to poly anion cathodes (such as lithium iron phosphate). Excess metal oxide cathodes will focus on the development of high-capacity, high-power cathode materials, as well as improvements in synthesis methods and the development of more compatible electrolyte additives.

7.5.1 Research on High Capacity NCA Cathodes

From current research, it can be seen that the Ni element in NCA materials still exists in the form of Ni^{3+} after being fully charged and stored at 45 °C for 2 years, and the surface structure has not changed to form NiO due to long-term storage. This shows that NCA has good storage and cycling performance. However, at higher temperatures (60 °C) or a wider range of charge and discharge voltages (0 ~ 100%ADOD), the surface microstructure of NCA will change (Fig. 7.21a), which will lead to the deterioration of NCA cycling performance [19].

Shoichiro and others studied the cycling performance of $LiNi_{0.76}Co_{0.14}Al_{0.1}O_2$ (NCA)/graphite batteries at 25 and 60 °C under different charge and discharge ranges. Using XPS, high-angle annular dark-field scanning transmission electron microscopy (HAADF-STEM), and scanning transmission electron energy loss spectroscopy (STEM-EELS), they studied the mechanism of cycle performance degradation under different charge and discharge ranges. After the ADOD 0 ~ 100% cycle test, many cracks were produced between the primary grain boundaries [19]. The electrolyte permeates through micro-cracks to the surface of primary particles and decomposes, forming a passivation film with poor conductivity (Fig. 7.21b). However, this is not the main reason for the poor cyclic performance. The main reason is that a NiO-like layer with an $Fm\bar{3}m$ rock salt structure is formed on the surface of each primary particle. The generation of cracks increases the contact resistance between particles, leading to poorer cyclic performance. The formation of micro-cracks and the rock salt NiO-like layer are the main reasons for the deterioration of cyclic performance. Therefore, controlling the depth of charge

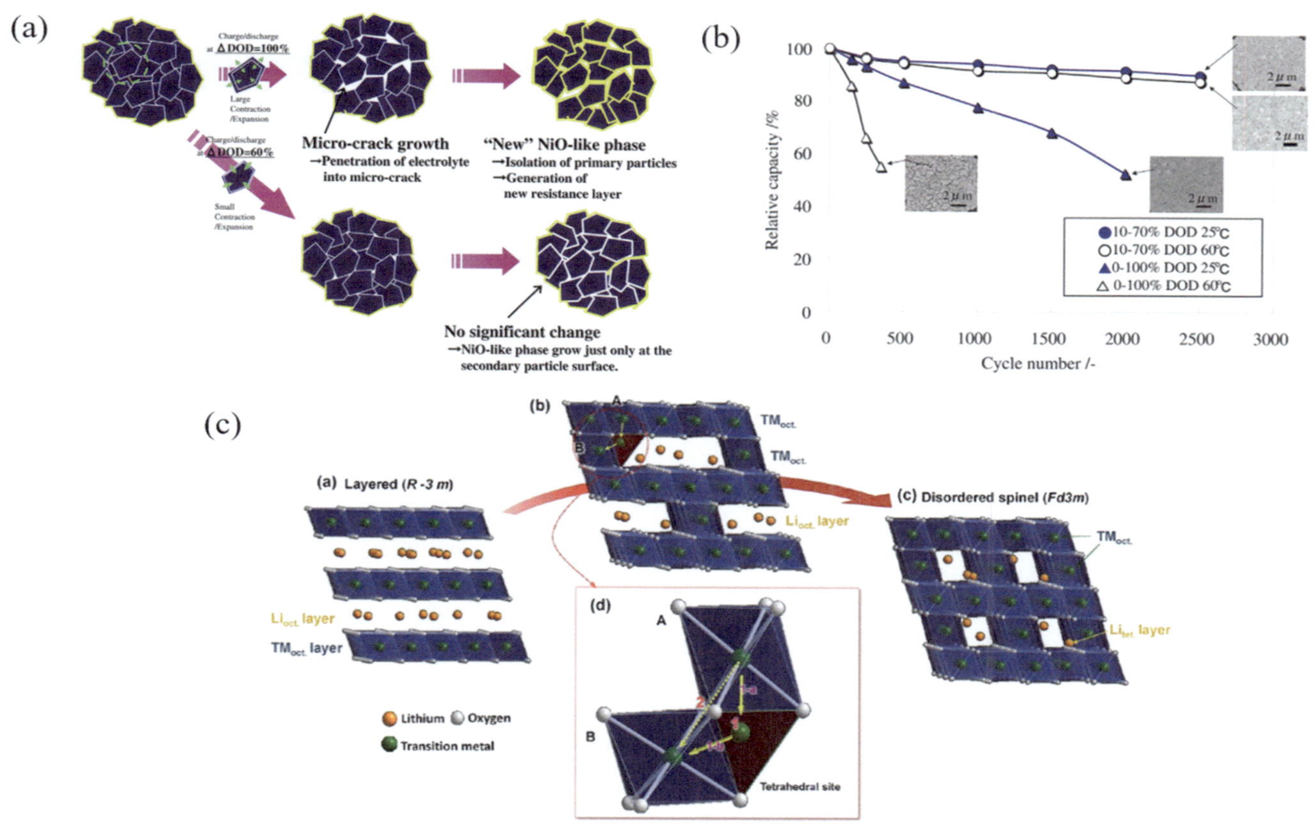

Fig. 7.21 **a** Schematic model of NCA particle degradation in cyclic tests; **b** cyclic performance of cylindrical model battery and cross-sectional SEM images of NCA cathode under two DDOD conditions at 25 °C and 60 °C: (●) 25 °C at 10–70%, (○) 60 °C at 10–70%, (▲) 25 °C at 0–100%, (Δ) 60 °C at 0–100% [19]; **c** during heating, $Li_xNi_{0.8}Co_{0.15}Al_{0.05}O_2$ Phase transition of the charging positive electrode: a layered structure, b cation migration during the phase transition from the layered structure to the disordered spinel structure, c disordered spinel structure, and d cation migration path from octahedron A to octahedron B [82]

and discharge can improve the cyclic performance of the battery.

In addition, the poor thermal stability of NCA batteries also severely hinders their development. It decomposes and undergoes exothermic reactions within a temperature range of 200 ~ 300 °C, exacerbating the safety risks of power batteries such as electric vehicles. Bak et al. used a combination of in-situ time-resolved X-ray diffraction (TRXRD) and mass spectrometry (MS) to study the thermal decomposition mechanism of battery cathode materials. They believe that the production of O_2 and CO_2 gases is closely related to the phase changes that occur during thermal decomposition. When x = 0.1, $LNi_{0.8}Co_{0.1}Al_{0.1}O_2$ shows severe oxygen release at about 175 °C, accompanied by a spinel phase structure change from layered to disordered. Since the released oxygen is highly reactive, it will cause the decomposition of PVDF binder and conductive carbon electrode carbon, thereby producing additional CO_2 at lower than expected temperatures. In the overcharged state, the material undergoes a phase change, and the phase change kinetics are related to the decay depth and cation migration path [82] (Fig. 7.21c) shows the transformation path of NCA from a layered structure to a disordered spinel structure during the delithiation process. After the disordered spinel structure is formed, it will continue to transform into a NiO-type rock salt structure and release more oxygen. In the transition process from disordered spinel to rock salt phase, TM ions immediately move from tetrahedral positions to octahedral positions. According to their conclusions, combined with Co K-edge XANES and EXAFS analysis, most of the Co cations occupy tetrahedral sites, increasing the phase transition temperature of the rock salt structure.

Based on the above analysis and discussion, the formation of rock salt phase NiO is the main reason for the deterioration of cyclic performance and safety performance. Delaying or inhibiting the formation of the rock salt phase can effectively improve the electrochemical performance and safety of NCA. From the above discussion, it is known that doping and surface modification of NCA can inhibit the formation of the rock salt phase. For example, if the doping element is stable at the tetrahedral position, it can delay or inhibit the formation of the rock salt phase. In addition, it is very important to screen alternative elements using computer simulation software.

7.5.2 Research on High-Power NCM or NCA Materials

High-power ternary materials have broad application prospects in the fields of electric tools, electric bicycles, electric vehicles, etc. As a lithium-ion power battery that has been a focus of development in recent years, the battery system with nickel cobalt manganese (NCM) and nickel cobalt aluminum (NCA) ternary as the positive electrode and graphite carbon as the negative electrode has exceeded 200 Wh kg^{-1}. However, the poor safety of electric vehicles has always restricted their vigorous development. That is to say, how to improve the safety of ternary materials is the focus of research on high-power ternary materials. In addition, improving the rate performance and specific capacity of the material is also the main research content of high-power ternary materials.

The specific capacity, thermal stability, and cycle performance of the ternary materials used in high-power batteries are closely related to the ratio of Ni–Co–Mn. In this regard, Surendra K and his research team compared and discussed the structure and electrochemical performance of $LiNi_{0.5}Mn_{0.5}O_2$, $LiNi_{1/3}Co_{1/3}Mn_{1/3}O_2$ and $LiNi_{0.4}Co_{0.2}Mn_{0.4}O_2$. The first material has a higher specific capacity (190 mAh g^{-1}) at a 0.1 C rate and within a voltage range of 2.5 ~ 4.5 V, but its rate performance is poor. When the discharge rate is 1 C, the specific capacity drops to below 100 mAh g^{-1}, and at a discharge rate of 3 C, the specific capacity is only about 50 mAh g^{-1} [83]. In contrast, the specific capacity of $LiNi_{0.4}Co_{0.2}Mn_{0.4}O_2$ at a 0.1 C rate is 170 mAh g^{-1}, but its rate performance is better. When the discharge rate is 3 C, the specific capacity is still above 110 mAh g^{-1}. For NCM442 material, the specific capacity is 180 mAh g^{-1} at a 0.1 C rate, higher than 110 mAh g^{-1} at a 1 C rate, and can maintain 80 mAh g^{-1} at a 3 C rate. The above capacity. This indicates that as the Co content increases from 0 to 0.33, the rate performance of the material improves. When the Ni content increases from 0.33 to 0.5, the specific capacity at 0.1 C rate increases from 170 to 190 mAh g^{-1}. In addition, Liu and his collaborators also compared the cycle performance and rate performance of ternary materials NCM523 and NCM433, and the results showed that NCM433 has better cycle performance and rate performance than NCM523. This is mainly due to the more stable structure of NCM433 and less cation mixing [84]. In addition, the ratio of the positive and negative electrodes of the fully charged battery will also have a certain impact on the cycle performance. The research of Sun's research group also emphasized the impact of Ni content on the electrochemical performance, structure, and thermal stability of NCM materials [10]. Comprehensive research results show that the electrochemical performance, structure, and thermal stability of the material are closely related to its components. As the Ni content increases, the discharge specific capacity increases, but the capacity retention rate and safety performance decrease. Finally, there are some noteworthy experimental phenomena. For example, based on the results of XPS experiments, the concentration of Ni on the surface is higher than the concentration of Mn in the electrolyte, and its concentration is twice that of Co. Therefore, how to inhibit the diffusion of Ni to the surface becomes a research problem, which involves the analysis of Ni diffusion channels. Another experimental phenomenon is the dissolution of Mn, which is mainly caused by the disproportionation reaction of Mn^{3+}–Mn^{2+}. Therefore, controlling the content of Mn^{3+}, especially the content of surface Mn^{3+}, is of great significance. Therefore, in the field of high-power ternary materials, it is necessary to consider multiple factors, including battery safety, specific capacity cycle life, calendar life, rate performance, cost, etc. This further requires in-depth research on the relationship between Ni/Co/Mn ratio and structure, and find the best conditions for optimizing comprehensive performance. These efforts will help achieve better overall performance.

7.5.3 Improvement of Synthesis Methods

In the current synthesis process of ternary material precursors, considering the cost and uniformity of the precursors, most of them are prepared by co-precipitation. However, due to the different solubility products of Ni, Co, and Mn, sometimes the synthesis conditions are not strictly controlled, which may cause Ni segregation, and Ni segregation will also affect the multiplicity and cycle performance of the material. In the above paper, we compared the NCM materials prepared by co-precipitation, sol-gel, solid phase, and hydrothermal assisted synthesis, and found that the materials prepared by hydrothermal assisted synthesis have the most uniform element distribution and better capacity retention rate, as shown in Table 7.4.

However, in industrial production, the co-precipitation method has higher efficiency and is easier to prepare. Therefore, Kim and others introduced the Couette-Taylor reactor in the process of synthesizing NCM precursors by co-precipitation. The precursor synthesized by this method has a narrow particle size distribution, is spherical, has a rough surface, and greatly improves the rate performance and cycle performance of NCM [85]. However, the Taylor flow reactors reported in the literature are mostly tubular structures, the reactor volume is generally small, the magnification is limited, and it is difficult to meet the needs of large-scale industrial production, so further research and improvement are needed (Fig. 7.22). One is the design of a new type of gas distribution device, which successfully induced the formation of Taylor flow in the stirred tank reactor and conducted experiments. Under the same experimental conditions, compared with the traditional stirred tank reactor, the conversion rate of sucrose hydrolysis increased by 26.7%. Ultrasonic vibration can also be added to the existing stirred tank to make the element distribution in the product more uniform. Another method is to modify the existing stirred tank reactor, design a new gas distribution device using the Taylor-Prudhomme theorem, adjust the two important parameters of speed and inlet flow in the stirring device, and successfully construct a complete Taylor vortex column structure in the stirred tank.

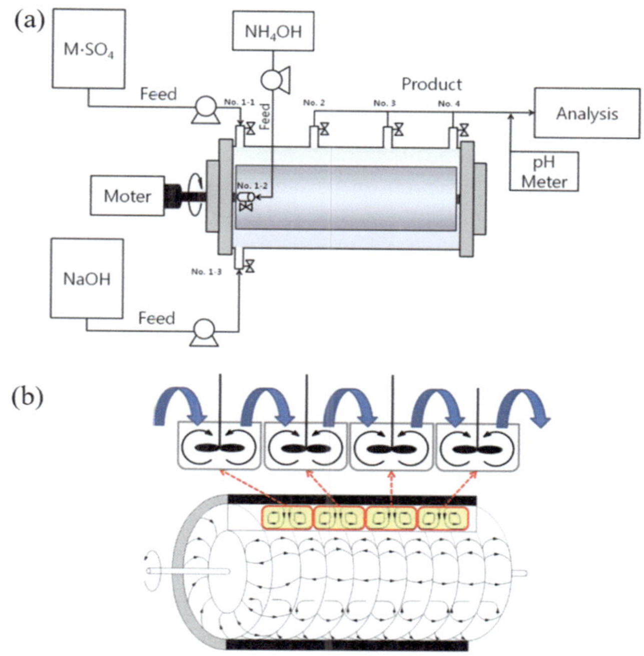

Fig. 7.22 **a** Experimental system of reaction crystallization of metal hydroxides in continuous Couette-Taylor crystallizer and **b** schematic diagram of Taylor vortex induced in Couette-Taylor crystallizer [85]

Table 7.4 Basic characteristics of ternary cathode materials prepared by different methods

Preparation method	Sol-gel method	Hydrothermal method	Solid phase method	Co-precipitation method
Capacity	Medium	High	Low	High
Cycle performance	Average	Good	Poor	Good
Cost	High	Medium	Low	Medium
Synthesis temperature	Medium	High	Medium	Low
Particle size	Uniform	Uniform	Non-uniform	Medium
Electrochemical performance	Average	Good	Poor	Good

7.5.4 Research on Electrolyte Additives in NCM Materials

In the development of lithium-ion batteries with ternary materials as the positive electrode, another very important task is the research on its electrolyte. In recent years, the research on the electrolyte of NCM materials has also received a lot of attention, but in industrial production, it is mainly based on $LiPF_6$ The main salt, with carbonate as the main solvent, additives are added to improve the performance of the battery. The role of additives is manifold, mainly to improve the surface state of the positive electrode and reduce the reaction of active substances with the electrolyte. Usually, by decomposing the additive at a lower voltage, a dense film can be formed to protect the active material. It also improves the applicability of the battery in low-temperature environments, etc.

Lithium metal batteries, composed of a lithium metal anode and a high-voltage nickel-rich cathode, have the potential to achieve high capacity and power density. However, there is an urgent need for suitable electrolytes to withstand the cathode oxidation under high cut-off voltages. Here, Yang et al. proposed a strategy for a helmet-like inorganic rich cathode electrolyte interface (CEI) with pentafluorophenylboronic acid (PFPBA) as an additive, to explore maintaining a 4.8 V Li||$LiNi_{0.6}Co_{0.2}Mn_{0.2}O_2$ (NCM622) battery's anti-oxidation electrolyte. In this CEI, the armored lithium borate surrounded by the upper layer of the CEI inhibits the dissolution of molecules inside the CEI, also improving the lithium-ion conductivity of the CEI, while a large amount of lithium is distributed in the CEI, compared to polymer molecules, improving mechanical stability and Li^+ conductivity (Fig. 7.23a). With such a powerful Li^+ conductive CEI, Li||NCM622 battery has excellent stability under a cut-off voltage of 4.6 V, with a capacity retention rate of 91.2% after 400 cycles. Even at a cut-off voltage of 4.8 V, excellent cycling performance can be achieved [86].

Lithium-ion batteries (LIB) that can operate at low temperatures (LT) are essential for orbital missions, seabed areas, and electric vehicle applications. Unfortunately, there is a significant loss of battery capacity due to the large kinetic barriers that occur at low temperatures. Zheng et al. proposed using Na^+ as a heteropositive ion additive, and further characterizing the Li-Na mixture and fluoride-rich interface by preferentially reducing Na^+ (solvent/anion) clusters, finding that it effectively promotes Li^+ migration through LiF/NaF grain boundaries (Fig. 7.23b). The graphite/nickel manganese cobaltate (NCM) full battery, under the condition of an N/P ratio of 1.1 (positive electrode loading of 18.5 mg cm^{-2}), can achieve a capacity of 125.1 mAh g^{-1} after 100 long-term cycles at −20 °C. Finally, we assembled a 270 Wh kg^{-1} graphite/NCM pocket battery, which discharged at −40 °C during the initial cycle, with a capacity of 108.7 mAh g^{-1}. This research is both fundamental and practical, and will lead to further progress, enabling lithium-ion batteries to operate at unprecedented extreme working temperatures [87].

It is easy to see that the use of additives can effectively control the surface properties of NCM. The impact of various electrolyte additives and combinations of additives on battery performance has received a lot of attention from researchers. They also concluded that electrolyte additives are necessary for improving the cycle performance of batteries. Additives should help form a passivation film at the positive and negative electrodes, have a low charge transfer resistance, and reduce side reactions. At present, there is still a lot of work to be done in the research of electrolyte matching ternary materials, such as the research of high-pressure ternary, high-nickel ternary electrolyte, the research of high-power battery electrolyte, and so on!

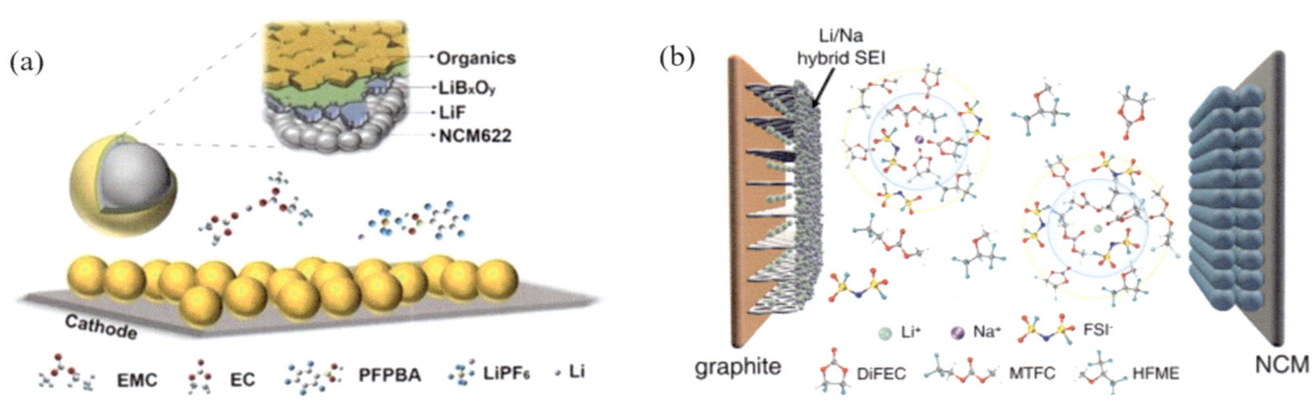

Fig. 7.23 **a** Schematic diagram of PFPBA additive in NCM622 battery [86]; **b** schematic diagram of the solvation environment of 1 M $LiPF_6$ EC/EMC FWSE + Na electrolyte [87]

In summary, due to the comprehensive advantages of NCM/NCA ternary materials in terms of specific capacity, working potential, safety, cost, and environmental friendliness, they are receiving more and more attention. Especially the low-Co, Ni-rich materials, due to their higher energy density and lower cost, show greater application prospects in automotive power batteries. Many organizations and companies, including the U.S. Department of Energy, are funding a large number of research groups. In order to achieve its large-scale application in electric vehicles, the following goals need to be met (driving distance: 300 miles, battery pack energy: 60–70 kWh, single cell energy density: ~300 Wh kg^{-1}, cathode material energy density: ~750 Wh kg−[1]); therefore, NCM/NCA cathode materials need to meet the following requirements [88]: (1) specific capacity ≥ 200 mAh g^{-1}, (2) cut-off potential ≥ 4.4 V (vs. Li/Li^{+}), (3) electrode density ≥ 3.9 g cm^{-3} (currently ≤ 3.6 g cm−[3]), (4) and cost ≤ 25% lower than $LiCoO_2$. However, the inherent defects of NCM/NCA materials themselves: severe capacity decay, poor rate performance, obvious voltage degradation, prominent safety issues, and low tap density, etc., seriously affect their commercial development. To achieve the expected goals, experts and scholars need to continue in-depth research to solve various problems. Finally, more extensive and closer cooperation between academia and industry is indispensable for the large-scale application of NCM/NCA cathode materials in high-performance lithium-ion batteries.

References

1. Liu Z, Yu A, Lee JY (1999) Synthesis and characterization of $LiNi_{1-x-y}Co_xMn_yO_2$ as the cathode materials of secondary lithium batteries. J Power Sour 81–82:416–419
2. Ohzuku TMY (2001) Layered lithium insertion material of $LiCo_{1/3}Ni_{1/3}Mn_{1/3}O_2$ for lithium-ion batteries. Chem Lett 7:642–643
3. Melot Bc TJ (2013) Design and preparation of materials for advanced electrochemical storage. Acc Chem Res 46(5):1226–1238
4. Saubanère M, Mccalla E, Tarascon JM et al (2016) The intriguing question of anionic redox in high-energy density cathodes for Li-ion batteries. Energy Environ Sci 9(3):984–991
5. Ding Y, Mu D, Wu B et al (2017) Recent progresses on nickel-rich layered oxide positive electrode materials used in lithium-ion batteries for electric vehicles. Appl Energy 195:586–599
6. Li W, Lee S, Manthiram A (2020) High-nickel NMA: a cobalt-free alternative to NMC and NCA cathodes for lithium-ion batteries. Adv Mater 32(33)
7. Venkatraman S, Choi J, Manthiram A (2004) Factors influencing the chemical lithium extraction rate from layered $LiNi_{1-y-z}Co_yMn_zO_2$ cathodes. Electrochem Commun 6(8):832–837
8. Liang M, Song D, Zhang H et al (2017) Improved performances of $LiNi_{0.8}Co_{0.15}Al_{0.05}O_2$ Material Employing $NaAlO_2$ as a new aluminum source. ACS Appl Mater Interfaces 9(44):38567–38574
9. Lei T, Li Y, Su Q et al (2018) High-voltage electrochemical performance of $LiNi_{0.5}Co_{0.2}Mn_{0.3}O_2$ cathode materials via Al concentration gradient modification. Ceramics Int 44(8):8809–8817
10. Noh H-J, Youn S, Yoon CS et al (2013) Comparison of the structural and electrochemical properties of layered $Li[Ni_xCo_yMn_z]O_2$ (x = 1/3, 0.5, 0.6, 0.7, 0.8 and 0.85) cathode material for lithium-ion batteries. J Power Sour 233:121–130
11. Kim H-G, Myung S-T, Lee JK et al (2011) Effects of manganese and cobalt on the electrochemical and thermal properties of layered $Li[Ni_{0.52}Co_{0.16+}Mn_{0.32-x}]O_2$ cathode materials. J Power Sour 196(16):6710–6715
12. Yoshizawa H, Ohzuku T (2007) An application of lithium cobalt nickel manganese oxide to high-power and high-energy density lithium-ion batteries. J Power Sour 174(2):813–817
13. Armand M, Tarascon JM (2008) Building better batteries. Nature 451:652
14. Gupta A, Manthiram A (2020) Designing advanced lithium-based batteries for low-temperature conditions. Adv Energy Mater 10(38)
15. Fly A, Kirkpatrick I, Chen R (2021) Low temperature performance evaluation of electrochemical energy storage technologies. Appl Thermal Eng 189
16. Hou S, Gao T, Li X et al (2020) Operando probing ion and electron transport in porous electrodes. Nano Energy 67
17. Sun Z, Li Z, Gao L et al (2018) Grafting Benzenediazonium Tetrafluoroborate onto $LiNixCoyMnzO_2$ materials achieves sub-zero-temperature high-capacity lithium-ion storage via a diazonium soft-chemistry method. Adv Energy Mater 9(6)
18. Sun Z, Li Z, Wu X-L et al (2019) A practical li-ion full cell with a high-capacity cathode and electrochemically exfoliated graphene anode: superior electrochemical and low-temperature performance. ACS Appl Energy Mater 2(1):486–492
19. Watanabe S, Kinoshita M, Hosokawa T et al (2014) Capacity fade of $LiAl_yNi_{1-x-y}Co_xO_2$ cathode for lithium-ion batteries during accelerated calendar and cycle life tests (surface analysis of $LiAl_yNi_{1-x-y}Co_xO_2$ cathode after cycle tests in restricted depth of discharge ranges). J Power Sour 258:210–217
20. Watanabe S, Kinoshita M, Nakura K (2011) Comparison of the surface changes on the cathode during long term storage testing of high energy density cylindrical lithium-ion cells. J Power Sour 196(16):6906–6910
21. Lu H, Zhou H, Svensson AM et al (2013) High capacity $Li[Ni_{0.8}Co_{0.1}Mn_{0.1}]O_2$ synthesized by sol-gel and co-precipitation methods as cathode materials for lithium-ion batteries. Solid State Ionics 249–250:105–111
22. Wu H, Pang X, Bi J et al (2020) Cellulose nanofiber assisted hydrothermal synthesis of Ni-rich cathode materials with high binding particles for lithium-ion batteries. J Alloys Compounds 829
23. Xiao Z-W, Zhang Y-J, Wang Y-F. Synthesis of high-capacity $LiNi_{0.8}Co_{0.1}Mn_{0.1}O_2$ cathode by transition metal acetates. Trans Nonferrous Metals Soc China 25(5):1568–1574
24. Huang Y, Wang Z-X, Li X-H et al Synthesis of $Ni_{0.8}Co_{0.1}Mn_{0.1}(OH)_2$ precursor and electrochemical performance of $LiNi_{0.8}Co_{0.1}Mn_{0.1}O_2$ cathode material for lithium batteries. Trans Nonferrous Metals Soc China 25(7):2253–2259
25. Myung S-T, Maglia F, Park K-J et al (2016) Nickel-rich layered cathode materials for automotive lithium-ion batteries: achievements and perspectives. ACS Energy Lett 2(1):196–223
26. Li W, Reimers JN, Dahn JR (1993) In situ X-ray diffraction and electrochemical studies of $Li_{1-x}NiO_2$. Solid State Ionics 67:123–130
27. Zhuang GV, Chen G, Shim J et al (2004) Li_2CO_3 in $LiNi_{0.8}Co_{0.15}Al_{0.05}O_2$ cathodes and its effects on capacity and power. J Power Sour 134(2):293–297

28. Noh Mijung YL, Jaephil C (2006) Water adsorption and storage characteristics of optimized $LiCoO_2$ and $LiNi_{1/3}Co_{1/3}Mn_{1/3}O_2$ composite cathode material for Li-ion cells. J Electrochem Soc 153(5):A935–A940
29. Shizuka K, Kiyohara C, Shima K et al (2007) Effect of CO_2 on layered $Li_{1+z}Ni_{1-x-y}Co_xM_yO_2$ (M = Al, Mn) cathode materials for lithium ion batteries. J Power Sour 166(1):233–238
30. Liang L, Hu G, Jiang F et al (2016) Electrochemical behaviours of SiO_2 -coated $LiNi_{0.8}Co_{0.1}Mn_{0.1}O_2$ cathode materials by a novel modification method. J Alloys Compounds 657:570–581
31. Hwang S, Chang W, Kim SM et al (2014) Investigation of changes in the surface structure of $Li_xNi_{0.8}Co_{0.15}Al_{0.05}O_2$ cathode materials induced by the initial charge. Chem Mater 26(2):1084–1092
32. Yano A, Aoyama S, Shikano M et al (2015) Surface structure and high-voltage charge/discharge characteristics of Al-oxide coated $LiNi_{1/3}Co_{1/3}Mn_{1/3}O_2$ cathodes. J Electrochem Soc 162(2):A3137–A3144
33. Sallis S, Pereira N, Mukherjee P et al (2016) Surface degradation of $Li_{1-x}Ni_{0.80}Co_{0.15}Al_{0.05}O_2$ cathodes: correlating charge transfer impedance with surface phase transformations. Appl Phys Lett 108(26)
34. Solchenbach S, Hong G, Freiberg ATS et al (2018) Electrolyte and SEI decomposition reactions of transition metal ions investigated by on-line electrochemical mass spectrometry. J Electrochem Soc 165(14):A3304–A3312
35. Hausbrand R, Becker D, Jaegermann W (2014) A surface science approach to cathode/electrolyte interfaces in Li-ion batteries: contact properties, charge transfer and reactions. Prog Solid State Chem 42(4):175–183
36. Hou P, Li G, Gao X (2016) Tailoring atomic distribution in micron-sized and spherical Li-rich layered oxides as cathode materials for advanced lithium-ion batteries. J Mater Chem A 4(20):7689–7699
37. Feng L, Liu Y, Wu L et al (2021) Enhancement on inter-layer stability on Na-doped $LiNi_{0.6}Co_{0.2}Mn_{0.2}O_2$ cathode material. Powder Technol 388:166–175
38. Liao P-Y, Duh J-G, Lee J-F (2009) Valence change and local structure during cycling of layer-structured cathode materials. J Power Sour 189(1):9–15
39. Xu Y-D, Zhang J, Wu Z-G et al (2019) Stabilizing the structure of nickel-rich lithiated oxides via Cr doping as cathode with boosted high-voltage/temperature cycling performance for Li-ion battery. Energy Technol 8(1)
40. Nurpeissova A, Choi MH, Kim J-S et al (2015) Effect of titanium addition as nickel oxide formation inhibitor in nickel-rich cathode material for lithium-ion batteries. J Power Sour 299:425–433
41. Kageyama M, Li D, Kobayakawa K et al (2006) Structural and electrochemical properties of $LiNi_{1/3}Mn_{1/3}Co_{1/3}O_{2-x}Fx$ prepared by solid state reaction. J Power Sour 157(1):494–500
42. Zhang P, Liu Z, Ma B et al (2021) Improving the single crystal $LiNi_{0.8}Co_{0.1}Mn_{0.1}O_2$ cathode material performance by fluorine doping. Ceramics Int 47(23):33843–33852
43. Xue L, Li Y, Xu B et al (2018) Effect of Mo doping on the structure and electrochemical performances of $LiNi_{0.6}Co_{0.2}Mn_{0.2}O_2$ cathode material at high cut-off voltage. J Alloys Compd 748:561–568
44. Sun H, Cao Z, Wang T et al (2019) Enabling high rate performance of Ni-rich layered oxide cathode by uniform titanium doping. Mater Today Energy 13:145–151
45. Lu Y, Mo Y, Chen Y et al (2021) Effects of various elements doping on $LiNi_{0.6}Co_{0.2}Mn_{0.2}O_2$ layered materials for lithium-ion batteries. Energy Technol 9(7)
46. Chu B, You L, Li G et al (2021) Revealing the role of W-doping in enhancing the electrochemical performance of the $LiNi_{0.6}Co_{0.2}Mn_{0.2}O_2$ cathode at 4.5 V. ACS Appl Mater Interfaces 13(6):7308–7316
47. Wu F, Li Q, Chen L et al (2019) Improving the structure stability of $LiNi_{0.8}Co_{0.1}Mn_{0.1}O_2$ by surface perovskite-like $La_2Ni_{0.5}Li_{0.5}O_4$ self-assembling and subsurface La^{3+} doping. ACS Appl Mater Interfaces 11(40):36751–36762
48. Liu S, Chen X, Zhao J et al (2018) Uncovering the role of Nb modification in improving the structure stability and electrochemical performance of $LiNi_{0.6}Co_{0.2}Mn_{0.2}O_2$ cathode charged at higher voltage of 4.5 V. J Power Sour 374:149–157
49. He Y-S, Pei L, Liao X-Z et al (2007) Synthesis of $LiNi_{1/3}Co_{1/3}Mn_{1/3}O_{2-z}F_z$ cathode material from oxalate precursors for lithium ion battery. J Fluorine Chem 128(2):139–143
50. Shin H-S, Shin D, Sun Y-K (2006) Improvement of electrochemical properties of $Li[Ni_{0.4}Co_{0.2}Mn_{(0.4-x)}Mg_x]O_{2-y}F_y$ cathode materials at high voltage region. Electrochimica Acta 52(4):1477–1482
51. Wu L, Tang X, Chen X et al (2020) Improvement of electrochemical reversibility of the Ni-Rich cathode material by gallium doping. J Power Sour 445
52. Cho D-H, Jo C-H, Cho W et al (2014) Effect of residual lithium compounds on layer Ni-rich $Li[Ni_{0.7}Mn_{0.3}]O_2$. J Electrochem Soc 161(6):A920–A926
53. Chen Y, Zhang Y, Wang F et al (2014) Improve the structure and electrochemical performance of $LiNi_{0.6}Co_{0.2}Mn_{0.2}O_2$ cathode material by nano-Al_2O_3 ultrasonic coating. J Alloys Compd 611:135–141
54. Lai Y-Q, Xu M, Zhang Z-A et al (2016) Optimized structure stability and electrochemical performance of $LiNi_{0.8}Co_{0.15}Al_{0.05}O_2$ by sputtering nanoscale ZnO film. J Power Sour 309:20–26
55. Kim HB, Park BC, Myung ST et al (2008) Electrochemical and thermal characterization of AlF3-coated $Li[Ni_{0.8}Co_{0.15}Al_{0.05}]O_2$ cathode in lithium-ion cells. J Power Sour 179(1):347–350
56. Sun YK, Cho SW, Lee SW, Yoon CS, Amine K (2007) AlF_3-coating to improve high voltage cycling performance of $LiNi_{1/3}Co_{1/3}Mn_{1/3}O_2$ cathode materials for lithium secondary batteries. J Electrochem Soc 154(3):A168–A172
57. Xia S, Li F, Chen F et al (2018) Preparation of FePO4 by liquid-phase method and modification on the surface of $LiNi_{0.80}Co_{0.15}Al_{0.05}O_2$ cathode material. J Alloys Compd 731:428–436
58. Jan SS, Nurgul S, Shi X et al (2014) Improvement of electrochemical performance of $LiNi_{0.8}Co_{0.1}Mn_{0.1}O_2$ cathode material by graphene nanosheets modification. Electrochimica Acta 149:86–93
59. Liu S, Zhang C, Su Q et al (2017) Enhancing electrochemical performance of $LiNi_{0.6}Co_{0.2}Mn_{0.2}O_2$ by lithium-ion conductor surface modification. Electrochimica Acta 224:171–177
60. Wang H, Ge W, Li W et al (2016) Facile fabrication of ethoxy-functional polysiloxane wrapped $LiNi_{0.6}Co_{0.2}Mn_{0.2}O_2$ cathode with improved cycling performance for rechargeable Li-Ion battery. ACS Appl Mater Interfaces 8(28):18439–18449
61. Li S, Fu X, Zhou J et al (2016) An effective approach to improve the electrochemical performance of $LiNi_{0.6}Co_{0.2}Mn_{0.2}O_2$ cathode by an MOF-derived coating. J Mater Chem A 4(16):5823–5827
62. Song G, Zhong H, Dai Y, Zhou X, Yang J (2019) WO_3 membrane-encapsulated $LiNi_{0.6}Co_{0.2}Mn_{0.2}O_2$ advanced Li-ion batteries cathode material. Ceramics Int 45(6):6774–6781
63. Chen Y, Zhang Y, Chen B et al (2014) An approach to application for $LiNi_{0.6}Co_{0.2}Mn_{0.2}O_2$ cathode material at high cutoff voltage by TiO_2 coating. J Power Sour 256:20–27
64. Huang Y, Chen J, Ni J et al (2009) A modified ZrO_2-coating process to improve electrochemical performance of $Li(Ni_{1/3}Co_{1/3}Mn_{1/3})O_2$. J Power Sour 188(2):538–545
65. Wang Y, Li Y, Zhang Y et al (2021) Influence of synergistic effect of $LiNi_{0.8}Co_{0.15}Al_{0.05}O_2@Cr_2O_5$ composite on the electrochemical properties. Chinese J Chem Eng 33:327–336

66. Liu T, Zhao S-X, Wang K et al (2012) CuO-coated $Li[Ni_{0.5}Co_{0.2}Mn_{0.3}]O_2$ cathode material with improved cycling performance at high rates. Electrochimica Acta 85:605–611
67. Kim CS, Cho J-H, Park YJ (2014) Electrochemical properties of FeF_3-coated $Li[Ni_{1/3}Co_{1/3}Mn_{1/3}]O_2$ cathode material. Mater Res Bull 58:49–53
68. Yun SH, Park K-S, Park YJ (2010) The electrochemical property of ZrFx-coated $Li[Ni_{1/3}Co_{1/3}Mn_{1/3}]O_2$ cathode material. J Power Sour 195(18):6108–6115
69. Cho J, Kim TJ, Kim J, Noh M, Park B (2004) Synthesis, thermal, and electrochemical properties of $AlPO_4$-coated $LiNi_{0.8}Co_{0.1}Mn_{0.1}O_2$ cathode materials for a li-ion cell. J Electrochem Soc 151(11):A1899–A1904
70. Song HG, Kim JY, Kim KT et al (2011) Enhanced electrochemical properties of $Li(Ni_{0.4}Co_{0.3}Mn_{0.3})O_2$ cathode by surface modification using Li_3PO_4-based materials. J Power Sour 196(16):6847–6855
71. Sim S-J, Lee S-H, Jin B-S et al (2020) Use of carbon coating on $LiNi_{0.8}Co_{0.1}Mn_{0.1}O_2$ cathode material for enhanced performances of lithium-ion batteries. Sci Rep 10(1)
72. Liang H, Wang Z, Guo H, Wang J, Leng J (2017) Improvement in the electrochemical performance of $LiNi_{0.8}Co_{0.1}Mn_{0.1}O_2$ cathode material by Li_2ZrO_3 coating. Appl Surf Sci 423:1045–1053
73. Meng K, Wang Z, Guo H, Li X, Wang D (2016) Improving the cycling performance of $LiNi_{0.8}Co_{0.1}Mn_{0.1}O_2$ by surface coating with Li_2TiO_3. Electrochimica Acta 211:822–831
74. Liu S, Wu H, Huang L, Xiang M, Liu H, Zhang Y (2016) Synthesis of $Li_2Si_2O_5$-coated $LiNi_{0.6}Co_{0.2}Mn_{0.2}O_2$ cathode materials with enhanced high-voltage electrochemical properties for lithium-ion batteries. J Alloys Compd 674:447–454
75. Liu X, Li H, Li D et al (2013) PEDOT modified $LiNi_{1/3}Co_{1/3}Mn_{1/3}O_2$ with enhanced electrochemical performance for lithium ion batteries. J Power Sour 243:374–380
76. Kosova NV, Devyatkina ET, Kaichev VV et al (2011) From 'core–shell' to composite mixed cathode materials for rechargeable lithium batteries by mechanochemical process. Solid State Ionics 192(1):284–288
77. Li Y, Makita Y, Lin Z et al (2011) Synthesis and characterization of lithium manganese oxides with core-shell $Li_4Mn_5O_{12}$@ Li_2MnO_3 structure as lithium battery electrode materials. Solid State Ionics 196(1):34–40
78. Koenig GM, Belharouak I, Deng H et al (2011) Composition-tailored synthesis of gradient transition metal precursor particles for lithium-ion battery cathode materials. Chem Mater 23(7):1954–1963
79. Yang X, Wang X, Zou G et al (2013) Spherical lithium-rich layered $Li_{1.13}[Mn_{0.534}Ni_{0.233}Co_{0.233}]_{0.87}O_2$ with concentration-gradient outer layer as high-performance cathodes for lithium ion batteries. J Power Sour 232:338–347
80. Noh H-J, Myung S-T, Lee YJ et al (2014) High-energy layered oxide cathodes with thin shells for improved surface stability. Chem Mater 26(20):5973–5979
81. Kim H, Kim MG, Jeong HY et al (2015) A new coating method for alleviating surface degradation of $LiNi_{0.6}Co_{0.2}Mn_{0.2}O_2$ cathode material: nanoscale surface treatment of primary particles. Nano Lett 15(3):2111–2119
82. Bak S-M, Nam K-W, Chang W et al (2013) Correlating structural changes and gas evolution during the thermal decomposition of charged $Li_xNi_{0.8}Co_{0.15}Al_{0.05}O_2$ cathode materials. Chem Mater 25(3):337–351
83. Martha SK, Sclar H, Szmuk Framowitz Z et al (2009) A comparative study of electrodes comprising nanometric and submicron particles of $LiNi_{0.50}Mn_{0.50}O_2$, $LiNi_{0.33}Mn_{0.33}Co_{0.33}O_2$, and $LiNi_{0.40}Mn_{0.40}Co_{0.20}O_2$ layered compounds. J Power Sour 189(1):248–255
84. Liu S, Xiong L, He C (2014) Long cycle life lithium ion battery with lithium nickel cobalt manganese oxide (NCM) cathode. J Power Sour 261:285–291
85. Kim J-M, Chang S-M, Chang JH et al (2011) Agglomeration of nickel/cobalt/manganese hydroxide crystals in Couette-Taylor crystallizer. Colloids Surf A 384(1–3):31–39
86. Yang Y, Wang H, Zhu C et al (2023) Armor-like Inorganic-rich Cathode Electrolyte Interphase Enabled by the Pentafluorophenylboronic Acid Additive for High-voltage Li||NCM622 Batteries. Angewandte Chemie International Edition 62(22)
87. Zheng X, Cao Z, Luo W et al (2023) Solvation and interfacial engineering enable −40 °C operation of graphite/NCM batteries at energy density over 270 Wh kg^{-1}. Adv Mater 35(10)
88. Myung S-T, Maglia F, Park K-J, Yoon CS, Lamp P, Kim S-J et al (2017) Nickel-rich layered cathode materials for automotive lithium-ion batteries: achievements and perspectives. ACS Energy Lett 2(1):196–223

8 Lithium Iron Phosphate and Lithium Iron Manganese Phosphate Cathode

Olivine-type phosphate cathode material $LiFePO_4$ has attracted great attention from the scientific community since it was first reported by Goodenough et al. [1] in 1997, and has gradually developed into one of the most widely used lithium-ion battery cathode materials in commercialization. Due to the non-toxicity of iron elements and their widespread distribution in the earth's crust, $LiFePO_4$ has the advantages of low cost and environmental friendliness. It also has a working voltage of 3.4 V (Li/Li^+) and a theoretical capacity of 170 mAh g^{-1}, and exhibits high safety and high cycle stability. These advantages make $LiFePO_4$ a high-potential cathode material. However, this material also has some shortcomings, such as low electronic conductivity and low lithium ion diffusion rate, which to some extent limit the larger scale application of this material. In response to these issues, researchers have optimized the performance of $LiFePO_4$ materials through various modification methods. The following will introduce the structure, preparation methods, and research progress of $LiFePO_4$ materials in detail, as well as the mechanism and electrochemical performance of $LiFePO_4$ materials.

8.1 Structure of $LiFePO_4$ Material

The olivine structure $LiFePO_4$ material belongs to the orthorhombic crystal system P_{nmb} space group. Its multi-atom framework crystal is composed of alternating FeO_6/ LiO_6 octahedra and PO_4 tetrahedra to form a spatial skeleton structure (as shown in Fig. 8.1). Each FeO_6 octahedron is connected to the four adjacent FeO_6 octahedra through the shared vertices on the bc face to form a transition metal layer for electron transmission; LiO_6 octahedra form shared edge chains in the b-axis direction. On the bc plane, one FeO_6 octahedron, two LiO_6 octahedra, and one PO_4 tetrahedron group share an edge, while one PO_4 tetrahedron group shares an edge with one FeO_6 octahedron and two LiO_6 octahedra. In the entire crystal structure, Li, Fe, and O all exist in the form of covalent bonds, and O is not easy to remove, so the deintercalation behavior of lithium ions during the charging and discharging process is not easy to cause material structural changes. In addition, the small volume change during the lithium ion deintercalation process makes the $LiFePO_4$ material have excellent cycle performance. However, there is no continuous FeO_6 The shared-edge octahedron network structure results in poor electronic conductivity; on the other hand, the oxygen atoms in the crystal are arranged in a close-packed hexagonal manner, providing only limited channels for lithium ions, which makes the migration rate of lithium ions in the charging and discharging process of the material low [2].

8.2 Preparation Method

Improving the performance of $LiFePO_4$ through controlled design of the cathode material is an effective means. The preparation methods of $LiFePO_4$ cathode material usually include solid phase method, carbothermal reduction method, solvothermal method, co-precipitation method, and sol-gel method.

8.2.1 Solid Phase Method

The solid phase method is one of the most common large-scale commercial synthesis methods for $LiFePO_4$, obtaining the desired substance through solid raw materials. The solid phase method generally requires high-temperature calcination to carry out the reaction, and the reaction temperature, heating rate, reaction environment, etc., will affect the performance of the product. Although the solid phase method has the advantages of simple operation and suitability for large-scale production, the particle size and uniformity of $LiFePO_4$ materials synthesized by the solid phase method need to be improved. Wang et al. [3] successfully

X. Li et al., *Electrode Development for Lithium-ion Batteries*, https://doi.org/10.1007/978-981-96-3021-9_8

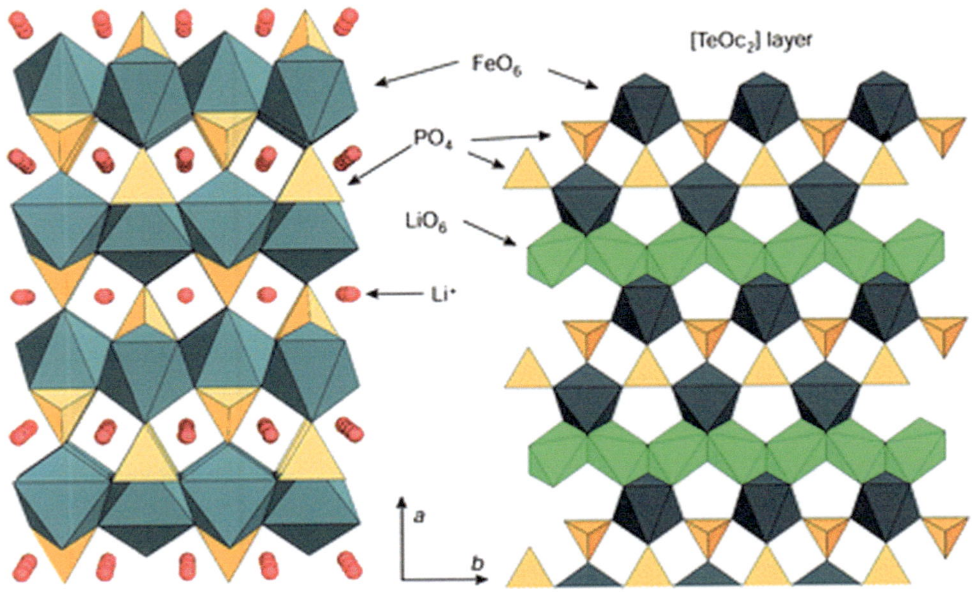

Fig. 8.1 Olivine $LiFePO_4$ crystal structure projected along [001] [2]

synthesized Y and F co-doped $LiFePO_4$ materials using the high-temperature solid phase method. Doping with F causes a rearrangement of the PO_4^{3-} electron cloud, while doping with Y introduces Li^+ vacancies, thereby reducing the diffusion resistance of Li ions. In addition, the results of material structure refinement show that Y and F doping weakens the Li–O bond and also widens the lithium ion diffusion channel, thereby increasing the lithium ion diffusion rate. Therefore, the modified sample has a super high discharge specific capacity of 135.8 mAh g^{-1} at a high rate of 10 C, and the discharge specific capacity at a rate of 5 C is as high as 148.6 mAh g^{-1}, and the capacity is almost unchanged after 700 cycles.

8.2.2 Carbothermal Reduction Method

When $LiFePO_4$ is synthesized by the carbothermal reduction method, trivalent iron, which is rich in raw materials, is usually used as the iron source, and an appropriate amount of carbon source is added. The carbon source is used as a reducing agent and as a carbon coating layer to improve the electronic conductivity of $LiFePO_4$. The choice and amount of different carbon sources have a significant impact on the morphology and electrochemical performance of the material. Currently, the most commonly used carbon sources are cheap organic carbon sources such as cane sugar, glucose, and citric acid. Zhu et al. [4] prepared $LiFePO_4$/(C+Cu) composite material by an improved carbothermal reduction method. Experiments have shown that after coating the surface of $LiFePO_4$ with a C+Cu layer, the $LiFePO_4$ The olivine structure has a particle size of about 100~200 nm. The $LiFePO_4$/(C+Cu) composite material exhibits an initial discharge specific capacity of 160.7 mAh g^{-1} at a 0.1 C rate, and the capacity retention rate is 98% after 200 cycles. At a 5 C rate, the initial discharge specific capacity is 105 mAh g^{-1}, and the capacity retention rate can reach 99% after 500 cycles, showing excellent cycle life and rate performance. Compared with pure carbon-coated materials, the introduction of copper particles and carbon materials further enhances the electronic conductivity of $LiFePO_4$, further improving its electrochemical performance. (3) Solvothermal method The solvothermal method is a widely used synthesis method. Through the solubility of raw materials in the solvent, the material crystallizes and grows under high temperature and pressure conditions. When water is generally used as the solvent, it can be called the hydrothermal method. The solvothermal method can prepare $LiFePO_4$ materials of different morphologies and crystallinities, and the generally adjusted parameters are reaction temperature and time, mixing order, solution pH value, etc. Although the solvothermal method can effectively adjust the material morphology and thereby improve the lithium ion diffusion rate, the process steps in this synthesis process need to be fully controlled, which is relatively complex. Li et al. [5] synthesized $LiFePO_4$ nanoparticles by the solvothermal method, which grow between graphene layers. Using the advantages of graphene's high conductivity and porous structure, the prepared nano $LiFePO_4$ has excellent cycle and rate performance. The discharge specific capacity is 160 mAh g^{-1} at a 0.2 C rate, and the discharge specific capacity is 107 mAh g^{-1} at a high rate of 60 C, and the capacity retention rate is 95% after more than 2000 cycles. (4) Co-precipitation method The co-precipitation method can prepare chemically homogeneous nanomaterials through chemical precipitation, where the key point is the uniform

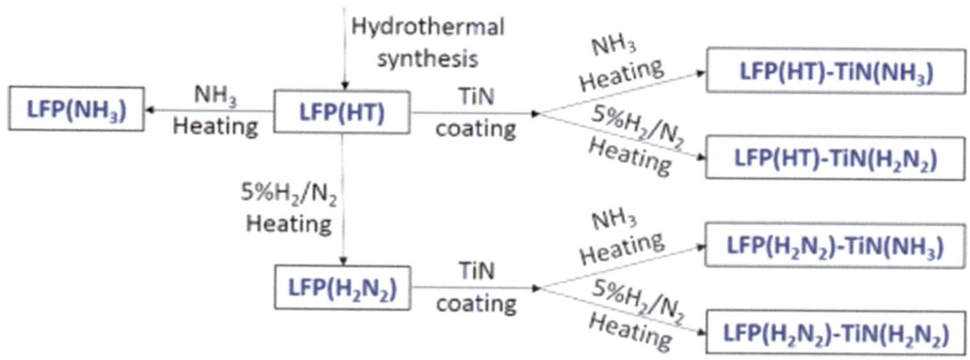

Fig. 8.2 Preparation conditions of TiN coated $LiFePO_4$ material [7]

mixing of materials. By adjusting the conditions such as the speed of raw material addition, the concentration of raw materials, the stirring speed, and the pH value of the solution, the growth kinetics of nanomaterials can be effectively controlled. Delacourt et al. [6] synthesized carbon-free $LiFePO_4$ by low-temperature chemical co-precipitation. The material has a narrow particle size distribution, small particle agglomeration, and a particle size of 100~200 nm. Although this study did not coat a conductive carbon layer, the prepared $LiFePO_4$ material still showed good electrochemical performance, with a discharge specific capacity of 146 mAh g^{-1} at 0.5 C, and a discharge specific capacity of 133 mAh g^{-1} at a higher rate of 5 C, and this material shows excellent volumetric energy density. (5) Sol–gel method The basis of the sol–gel method is colloid chemistry. A stable and uniform sol system is formed by controlling the liquid precursor, which is then aged and polymerized into a network gel, and then dried to form $LiFePO_4$ material. Although the $LiFePO_4$ material prepared by this method… The material has a small particle size and the product is uniform, but the shrinkage after gel drying is large, making it difficult to apply on a large scale. Zhang et al. [7] developed a high conductivity nano TiN material using the sol-gel synthesis route and used it as a conductive coating for $LiFePO_4$ battery material (Fig. 8.2). Compared with the uncoated $LiFePO_4$, the modified material has a higher specific capacity and better rate performance. At a 0.1 C rate, the material with a TiN mass fraction of 10% performs best, with a discharge specific capacity of 159 mAh g^{-1}, which is about 93% of the theoretical specific capacity, an increase of 38 mAh g^{-1} compared to pure $LiFePO_4$, and the initial discharge specific capacity at 5 C can reach 80 mAh g^{-1}, an increase of 40 mAh g^{-1} compared to pure $LiFePO_4$.

8.3 Research Progress

The low cost, high safety, and high cycle stability of $LiFePO_4$ material make it one of the widely used cathode materials in the field of power batteries and energy storage. However, the low intrinsic electron conductivity and relatively low lithium ion diffusion rate caused by the crystal structure characteristics of lithium iron phosphate batteries limit their energy density and high rate performance, and the discharge performance of $LiFePO_4$ material is poor in low temperature environments. In response to these problems, researchers mainly modify the material through surface coating, ion doping, nanosizing, and electrolyte, and have achieved significant results.

8.3.1 Surface Coating

The conductivity of $LiFePO_4$ material is extremely poor. By coating the material surface with conductive and structurally stable materials, the electron and ion conduction between materials can be improved. The commonly used coating materials include carbon materials, metals or metal oxides, and ion-conductive materials.

Carbon material is the simplest and cheapest coating material, with the following main functions: carbon material as a reducing agent can prevent the oxidation of Fe^{2+} in $LiFePO_4$ material; improve material conductivity; hinder direct contact between particles, effectively inhibiting particle growth. Liu et al. [8] research shows that the formation of $LiFePO_4$ composite material is related to particle size, temperature, atmosphere, and thermodynamic mechanism during the carbon coating process. By in-situ polymerization, the organic carbon source is coated on the surface of $LiFePO_4$, and after high-temperature carbonization and annealing, the composite material of $LiFePO_4$ and carbon is obtained. This method can not only obtain a complete and uniform carbon layer, but also the consistency of the coated carbon layer is good, the thickness is controllable, and it is compatible with $LiFePO_4$ The close contact effectively suppresses the excessive growth and agglomeration of $LiFePO_4$ particles. Xu et al. [9] developed a binary conductive additive containing only 1% porous graphene and 1% carbon black. The porous graphene with a large number of pores and a large specific surface area enhances the electronic conductivity of $LiFePO_4$, but does not affect the efficient transmission of ions. A small amount of carbon black, as a supplement to the conductive network formed by graphene, can fully contact with $LiFePO_4$, thereby achieving excellent conductivity throughout the $LiFePO_4$ electrode. The double carbon-coated $LiFePO_4$ material has an initial discharge specific capacity of up to 160 mAh g^{-1} at a rate of 0.1 C, and the initial capacity at 1 C can reach 140 mAh g^{-1}, with no decay after 50 cycles.

Some researchers have prepared $LiFePO_4$ cathode materials modified by metal or metal oxide and carbon composites. Lu et al. [10] proposed a simple method to uniformly disperse a small amount of highly conductive silver (mass fraction about 1.37%) throughout the $LiFePO_4$ surface to improve the conductivity of the $LiFePO_4$ material. The results show that the $LiFePO_4$ cathode coated with metal Ag exhibits superior electrochemical performance, with higher specific capacity and good cycle stability, and most importantly, the capacity of the material is significantly improved at high rates. The $LiFePO_4$–Ag composite has an initial specific capacity of 155 mAh g^{-1} at a rate of 0.7 C, the discharge specific capacity does not decay after 200 cycles, and the initial discharge specific capacity can reach 85 mAh g^{-1} at 14 C. Liu et al. [11] prepared WO_2 coated modified $LiFePO_4$/C cathode material by in-situ synthesis method. WO_2 can uniformly coat the surface of $LiFePO_4$ particles without changing the crystal structure of $LiFePO_4$. The $LiFePO_4$/C sample coated with WO_2 shows a better initial discharge specific capacity, at a rate of 1C, the initial discharge specific capacity is 130 mAh g^{-1}, which is about 20 mAh g^{-1} higher than the unmodified $LiFePO_4$ cathode material.

Ionic conductive materials have high ionic conductivity and Li^+ storage capacity, and can provide extra Li^+ for the battery during the charging and discharging process. In addition, the introduction of ionic conductive materials can effectively suppress Li/Fe antisite defects and improve the performance of $LiFePO_4$. The electrochemical performance of positive electrode materials. Among them, the most common is the three-dimensional framework material $Li_3V_2(PO_4)_3$ with a high Li^+ diffusion rate and working potential. Zhang et al. [12] prepared $LiFePO_4$ powder coated with fast ion conductor $Li_3V_2(PO_4)_3$ using the sol-gel method. The improved sample contains olivine $LiFePO_4$ phase and monoclinic $Li_3V_2(PO_4)_3$ phase. Some V^{3+} and Fe^{2+} elements are doped into the lattices of $LiFePO_4$ and $Li_3V_2(PO_4)_3$, respectively, causing lattice shrinkage and defect formation. Compared with commercial $LiFePO_4$, the lithium ion diffusion coefficient of modified $LiFePO_4$ increased by an order of magnitude, reaching 1.2×10^{-10} $cm^2\ s^{-1}$. Electrochemical test results show that the rate performance and cycle stability of modified $LiFePO_4$ have been significantly improved. The initial capacities of the modified $LiFePO_4$ at 5 and 10 C rates are 143.78 mAh g^{-1} and 111.06 mAh g^{-1}, respectively, and after 190 cycles, they show discharge specific capacities of 134.85 mAh g^{-1} and 104.99 mAh g^{-1}, and high capacity retention rates of 93.79% and 94.53%, respectively. (2) Ion doping Ion doping is considered an important method to improve the internal electronic conductivity and ion diffusion rate of $LiFePO_4$ material. Doping a small amount of ions/elements at Li sites, Fe sites, or other sites can effectively improve the charge–discharge performance and cycle stability of $LiFePO_4$ material at high rates. Li-site doping usually uses some smaller radius metal ions, such as Zr, Cr, Nb, Mg, Na, etc. Chung et al. [13] first reported $LiFePO_4$ materials doped with Ti, Al, Mg, etc. at Li sites. The doping of multivalent positive ions will produce positive ion defects, Fe^{3+}/Fe^{2+}. As the concentration increases, the mixed valence promotes the formation of P-type semiconductors, thereby increasing the electronic conductivity of lithium iron phosphate to 10^{-2} S cm^{-1}, an increase of about 108 times. Johnson et al. [14] found that Nb doping $LiFePO_4$ material can cause lattice distortion of the material, further generating more Li vacancies, and the transmission rate of Li^+ increases. The cycling performance and rate performance of $LiFePO_4$ material improve with the increase of Nb doping. The doped sample has a discharge specific capacity of up to 96.7 mAh g^{-1} at 10 C, and the capacity retention rate is 96.7 mAh g^{-1} after 200 cycles.

8.3.2 Doping

Common elements doped at the Fe site include Ni, Mn, Mg, Co, V, Zn, Mo, Cu, Cr, etc. Doping at the Fe site usually weakens the Li–O bond, increases the lattice volume, obtains a higher lithium ion migration rate and diffusion coefficient, and reduces the lattice distortion of lithium iron phosphate; at the same time, doping at the Fe site can inhibit the generation of Li/Fe antisite defects, thereby making Li^+ in the crystal diffusion channel easier to migrate without being blocked, all of which contribute to the improvement of electrochemical performance. Kim et al. [15] used a solid-phase reaction to dope Ti into $LiFePO_4$ at 700 °C, and prepared the target product $LiTi_{0.08}Fe_{0.92}PO_4$ material. By replacing a small amount of Fe^{2+} ions with Ti^{4+} ions, the material belongs to P-type semiconductors, its conductivity is enhanced, and the electrochemical performance of the material is ultimately improved. This cathode $LiTi_{0.08}Fe_{0.92}PO_4$ has a high charge and discharge specific capacity and good capacity retention rate. The initial discharge specific capacity at 0.2 C rate is 160 mAh g^{-1}, and the capacity loss is less than 2% after 50 cycles. The discharge specific capacity of this material at 11.4 C rate is 110 mAh g^{-1}, and the rate performance is good. Zhang et al.

[16] conducted a study on Mo-doped $LiFe_{1-x}Mo_xPO_4/C$ material. XRD results show that the doping of Mo does not change the olivine structure of $LiFePO_4/C$, but it also forms Li^+ vacancies, which helps to improve the diffusion rate of Li^+. The $LiFe_{0.98}Mo_{0.02}PO_4/C$ material has a discharge specific capacity of 143.2 mAh g^{-1} after 100 cycles at a 0.1 C rate, and when the rate reaches 5 C, $LiFe_{0.98}Mo_{0.02}PO_4/C$'s discharge capacity can still maintain above 90 mAh g^{-1}, and the capacity retention rate is 93.2% after 100 cycles.

In addition to the above-mentioned cation doping, anion doping is also an effective way to improve the conductivity of $LiFePO_4$ material. Some studies choose Cl^- and F^- as substitutes for O^{2-} anions. Lu et al. [17] reported F-doped $LiFePO_4/C$ nanoparticles synthesized through low-temperature hydrothermal reaction and high-temperature treatment. The discharge capacities at different rates are 120.4 (5 C), 101.3 (10 C), and 90.5 (15 C) mAh g^{-1} respectively. Sun et al. [18] synthesized Cl-doped $LiFePO_4/C$ cathode material through a carbothermal reduction route, which exhibited a high discharge capacity of about 90 mAh g^{-1} at 20 C. Cl doping optimized the electrochemical reactions and lithium ion diffusion in $LiFePO_4$. More and more researchers are modifying $LiFePO_4$ materials through multi-element doping, mainly regulating with two or more metal elements. Gao et al. [19] studied the mechanism of Zr, Co co-doping $LiFePO_4$ through first-principles calculations. The results showed that Zr, Co co-doping can reduce the band gap of $LiFePO_4$ and the diffusion barrier of Li^+, and improve the diffusion coefficient of Li^+. Among them, $Li_{0.99}Zr_{0.0025}Fe_{0.99}Co_{0.01}PO_4$ material showed excellent rate performance and cycle stability. After 50 cycles at 0.1 C rate, the discharge capacity is 139.9 mAh g^{-1}, and the capacity retention rate is 85%.

8.3.3 Nanosizing

Because the diffusion channel of Li^+ in the olivine-type $LiFePO_4$ lattice is one-dimensional and its transmission has high anisotropy, this to some extent limits the transmission efficiency of Li^+. The nanosizing modification method of materials, on the one hand, is beneficial to shorten the diffusion path of ions and improve the cycling performance of the material; on the other hand, it is beneficial to increase the specific surface area of the material and provide more diffusion channels for interface reactions. Therefore, the nanosizing method is widely used in the modification work of $LiFePO_4$ materials. Wang et al. [20] Ultrafine $LiFePO_4$ quantum dots co-modified with amorphous carbon and graphitized conductive carbon were successfully synthesized through a novel microreactor strategy (as shown in Fig. 8.3). This material, with a size of about 6.5 nm, exhibits ultra-fast lithium storage behavior, possessing extraordinary rate performance (providing 78 mAh g^{-1} capacity at 200 C) and excellent cycle stability (retention rate of about 99% after 1000 cycles at 20 C). The excellent performance is due to the large charge adsorption specific surface area and abundant Faraday reaction active sites of this nanocomposite material, as well as the ideal electronic and ionic transport kinetics. Jiang et al. [21] prepared $LiFePO_4$ of different morphologies and sizes by changing the concentration of glycerol using the solvothermal method. The results showed that when the ratio of water to glycerol concentration was 1:4, the synthesized nanorod-shaped $LiFePO_4$ had the smallest size in the [001] direction and exhibited excellent electrochemical performance, with capacity retention rates still as high as 99.5% and 93.2% after 500 cycles at 5 C and 20 C rates, respectively. Zhou et al. prepared layered

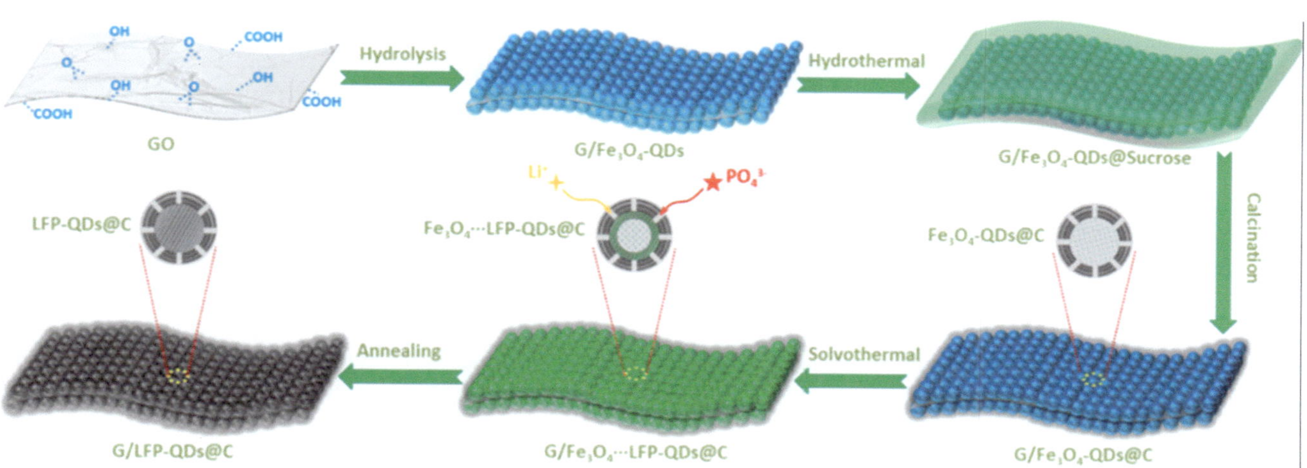

Fig. 8.3 Schematic diagram of the preparation of G/LFP-QDs@C by a novel microreactor strategy [20]

dumbbell-shaped $LiFePO_4$ and layered flower-shaped $LiFePO_4/C$ composite materials using the solvothermal method. The pure phase dumbbell-shaped $LiFePO_4$ had a discharge capacity of 140 mAh g^{-1} at a 0.1 C rate, mainly due to the larger specific surface area and shorter diffusion distance of the dumbbell-shaped structure composed of $LiFePO_4$ nanorods; while the carbon-coated flower-shaped $LiFePO_4$ crystals had a higher capacity, with a discharge specific capacity of 161 mAh g^{-1} [22, 23] at a 0.1 C rate.

8.3.4 Electrolyte

The cycle stability and safety of a battery largely depend on the stability and safety of the electrolyte. As the carrier of ion transport in $LiFePO_4$ batteries, the electrolyte plays a key role in conducting lithium ions between the positive and negative electrodes. Therefore, the modification of the electrolyte is particularly important, which has a significant impact on the specific capacity, operating temperature, cycle performance, and safety performance of $LiFePO_4$ batteries, and researchers have conducted a lot of research in this area. Zhang et al. [24] studied the synergistic effect of lithium bis(fluorosulfonyl)imide (LiFSI) and lithium bis(oxalato)borate (LiBOB) in $LiPF_6$/ethyl carbonate (EC)/ethyl methyl carbonate (EMC) electrolyte. In 1.0 mol L^{-1} $LiPF_6$ based electrolyte, 0.2 mol L^{-1} LiFSI can effectively improve the cycle stability and rate performance of the graphite anode. However, due to the severe corrosion of LiFSI on the aluminum current collector, it is difficult for the $LiFePO_4$ cathode to cycle in this electrolyte. On the one hand, the introduction of 0.2 mol L^{-1} LiBOB can effectively passivate the aluminum current collector and inhibit the corrosion caused by the bis(fluorosulfonyl) imide anion. On the other hand, LiFSI helps to reduce the high impedance at the electrode/electrolyte interface caused by LiBOB salt. In the electrolyte based on the $LiPF_6$/LiFSI/LiBOB ternary salt system, both the $LiFePO_4$ cathode and the graphite anode show high rate performance and improved cycle stability. In addition, full cell testing also confirmed the good compatibility between the graphite anode and the $LiFePO_4$ cathode in the $LiPF_6$/LiFSI/LiBOB ternary salt electrolyte system. Chen et al. [25] used a mixture of lithium bis(trifluoromethanesulfonyl)imide (LiTFSI) and LiBOB to replace the $LiPF_6$ salt in non-aqueous electrolytes in order to give $LiFePO_4$ based batteries excellent long-term cycle stability at high temperatures. It was found that the addition of LiBOB to the LiTFSI-based electrolyte can effectively prevent the severe corrosion of the Al current collector often observed in the thermally stable LiTFSI-based electrolyte. Compared with batteries using $LiPF_6$-based electrolytes, $LiFePO_4$ batteries using $LiTFSI_{0.6}$-$LiBOB_{0.4}$-based electrolytes showed excellent high-temperature stability (less than 5.8% capacity decay after more than 1000 cycles at 60 °C). Fan et al. [26] successfully designed a dual-salt (LiTFSI-$LiPF_6$) gel polymer electrolyte with a 3D (3 Dimensions) cross-linked polymer network (as shown in Fig. 8.4), achieving a capacity retention rate of 87.93% after 300 cycles of the $LiFePO_4$|Li battery. This 3D gel polymer electrolyte effectively suppressed lithium dendrites and ensured excellent ionic conductivity of the material.

In the application of electric vehicles, ternary materials and lithium iron phosphate materials have always been the focus of attention. Among them, the olivine-structured lithium iron phosphate battery has gradually become the main force of power lithium-ion batteries due to its low cost, high cycle stability, and high safety. However, the low lithium-ion diffusion rate and low conductivity caused by the olivine structure limit the development of the energy density of its battery, and at the same time, the $LiFePO_4$ Batteries also face the problem of poor low-temperature performance. Researchers have conducted a large amount of modification work and achieved significant results in surface coating, ion doping, nanoscale, and electrolyte. However, further improving the energy density and high-rate performance, and low-temperature performance of $LiFePO_4$ batteries in large-scale applications is still an urgent challenge. In the design of cathode materials, the microstructure and morphological characteristics of $LiFePO_4$ materials need to be regulated by low-cost and effective synthesis methods to enhance the lithium storage characteristics of the materials; at the same time, the compatibility of the electrolyte needs to be considered, and diversified construction should be carried out in the functional design of liquid electrolyte and solid electrolyte; in addition, the consistency problem of the whole battery also needs to overcome the process problems, and the negative electrode materials and separators and other components should be optimized.

8.4 Lithium Iron Manganese Phosphate Cathode

With the high demand for energy density of power batteries, the 3.4 V working voltage (Li/Li^+) and 170 mAh g^{-1} theoretical capacity of $LiFePO_4$ materials are no longer satisfactory, and its energy density (578 Wh kg^{-1}) can no longer meet the demand. The energy density of $LiMnPO_4$ cathode material with the same olivine structure is 700 Wh kg^{-1}, which is 20% higher than that of $LiFePO_4$, and it has high safety, abundant raw materials, but the inherent Jahn-Teller effect of Mn^{3+} in $LiMnPO_4$ materials makes it easy to dissolve in the electrolyte, affecting the cycle performance of the battery, and its low conductivity and poor thermal stability lead to the need to improve the electrochemical

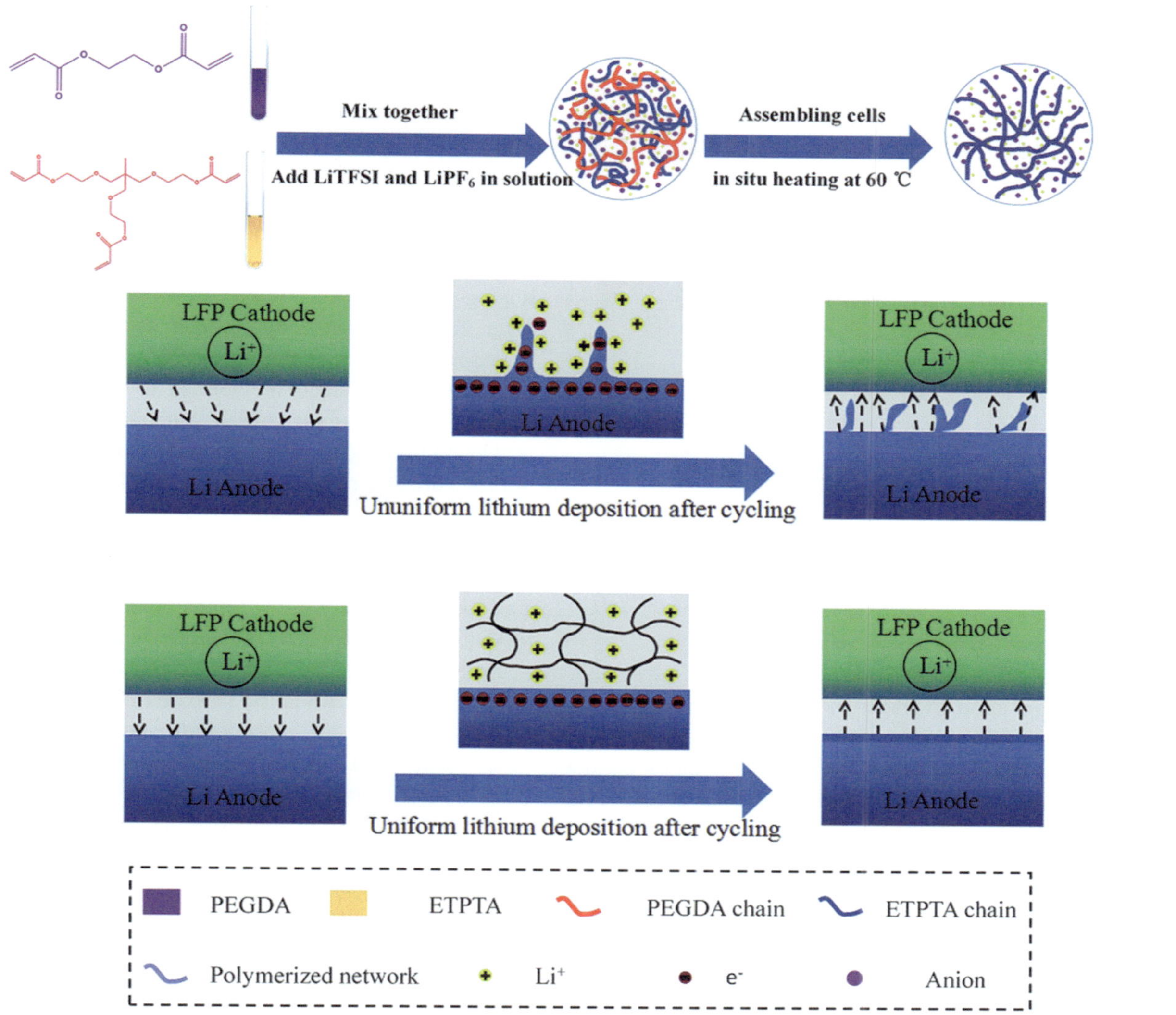

Fig. 8.4 In-situ polymerization process of gel polymer electrolyte [26]

performance of the battery. Therefore, researchers have extensively studied the solid solution lithium iron manganese phosphate $LiMn_{1-x}Fe_xPO_4$ material to combine the advantages of both $LiFePO_4$ and $LiMnPO_4$ materials.

8.4.1 Structure of $LiMn_{1-x}Fe_xPO_4$ Material

The olivine-type lithium iron manganese phosphate material $LiMn_{1-x}Fe_xPO_4$ belongs to the same olivine-type orthorhombic crystal system Pnmb space group as $LiFePO_4$ and $LiMnPO_4$ (as shown in Fig. 8.5). The oxidation-reduction potential of Mn is 4.1 V (Li/Li^+), and the reduction potential of Fe is 3.4 V (Li/Li^+). The radius of Mn^{2+} (0.8 Å) is larger than that of Fe^{2+} (0.74 Å), so in $LiMn_{1-x}Fe_xPO_4$ In the solid solution, lattice defects can be formed, which expands the lithium ion transmission channel and can improve the electronic conductivity and ionic conductivity of the material. Moreover, the interaction between Fe and Mn can enhance the stability of Mn ions during the charging and discharging process, thereby improving the cycle stability of the material. The olivine-type manganese iron phosphate lithium material $LiMn_{1-x}Fe_xPO_4$ is a single solid solution, and there is no clear conclusion on the optimal ratio of Mn and Fe. Its working voltage is 3.4–4.1 V (Li/Li^+), corresponding to the oxidation-reduction potentials of Mn and Fe ions. During the charging and discharging process of $LiMn_{1-x}Fe_xPO_4$, the two discharge platforms appearing at 3.4 and 4.1 V correspond to the oxidation-reduction reactions of Mn^{2+}/Mn^{3+} and Fe^{2+}/Fe^{3+}, and the lengths of the two platforms are proportional to the ratio of Mn and Fe ion content [27].

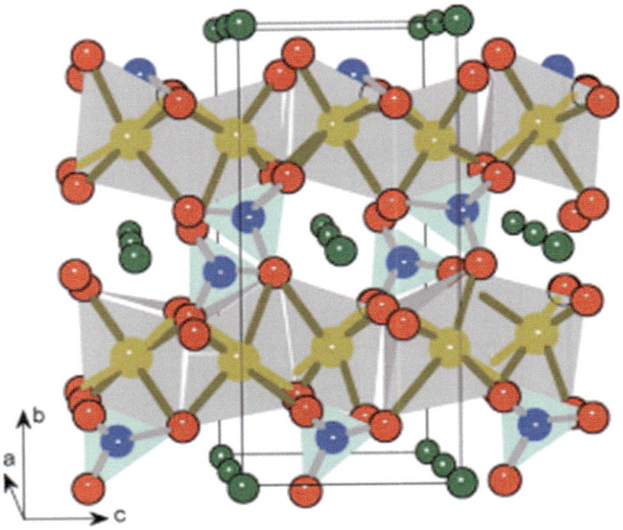

Fig. 8.5 Olivine $LiFe_{1-y}Mn_yPO_4$ structure [27]

8.4.2 Preparation Methods

Different preparation methods have a huge impact on the electrochemical performance of lithium manganese iron phosphate materials. General preparation methods include co-precipitation, sol-gel, solvent thermal, solid phase, spray drying, etc.

Co-precipitation is an effective liquid phase reaction method for synthesizing uniform products. This method does not require high equipment, has good repeatability, and produces uniform particles. Zong et al. [28] used co-precipitation with ethanol as the initiator to prepare $LiMn_{0.5}Fe_{0.5}PO_4$/C material. After 25 cycles at 0.05 C, it showed an excellent reversible capacity of 131 mAh g^{-1}, with a retention rate of about 91%. In addition, the sample also showed good rate performance, with a reversible capacity of 119 mAh g^{-1} at 0.2 C, and a capacity of 105 mAh g^{-1} at 1 C.

The lithium manganese iron phosphate material synthesized by the sol-gel method has characteristics of high purity and small particles, but the cost of raw materials for this method is high, making it difficult to achieve large-scale production. Budumuru et al. [29] prepared single-crystal carbon-coated nanotube and nanoparticle $LiFe_{1-x}Mn_xPO_4$ positive electrode material. The open microscopic structure of nanotubes and the oriented nanocrystals allow lithium ions to easily diffuse through the surface of the nanotubes along the main diffusion path b-axis [010] in the olivine structure. This promotes faster and easier diffusion of lithium ions. After 1000 cycles at 5 C, the capacity retention rate is 88%, showing excellent cycle performance.

Solvothermal method is one of the most common synthesis methods. It effectively avoids side reactions caused by high temperature and oxidation by reducing the reaction temperature and shortening the reaction time. However, the reaction conditions of this method are not easy to control, and the process is complex. Fan et al. [30] used sucrose as the carbon source and melamine as the nitrogen source to synthesize $LiMn_{0.8}Fe_{0.2}PO_4$ (LMFP-7) nanocrystals (as shown in Fig. 8.6) using a simple solvothermal method, and studied the doping mechanism of nitrogen atoms in the pyrolytic carbon layer. The results show that nitrogen doping does not change the crystal structure of the composite material. A large number of extrinsic defects are formed in the pyrolytic carbon of the LMFP-7 sample, which generate active defect sites, accelerate the migration of lithium ions, and are beneficial to improve the electronic conductivity and diffusion rate of lithium ions. The LMFP-7 sample showed excellent cycle performance, with a specific capacity of 131.8 mAh g^{-1} after 200 cycles at 1 C, and a capacity retention rate of up to 93.81%. The specific capacity of the sample without N doping is 99.3 mAh g^{-1}, and the capacity retention rate is 88.34%.

Solid phase method is the most conventional preparation method. Although it can be prepared on a large scale, it is difficult to adjust the morphology and particle uniformity of the product. Li et al. [31] prepared $LiMn_{0.5}Fe_{0.5}PO_4$/C (LMFP-2) composite material using an improved high-temperature solid phase method. The secondary carbon coating plays a role in refining particles and enhancing electron and ion transport, making the composite material show good electrochemical performance. The initial discharge specific capacity of the LMFP-2 sample at 0.1 C is 165 mAh g^{-1}, the discharge specific capacity at 20 C is 85 mAh g^{-1}, and the capacity retention rate after 100 cycles at 1 C is 90.1%. It has good rate performance and cycle stability. Through cyclic voltammetry and electrochemical impedance spectroscopy analysis, it is known that the lithium ion diffusion coefficient of the LMFP-2 sample has increased, and the charge transfer impedance has decreased, indicating that the secondary carbon coating effectively improves the carrier transport process in $LiMn_{0.5}Fe_{0.5}PO_4$, which plays an important role in improving material performance.

The spray drying method can obtain ultrafine powder of lithium iron manganese phosphate material, but this method has high requirements for equipment. Zhang et al. [32] first proposed a simple improved spray drying method using terephthalic acid and sucrose as carbon sources to synthesize carbon-coated nano-level $LiMn_{0.8}Fe_{0.2}PO_4$(LMFP) microspheres (as shown in Fig. 8.7). The diameter of these microsphere materials is 10–20 μm, the primary particle size is 100 nm, and there is a 5 nm carbon coating on the particles. This material shows very good rate performance and excellent cycle stability. After 200 cycles at 25 °C and

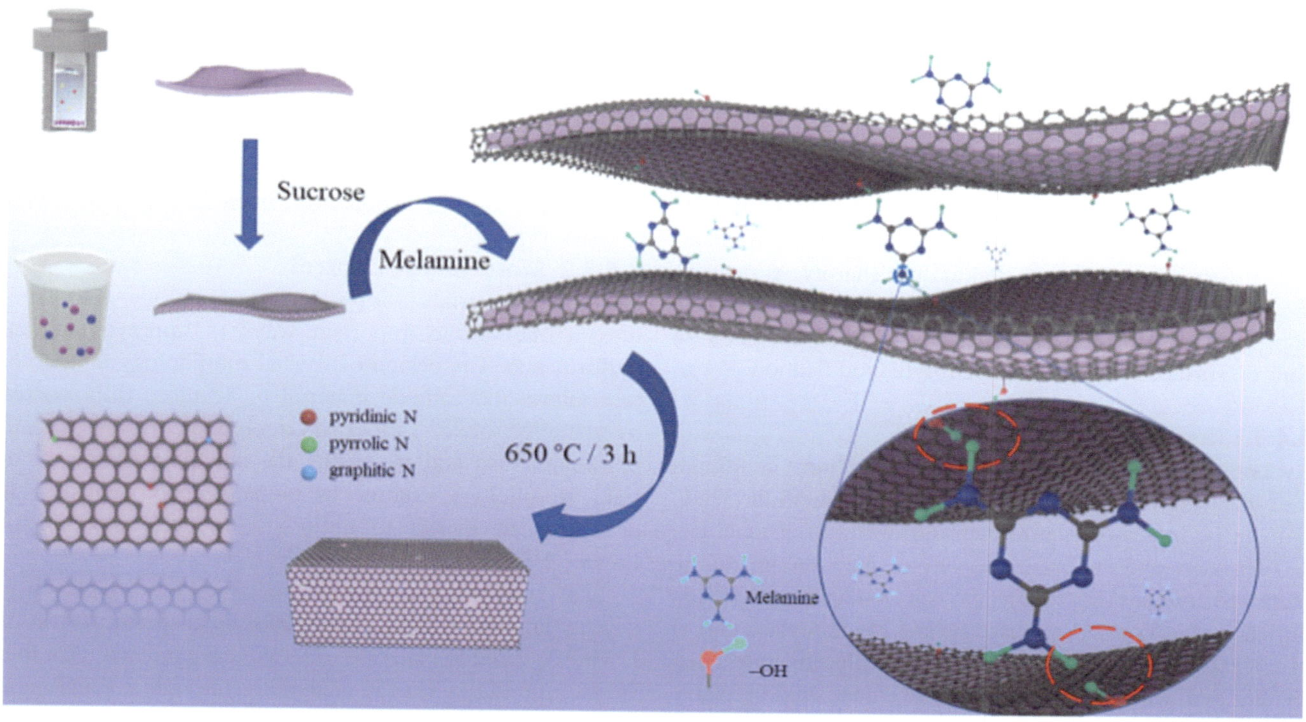

Fig. 8.6 Surface morphology of carbon-coated $LiMn_{0.8}Fe_{0.2}PO_4/C$ doped with nitrogen and the transition mechanism of nitrogen doping [30]

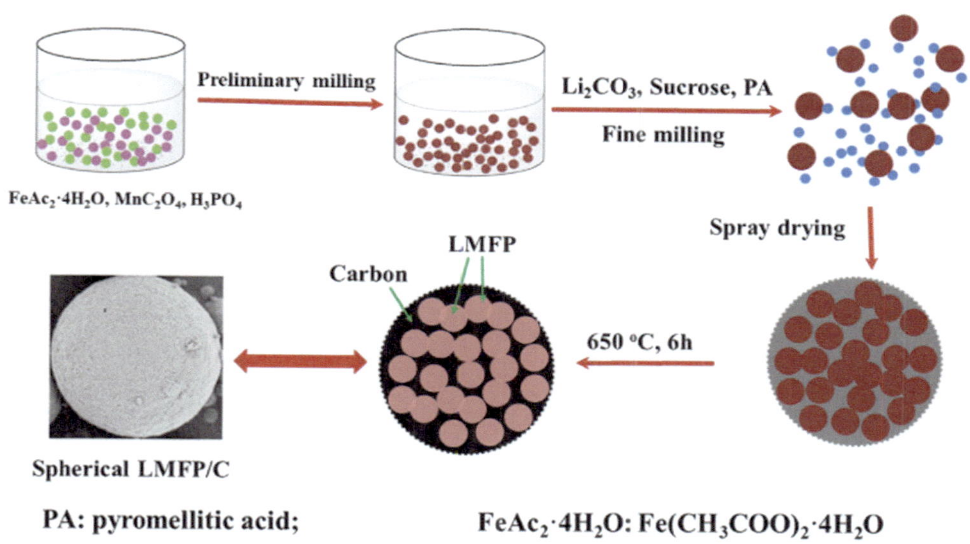

Fig. 8.7 Schematic diagram of the synthesis of LMFP nanoporous microspheres [32]

0.2 C, the capacity retention rate is 90.2%, and after 100 cycles at 55 °C and 1 C rate, the capacity retention rate is 95.0%. The research results show that the primary particles of LMFP are wrapped in a continuous carbon coating, which can prevent the positive electrode particles from being eroded by the electrolyte and ensure the excellent cycle performance of the sample.

8.4.3 Research Progress

Olivine-structured $LiMn_{1-x}Fe_xPO_4$ as a new type of positive electrode material for lithium-ion batteries, has the advantages of environmental friendliness, wide source of raw materials, low cost, high energy density and theoretical specific capacity, stable discharge voltage, good thermal stability and cycle performance, and has received widespread attention in the field of lithium-ion batteries. However, this material faces inherent disadvantages of olivine-structured phosphate compounds, such as low electronic conductivity and slow one-dimensional lithium ion diffusion rate, and the Jahn-Teller effect produced by Mn^{3+} ions in the reaction process, as well as the dissolution of Mn in the reaction process, will also cause the electrochemical performance of the material to drop sharply. At present, the electrochemical performance of lithium iron manganese phosphate is mainly improved by ion doping, morphology control, surface coating, and electrolyte modification.

8.4.3.1 Ion Doping

By doping some metal ions or non-metal ions in lithium iron manganese phosphate material, the defects in the material can be increased, the lithium ion diffusion channel can be expanded, and the carrier density of the material can be increased, thereby improving the ion conductivity and ion diffusion rate. Yi et al. [33] used Fe and Mg to replace part of Mn, and the results showed that double replacement has better rate performance and cycle stability than single replacement, and the discharge specific capacity can reach more than 140 mAh g^{-1} at 0.1 C. Wang et al. [34] prepared Fe^{2+}, Ni^{2+} co-doped $LiMn_{0.8}Fe_{0.2-x}Ni_xPO_4$ by an improved solvothermal method. ($x=0$, 0.01, 0.03, 0.05). Studies have shown that the introduction of Ni^+ can significantly reduce the particle size of the material, among which $LiMn_{0.8}Fe_{0.19}Ni_{0.01}PO_4$ has a capacity retention rate of 94.1% after 200 cycles at 0.5 C, and still has a discharge specific capacity of over 100 mAh g^{-1} at 20 C, demonstrating excellent electrochemical performance. Ni^{2+} has a higher electronegativity, and the introduction of Ni^{2+} changes the adsorption state of various metal ions on the surface of the intermediate product (Li_3PO_4), reducing the particle size of the material. At the same time, the doping of Ni^{2+} increases the lattice parameters of the material, expands the transmission channel of Li^+, and is conducive to the diffusion of Li^+ in the lattice.

8.4.3.2 Morphology Control

Designing materials with special morphology and structure is one of the important ways to improve material performance. The design of special morphology and structure allows the cathode material to be more fully wetted by the electrolyte, further shortening the diffusion path of lithium ions and increasing the reactivity of the material. Morphology control to prepare nanocomposites is an important means to optimize the cycle stability of $LiMn_{1-x}Fe_xPO_4$ cathode materials.

Liao et al. [35] discussed the effect of different iron doping amounts on the electrochemical performance of $LiMnPO_4$ cathode materials. The authors prepared ultra-thin $LiMn_{1-x}Fe_xPO_4(x \leq 0.15)$ nanoplates through a simple and controllable solvothermal method. The $LiFe_{0.15}Mn_{0.85}PO_4/C$ composite cathode material has a discharge specific capacity of 96.2 mAh g^{-1} at a 50 C rate, and maintains excellent cycle stability, with a capacity retention rate of nearly 70% after 1000 cycles at a 10 C rate. Ruan et al. [36] prepared $LiMn_{0.5}Fe_{0.5}PO_4$ with a concentration gradient through a simple and controllable two-step solvothermal method. Hollow core cathode material (HCG-LMFP/C) (as shown in Fig. 8.8). This material is rich in iron on the outside and

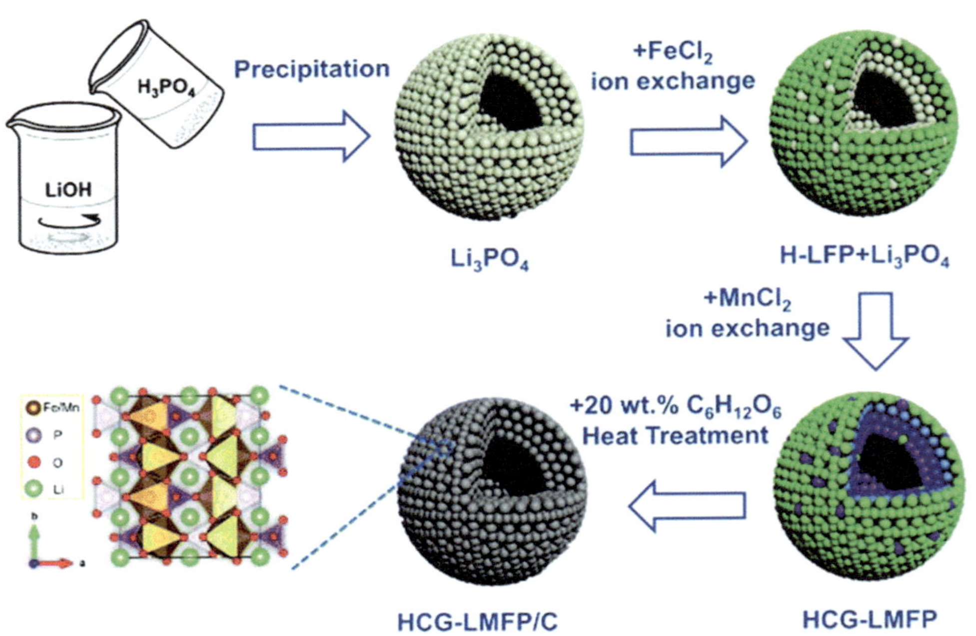

Fig. 8.8 Schematic diagram of the preparation of HCG-LMFP/C by a two-step hydrothermal method [36]

rich in manganese on the inside, with an optimized elemental gradient distribution as a shell around the hollow structure. The manganese-rich inner shell optimizes the material's energy density, while the surface iron-rich layer enhances the material's electrochemical activity and overcomes the inherent chemical instability of lithium manganese phosphate. In addition, the internal hollow hierarchical porous structure helps to reduce the level of anisotropic internal strain, thereby providing enhanced structural stability during the lithium insertion/extraction process. Therefore, the synthesized concentration gradient HCG-LMFP/C cathode displays high energy density (81 Wh kg^{-1}, power density of 10.6 kW kg^{-1}) and good rate performance (20 C and 60 C capacities are 111 and 78 mAh g^{-1} respectively) without affecting cycle life (capacity retention rate is 96% after 1000 cycles at 10 C). Hou et al. [37] prepared $LiMn_{0.8}Fe_{0.2}PO_4$ nanospheres using yeast cells as nucleating agents through biomimetic mineralization technology, and coated them with N-doped graphene nanoribbons. The network of N-doped graphene nanoribbons between the $LiMn_{0.8}Fe_{0.2}PO_4$/C nanospheres facilitates rapid electron migration and good electrolyte penetration. The prepared composite material has excellent long-term cycle stability over 2000 cycles.

8.4.3.3 Surface Coating

The cathode material is prone to fracture during the cycle, and there are persistent side reactions with the electrolyte, which severely damage the battery's cycle life and rate performance. Surface coating can inhibit the direct contact between the cathode material and the electrolyte, reduce side reactions, and lower the interface charge transfer resistance. As the electronic conductivity of $LiMn_{1-x}Fe_xPO_4$ material is relatively low, evenly coating conductive material on the particle surface can effectively improve the material's cycle performance.

Ma et al. [38] synthesized a $LiMn_{0.9}Fe_{0.1}PO_4$ cathode material co-doped with pyrolytic carbon and graphene oxide using a novel amine-assisted coating method. It has excellent electrochemical performance, with a specific capacity of 158.0 mAh g^{-1} at a 0.1 C rate, a specific capacity of 124.6 mAh g^{-1} at a 20 C rate, and still retains about 95% of its capacity after 140 cycles at 0.2 C. The initial capacity is %. Further analysis shows that the excellent electrochemical performance can be attributed to the combination of uniformly coated pyrolytic carbon and closely connected graphene oxide, which has excellent electronic conductivity. Li et al. [39] synthesized $LiMn_{0.6}Fe_{0.4}PO_4$/carbon aerogel (CA) composite material with high tap density and excellent electronic conductivity through wet ball milling, spray drying, and carbothermal reduction. The CA coating on the surface of the primary particles of the $LiMn_{0.6}Fe_{0.4}PO_4$ material can form a unique 3D woven network structure, which is conducive to improving the electronic conductivity and can absorb a large amount of electrolyte, which helps to improve the Li^+ migration rate between the electrode and the electrolyte. The appropriate addition of CA enhances the cycling performance of the material. The composite material also shows high tap density and excellent electronic conductivity. Electrochemical results show that the composite material containing 10% CA exhibits the best cycle stability and specific capacity, with a discharge capacity of 159.1 mAh g^{-1} at 0.2 C, and a capacity retention rate of 96.9% after 500 cycles at 1 C. This additive process is simple and environmentally friendly, providing a broad application prospect for large-scale commercial production. Ding et al. [40] provided a new green one-step wet ball milling graphene to produce porous graphene oxide (h-GO), and combined with carbon nanotubes (CNT) and $LiMn_{0.7}Fe_{0.3}PO_4$(LMFP), without using any catalyst to prepare lithium-ion battery composite cathode. The LMFP/h-GO/CNT composite cathode shows stable cycling performance, with a discharge capacity of 112 mAh g^{-1} at 20 C, and about 120 mAh g^{-1} after 200 cycles at 1 C. Yi et al. [41] successfully synthesized $LiMn_{0.5}Fe_{0.5}PO_4$(LMFP)@C and LMFP@$LiAlO_2$@C nanorods by solvent thermal method. This is the first time that this composite material is used as a cathode to improve the cycle stability and rate performance of lithium-ion batteries. LMFP@$LiAlO_2$ (5 wt.%)@C shows excellent rate performance, with reversible discharge capacities of 137.6 and 113.2 mAh g^{-1} at 0.05 C and 5 C, respectively. The composite material also shows good cycling performance, with a capacity of 107 mAh g^{-1} after 100 cycles at 5 C, and a capacity retention rate of 86.4%. The results show that the $LiAlO_2$ layer further avoids the direct contact of the cathode with the electrolyte, thus protecting the LMFP cathode from hydrofluoric acid corrosion during charging and discharging. Therefore, $LiAlO_2$ coating modification can effectively improve the cycle stability of the LMFP cathode.

8.4.3.4 Electrolyte Modification

Currently, the most commercially used is liquid electrolyte, which has the advantage of high ionic conductivity and good contact between the electrolyte and the porous electrode. At the same time, liquid electrolytes have the disadvantages of being flammable, prone to leakage, and volatile, which pose certain safety risks during battery use. The development of solid electrolytes overcomes the above shortcomings of liquid electrolytes and has good mechanical properties, which can to some extent inhibit the diffusion of lithium dendrites and intermediate products during discharge. Different electrolyte systems have different effects on the electrochemical performance of $LiMn_{1-x}Fe_xPO_4$ cathode materials.

Zhao et al. [42] chose lithium bis(oxalate)borate (LiBOB) as the lithium salt, ethylene carbonate (EC), dimethyl carbonate (DMC), and dimethyl sulfite (DMS) as the electrolyte solvents, and studied the compatibility between $LiFe_{0.6}Mn_{0.4}PO_4$ and the electrolyte. The combination of LiBOB and DMS solvent can promote the formation of an effective SEI film (Solid Electrolyte Interface), thereby showing better cycle and rate performance. After 50 cycles at 0.2 C, the capacity retention rate is 70.4%, which is higher than the traditional electrolyte $LiPF_6$ (56.1%). Tris(trimethylsilyl) phosphate (TMSP) is widely used as a functional additive in the electrolyte of lithium-ion batteries. Its decomposition products are conducive to forming a dense interface on the surface of the positive electrode, alleviating the erosion of the electrolyte on the positive electrode material, thereby improving the cycle stability of the battery. Kim et al. [43] prepared a porous structure $LiMn_{0.6}Fe_{0.4}PO_4$ with TMSP additive by an improved sol-gel method, showing a high capacity of 166 mAh g^{-1} at 0.1 C, and a capacity of 118 mAh g^{-1} after 100 cycles at 1 C, with an excellent capacity retention rate of 100%. And after 10 cycles at all rates, the capacity retention rate is 100%. The results show that the oxidized TMSP additive reacts with hydrofluoric acid in the electrolyte to form a product with a P-F group. The F-P-O and F-P compounds produced by the P-F-containing molecules are SEI compounds, which not only remove hydrofluoric acid but also facilitate the formation of the SEI film. Zhao et al. [44, 45] studied the electrochemical performance of $(LiMn_{0.6}Fe_{0.4}PO_4/C)//LiV_3O_8$ in different concentrations of Li_2SO_4, $LiNO_3$ aqueous electrolyte. The results show that in 5 M $LiNO_3$ or pH7 saturated $LiNO_3$ aqueous electrolyte, it has good specific capacity and rate performance, among which pH7 saturated $LiNO_3$ The energy density of the electrolyte system is higher. The authors then used LiTFSI and $ZnSO_4$ aqueous electrolyte to assemble a new type of $Zn/LiMn_{0.8}Fe_{0.2}PO_4$ aqueous mixed-ion battery. It was proven that the electrochemical characteristics of $LiMn_{0.8}Fe_{0.2}PO_4$ heavily depend on the pH value of the electrolyte. Ju et al. [46] successfully prepared polyvinyl carbonate (PVC) based composite solid electrolyte $PVCA\text{-}Li_{10}SnP_2S_{12}$(LSnPS) through the free radical polymerization method. The results showed that the electrolyte battery $LiFe_{0.2}Mn_{0.8}PO_4/Li$ has stable cycling performance compared to liquid electrolyte, with a capacity retention rate of 88% after 140 cycles at 0.5 C, and a Coulombic efficiency exceeding 99%. In addition, the elemental analysis of PVCA-LSnPS after cycle testing showed that there is no Mn element in the composite electrolyte. The in-situ generated PVCA-LSnPS can effectively inhibit the dissolution of Mn, promoting the cycling stability of the material.

8.5 Conclusion

The above content systematically summarizes the structure, preparation methods, modification methods, physical and chemical properties, and research progress of $LiMn_{1-x}Fe_xPO_4$ materials. Although the above strategies of ion doping, morphology control, surface coating, and electrolyte modification can effectively solve existing problems, there is still a certain distance from commercialization. There is still room for development in improving the energy density and cycling stability of lithium iron manganese phosphate materials, which will inevitably put higher requirements on the design and development of composite materials for this material. This can be combined with theoretical calculations and in-situ characterization methods for deep mechanism analysis; at the same time, the development of the electrolyte system can also be combined with various specific application scenarios for comprehensive design; on the other hand, the design of the N/P ratio in the battery matching work, the addition of lithium supplements and other work also need to be studied simultaneously, to overcome key technical difficulties, and lay a theoretical and technological foundation for the large-scale production of high energy density and high cycling stability lithium iron manganese phosphate materials.

References

1. Padhi AK, Nanjudaswamy KS, Goodenough JB (1997) Phospho-olivines as positive-electrode materials for rechargeable lithium batteries. J Electrochem Soc 144(4):1188–1194
2. Yuan LX, Wang ZH, Zhang WX et al (2011) Development and challenges of $LiFePO_4$ cathode material for lithium-ion batteries. Energy Environ Sci 4(2):269–284
3. Wang H, Lai A, Huang D et al (2021) Y-F co-doping behavior of $LiFePO_4/C$ nanocomposites for high-rate lithium-ion batteries. New J Chem 45(12):5695–5703
4. Zhu M, Cheng L, Liu Y et al (2018) $LiFePO_4/(C+Cu)$ composite with excellent cycling stability as lithium ion battery cathodes synthesized via a modified carbothermal reduction method. Ceram Int 44(11):12106–12111
5. Li F, Tao R, Tan X et al (2021) Graphite-embedded lithium iron phosphate for high-power-energy cathodes. Nano Lett 21(6):2572–2579
6. Delacourt C, Poizot P, Levasseur S et al (2006) Size effects on carbon-free $LiFePO_4$ powders: the key to superior energy density. Electrochem Solid-State Lett 9(7):A352–A355
7. Zhang M, Garcia-Araez N, Hector AL et al (2017) A sol–gel route to titanium nitride conductive coatings on battery materials and performance of TiN-coated $LiFePO_4$. J Mater Chem A 5(5):2251–2260
8. Liu Y, Wang J, Liu J et al (2018) Origin of phase inhomogeneity in lithium iron phosphate during carbon coating. Nano Energy 45:52–60
9. Xu L, Lv W, Shi K et al (2019) Holey graphenes as the conductive additives for $LiFePO_4$batteries with an excellent rate performance. Carbon 149:257–262

10. Lu ZG, Cheng H, Lo MF et al (2007) Pulsed laser deposition and electrochemical characterization of $LiFePO_4$–Ag composite thin films. Adv Func Mater 17(18):3885–3896
11. Liu S, Wang H (2014) WO_2 modified $LiFePO_4$/C cathode materials with improved electrochemical performance synthesized by in-situ synthesis method. Mater Lett 122:151–154
12. Zhang X, Guo H, Li X et al (2012) Studies of fast-ion conducting $Li_3V_2(PO_4)_3$ coated $LiFePO_4$ via sol–gel method. Solid State Ionics 212:106–111
13. Chung S-Y, Bloking JT, Chiang Y-M (2002) Electronically conductive phospho-olivines as lithium storage electrodes. Nat Mater 1(2):123–128
14. Johnson ID, Blagovidova E, Dingwall PA et al (2016) High power Nb-doped $LiFePO_4$Li-ion battery cathodes; pilot-scale synthesis and electrochemical properties. J Power Sour 326:476–481
15. Kim S, Mathew V, Kang J et al (2016) High rate capability of $LiFePO_4$ cathodes doped with a high amount of Ti. Ceram Int 42(6):7230–7236
16. Zhang Y, Shao Z, Zhang Y (2020) Preparation of Mo-doping $LiFePO_4$/C by carbon reduction method. Mater Manuf Processes 36(4):419–425
17. Lu F, Zhou Y, Liu J et al (2011) Enhancement of F-doping on the electrochemical behavior of carbon-coated $LiFePO_4$ nanoparticles prepared by hydrothermal route. Electrochim Acta 56(24):8833–8838
18. Sun CS, Zhang Y, Zhang XJ et al (2010) Structural and electrochemical properties of Cl-doped $LiFePO_4$/C. J Power Sour 195(11):3680–3683
19. Gao L, Xu Z, Zhang S (2018) The co-doping effects of Zr and Co on structure and electrochemical properties of $LiFePO_4$ cathode materials. J Alloy Compd 739:529–535
20. Wang B, Xie Y, Liu T et al (2017) $LiFePO_4$ quantum-dots composite synthesized by a general microreactor strategy for ultra-high-rate lithium ion batteries. Nano Energy 42:363–372
21. Jiang Y, Tian R, Liu H et al (2015) Synthesis and characterization of oriented linked $LiFePO_4$ nanoparticles with fast electron and ion transport for high-power lithium-ion batteries. Nano Res 8(12):3803–3814
22. Zhou N, Wang H-Y, Uchaker E et al (2013) Additive-free solvothermal synthesis and Li-ion intercalation properties of dumbbell-shaped $LiFePO_4$/C mesocrystals. J Power Sour 239:103–110
23. Zhou N, Uchaker E, Wang H-Y et al (2013) Additive-free solvothermal synthesis of hierarchical flower-like $LiFePO_4$/C mesocrystal and its electrochemical performance. RSC Adv 3(42):19366–19374
24. Zhang L, Chai L, Zhang L et al (2014) Synergistic effect between lithium bis(fluorosulfonyl)imide (LiFSI) and lithium bis-oxalato borate (LiBOB) salts in $LiPF_6$-based electrolyte for high-performance Li-ion batteries. Electrochim Acta 127:39–44
25. Chen X, Xu W, Engelhard MH et al (2014) Mixed salts of LiTFSI and LiBOB for stable $LiFePO_4$-based batteries at elevated temperatures. J Mater Chem A 2(7):2346–2352
26. Fan W, Li NW, Zhang X et al (2018) A dual-salt gel polymer electrolyte with 3D cross-linked polymer network for dendrite-free lithium metal batteries. Adv Sci 5(9):1800559
27. Deng Y, Yang C, Zou K et al (2017) Recent advances of Mn-rich $LiFe_{1-y}Mn_yPO_4$($0.5 \leq y < 1.0$) cathode materials for high energy density lithium ion batteries. Adv Energy Mater 7(13):1601958
28. Zong J, Peng Q, Yu J et al (2013) Novel precursor of $Mn(PO_3(OH))_3H_2O$ for synthesizing $LiMn_{0.5}Fe_{0.5}PO_4$ cathode material. J Power Sour 228:214–219
29. Budumuru AK, Viji M, Jena A et al (2018) Mn substitution controlled Li-diffusion in single crystalline nanotubular $LiFePO_4$ high rate-capability cathodes: experimental and theoretical studies. J Power Sour 406:50–62
30. Fan R-z, Fan C-l, Hu Z et al (2021) Construction of high performance N-doped carbon coated $LiMn_{0.8}Fe_{0.2}PO_4$ nanocrystal cathode for lithium-ion batteries. J Alloys Compd 876:160090
31. Li J, Qin Y, Ning X (2020) Improved high-temperature solid-phase method for preparing lithium manganese iron phosphate cathode material. Mater Guide 34(16):16001–16005
32. Zhang X, Hou M, Tamirate AG et al (2020) Carbon coated nano-sized $LiMn_{0.8}Fe_{0.2}PO_4$ porous microsphere cathode material for Li-ion batteries. J Power Sour 448:227438
33. Yi H, Hu C, Fang H et al (2011) Optimized electrochemical performance of $LiMn_{0.9}Fe_{0.1-x}Mg_xPO_4$/C for lithium ion batteries. Electrochimica Acta 56(11):4052–4057
34. Wang Y, Yang H, Wu CY et al (2017) Facile and controllable one-pot synthesis of nickel-doped $LiMn_{0.8}Fe_{0.2}PO_4$ nanosheets as high performance cathode materials for lithium-ion batteries. J Mater Chem A 5(35):18674–18683
35. Liao LH, Wang HT, Guo H et al (2015) Facile solvothermal synthesis of ultrathin $LiFe_xMn_{1-x}PO_4$ nanoplates as advanced cathodes with long cycle life and superior rate capability. J Mater Chem A 3(38):19368–19375
36. Ruan TT, Wang B, Wang F et al (2019) Stabilizing the structure of $LiMn_{(0.5)}Fe_{(0.5)}PO_4$ via the formation of concentration-gradient hollow spheres with Fe-rich surfaces. Nanoscale 11(9):3933–3944
37. Hou YK, Pan GL, Sun YY et al (2018) $LiMn_{0.8}Fe_{0.2}PO_4$/carbon nanospheres@graphene nanoribbons prepared by the biomineralization process as the cathode for lithium-ion batteries. ACS Appl Mater Interfaces 10(19):16500–16510
38. Ma F, Zhang XY, He P et al (2017) Synthesis of hierarchical and bridging carbon-coated $LiMn_{0.9}Fe_{0.1}PO_4$ nanostructure as cathode material with improved performance for lithium ion battery. J Power Sour 359:408–414
39. Li ZF, Ren X, Tian WC et al (2020) $LiMn_{0.6}Fe_{0.4}PO_4$/CA cathode materials with carbon aerogel as additive synthesized by wet ball-milling combined with spray drying. J Electrochem Soc 167(9):090516
40. Ding D, Maeyoshi Y, Kubota M et al (2020) Holey reduced graphene oxide/carbon nanotube/$LiMn_{0.7}Fe_{0.3}PO_4$ composite cathode for high-performance lithium batteries. J Power Sour 449:227553
41. Yi TF, Li Y, Fang ZK et al (2020) Improving the cycling stability and rate capability of $LiMn_{0.5}Fe_{0.5}PO_4$/C nanorod as cathode materials by $LiAlO_2$ modification. J Materiomics 6(1):33–44
42. Zhao Q, Li X, Tang F et al (2017) Compatibility between lithium bis(oxalate)borate-based electrolytes and a $LiFe_{0.6}Mn_{0.4}PO_4$/C cathode for lithium-ion batteries. Energy Technol 5(3):406–413
43. Kim J-K, Vijaya R, Zhu L et al (2015) Improving electrochemical properties of porous iron substituted lithium manganese phosphate in additive addition electrolyte. J Power Sour 275:106–110
44. Zhao MS, Huang GL, Qu F et al (2015) Electrochemical performances of ($LiMn_{0.6}Fe_{0.4}PO_4$/C)//LiV_3O_8 in different aqueous solution electrolyte. Electrochimica Acta 151:50–55
45. Zhao JW, Li YQ, Peng X et al (2016) High-voltage Zn/$LiMn_{0.8}Fe_{0.2}PO_4$ aqueous rechargeable battery by virtue of "water-in-salt" electrolyte. Electrochem Commun 69:6–10
46. Ju J, Wang Y, Chen B et al (2018) Integrated interface strategy toward room temperature solid-state lithium batteries. ACS Appl Mater Interfaces 10(16):13588–13597

Lithium-Rich Manganese-Based Cathode

9

Rechargeable lithium-ion batteries have become the main power source for portable electronic devices and are considered the preferred battery technology for electric vehicles and potential candidates for grid-scale storage. After thirty years of battery technology development, the energy density of commercial lithium-ion batteries has approached its limit. In order to continuously improve energy density and reduce costs, Lithium-rich manganese-based layered oxides (LMLO) cathodes are receiving increasing attention due to their high discharge capacity and low cost. LMLO cathodes have a very high reversible discharge capacity (>250 mAh g^{-1}) and extremely high energy density (>800 Wh kg^{-1}), and are considered one of the most promising cathode candidate materials for the next generation of lithium-ion batteries. However, before its actual application, there are many problems to be solved, such as microscopic material defects (structural instability, poor diffusion kinetics, etc.) and the resulting macroscopic electrochemical performance decay (low initial Coulombic efficiency, rapid capacity/voltage decay, etc.). This section discusses the structure, preparation methods, and existing problems of LMLO cathodes, summarizes various modification methods and strategies to promote performance, and looks forward to the future research direction of LMLO.

9.1 Structure of Lithium-Rich Manganese-Based Cathode

Lithium-rich manganese-based cathode materials (chemical formula: $xLi_2MnO_3 \cdot (1-x)LiTMO_2$, TM is one or more elements among Ni, Co, Mn, etc.; abbreviated as LMLO) have received increasing attention due to their higher specific capacity (250–400 mA h/g). As shown in Fig. 9.1, in the Lithium-rich manganese-based cathode materials, Li_2MnO_3 belongs to the monoclinic system, and $LiTMO_2$ belongs to the hexagonal system. The structure of LMLO is very complex, and there are still some disputes about its microscopic structure. In the currently established structural models, there are mainly two views, one believes that there is only a single phase in the Lithium-rich manganese-based cathode materials, which is a single-phase solid solution model; the other view believes that it is composed of two different phases, which is a two-phase composite model. The single-phase solid solution model was proposed by Dahn et al. Because the structures of the monoclinic Li_2MnO_3 phase and the tetragonal $LiTMO_2$ phase are very similar, it is believed that the Li_2MnO_3 phase and the $LiTMO_2$ phase exist in the material in the form of a solid solution, and its component symbol is represented as $LiTMO_2$. The two-phase composite model was proposed by Thackeray et al. based on previous research, believing that the Lithium-rich manganese-based cathode material is composed of two solid solutions, and its component symbol is represented as $xLi_2MnO_3 \cdot (1-x)LiTMO_2$. In summary, the special two-phase structure of Lithium-rich manganese-based materials endows it with a unique reaction mechanism. The first charge is divided into two stages, when the voltage is below 4.5 V, $LiTMO_2$ The transition metal in the selected phase undergoes redox reactions; the second stage is when the voltage is higher than 4.5 V, the Li_2MnO_3 component is activated and shows electrochemical activity, Li^+ is released from Li_2MnO_3 and is accompanied by the oxidation of O^{2-}. LMLO material has a lower ionic conductivity, lower Coulombic efficiency, and rapid capacity decay caused by phase changes during the cycle. Therefore, there is still a distance from the commercialization of LMLO materials, and it is expected to be commercially used in high-energy lithium-ion batteries in the future.

X. Li et al., *Electrode Development for Lithium-ion Batteries*, https://doi.org/10.1007/978-981-96-3021-9_9

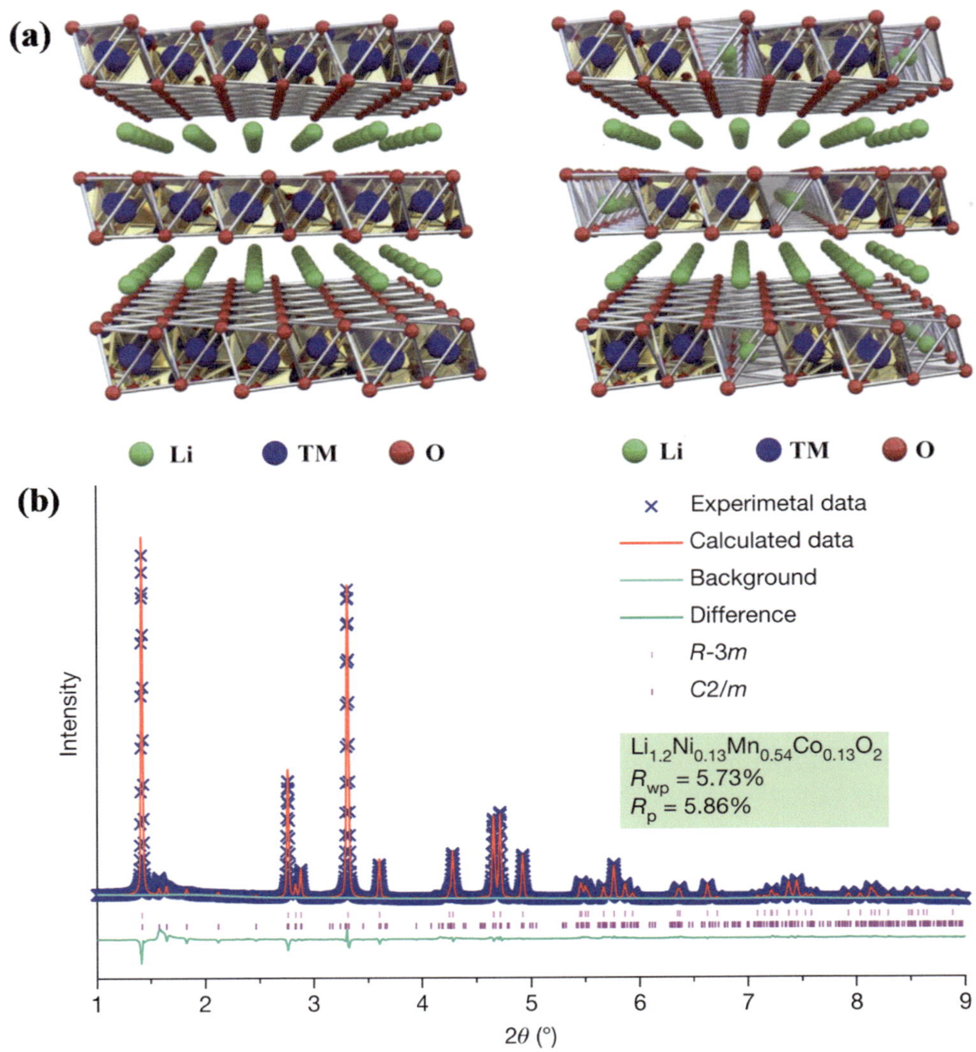

Fig. 9.1 **a** (left) $LiTMO_2$ phase (space group: R-3m, TM is elements such as Ni, Co, and Mn) and (right) monoclinic Li_2MnO_3 phase (space group: C^2/m) crystal structure [1]. **b** X-ray diffraction pattern and Rietveld refinement results of lithium-rich cathode [2]

9.2 Preparation of Lithium-Rich Manganese-Based Cathode

The preparation method plays a decisive role in the overall morphology, crystal structure, and other characteristics of the LMLO cathode, and also determines the electrochemical performance of the material. Since lithium-rich manganese-based materials contain a variety of metal elements, the choice of preparation method usually considers how to make multi-metal evenly distributed in the material, and common preparation methods include solid-phase method, sol-gel method, hydrothermal method, and co-precipitation method (Fig. 9.2).

(1) The high-temperature solid-phase method is to obtain the final material product by pre-treating and mixing the raw materials according to the corresponding chemical raw material ratio, and then calcining at high temperature. The advantages of this method are a simple preparation process, low equipment requirements, low cost, and it can realize large-scale industrial production. The current mechanochemical activation solid-phase method is also an important preparation method, which can prepare high-purity, uniform particle size, and more excellent electrochemical performance LMLO cathode materials compared with the traditional high-temperature solid-phase method.

(2) The co-precipitation method is to add a precipitant of the corresponding concentration to a salt solution of transition metals Mn, Ni, and Co in a certain chemical stoichiometric ratio, control the reaction temperature, pH value, and obtain a solid precipitate mixture through a series of chemical reactions. After washing and drying, the cathode precursor is obtained. The precursor is mixed evenly with lithium salt in a certain proportion, and the LMLO cathode material is prepared

Fig. 9.2 SEM images of lithium-rich manganese-based cathode materials prepared by different methods: **a** high-temperature solid-phase method [3], **b** co-precipitation method [4], **c** hydrothermal method [5], **d** sol-gel method [6]

by high-temperature calcination. This method can achieve atomic-level uniform mixing, avoid the generation of impurity phases during the reaction process, help to improve the electrochemical performance of the cathode material, and is currently the most widely used and mature preparation method.

(3) The hydrothermal method is to dissolve the raw materials of the corresponding chemical stoichiometric ratio in a lithium salt solution, complete the reaction under specific high temperature and high pressure, and prepare the LMLO cathode material after centrifugation, washing, drying, and high-temperature calcination. The cathode material prepared by this method has uniform particle size and phase, but it requires strict equipment, high cost, and certain safety hazards.

(4) The sol-gel method involves dissolving a certain proportion of transition metal salt compounds and chelating agents to form a uniform and transparent sol system. After heating and evaporation, thermal decomposition occurs between the material ions, forming a dry gel with a three-dimensional spatial structure. After high-temperature calcination, the LMLO cathode material is prepared. The sol-gel method has advantages such as low processing temperature, good product particle dispersion, simple experimental equipment, and the ability to easily modify metal elements during the preparation process. However, there are many variable process parameters during the preparation process.

(5) In addition, there are unconventional synthesis methods such as self-propagating combustion, ion exchange, electrospinning, and spray drying.

9.3 Problems with Lithium-Rich Manganese-Based Cathodes

Despite the excellent specific capacity of the LMLO cathode, it has problems such as low ion conductivity and Coulombic efficiency, poor structural stability, and phase transition behavior during the cycle, resulting in short cycle life and poor safety of the LMLO battery, which has not yet been realized on a large scale. Commercial applications. The large specific surface area inevitably promotes the release of oxygen during the cycle, leading to the collapse of the material structure [7]. The unique redox process of LMLO will trigger serious lattice oxygen release and irreversible migration of transition metal (TM) ions to the lithium layer during the cycle, leading to continuous voltage decay and potential safety hazards.

9.3.1 Low First-Cycle Coulombic Efficiency and Poor Diffusion Kinetics

Due to the unique first charge–discharge mechanism, lithium-rich manganese-based materials exhibit a low first-cycle Coulombic efficiency. Currently, it is generally believed that the irreversible capacity loss is due to the release of oxygen and corrosion of the material. A unique 4.5 V voltage plateau appears in the first charge curve. When the voltage is less than 4.5 V, $LiMnO_2$ in LMLO undergoes lithium ion deintercalation, and Li_2MnO_3 does not react, playing a role in stabilizing the material structure. When the charging process voltage is greater than 4.5 V, Li_2O deintercalation occurs in Li_2MnO_3, but during discharge, this part of lithium cannot be reintercalated into the layered lattice, resulting in irreversible first-cycle capacity loss. In addition, the large amount of Li^+ extraction will also cause oxygen to be released from the material surface. The loss of oxygen due to the release of oxygen will cause a highly disordered irreversible phase transition on the surface of LMLO. Li^+ cannot be intercalated into the phase transition lattice in the subsequent discharge process, which also causes irreversible first-cycle capacity loss [8].

Compared to traditional layered $LiNi_xMn_yCo_zO_2$ or spinel $LiMn_2O_4$, lithium-rich manganese-based materials generally have a relatively low conductivity due to the presence of the poorly conductive Li_2MnO_3 component. The rate performance of the cathode largely depends on its kinetic factors, including the Li^+ diffusion coefficient within the material structure and the charge transfer reaction at the electrode/electrolyte interface. It is generally believed that low conductivity and poor Li^+ diffusion kinetics are the main reasons for the poor rate performance of lithium-rich manganese-based cathode materials. Due to the large interface impedance of the Li_2MnO_3 component and poor Li^+ diffusion kinetics, its activity controls the rate capability of the lithium-rich manganese-based cathode, as it disrupts the electrode surface structure, resulting in higher impedance. In addition, the formation of spinel-like phases may also be an important reason for the poor kinetics of lithium-rich manganese-based cathodes (Fig. 9.3).

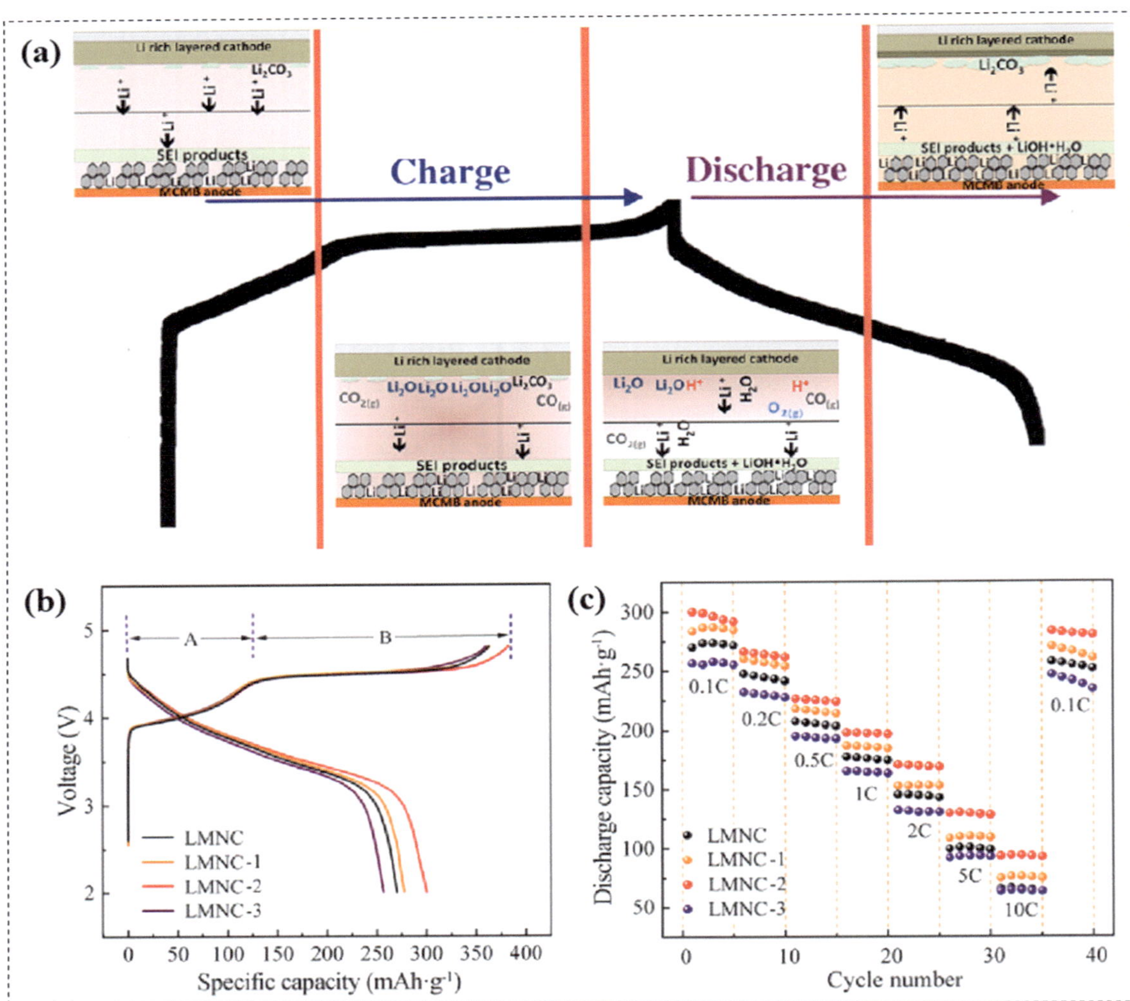

Fig. 9.3 **a** Surface reaction mechanism (top) and charge–discharge curves of different processes (bottom) of layered lithium-rich manganese-based cathode during oxygen activation component charge–discharge process [9]. **b**, **c** Charge–discharge curves and rate performance of original and modified lithium-rich cathodes at 0.1 C [10]

9.3.2 Rapid Capacity and Voltage Decay

Currently, the rapid capacity and voltage decay of LMLO materials is one of the most concerning issues. In the long cycle of LMLO batteries, layered structure (R-3m) LMLO will produce a spinel-like phase (Fd-3m) and even rock salt phase (Fm-3m), which are the main factors causing continuous voltage decay and capacity loss (as shown in Fig. 9.4). The transition metal cations (TM) in LMLO migrating from the TM layer to the Li layer will cause significant lattice strain, leading to nanoscale spinel nucleation and growth. The loss of oxygen during the first charge–discharge process causes TM ions to migrate from the surface to the crystal interior, forming a surface defect spinel layer, hindering the diffusion of Li ions, resulting in poor rate capability. In subsequent cycles, more TM ions migrate into the lithium layer, and the reconstructed layer at the crystal edge becomes thicker, forming a defective spinel structure. In this defective spinel structure, only the 16c octahedral sites are available for Li^+ to reversibly intercalate and deintercalate, while the tetrahedral vertices are electrochemically inert, thus causing a voltage drop. After long cycles, TM ions further occupy some of the 16c octahedral sites in the structure, transforming the defective spinel structure into a disordered rock salt phase, forming an irregular rock salt structure region on the surface. The next is the dissolution of metal ions in the electrolyte during the cycle, causing the capacity and voltage decay of LMLO materials.

9.4 Modification Research of Lithium-Rich Manganese-Based Cathode

To solve the problems of Coulombic efficiency, capacity decay, and structural stability of the LMLO cathode, and to promote the commercial development of LMLO materials, researchers have made many modification strategies for LMLO, mainly including element doping, morphology control, and surface coating, all of which have shown significant performance improvements. At the same time, other modification methods such as surface chemical treatment, electrolyte modification, oxygen vacancy and lithium regulation, etc., can also enhance the electrochemical performance of the LMLO cathode.

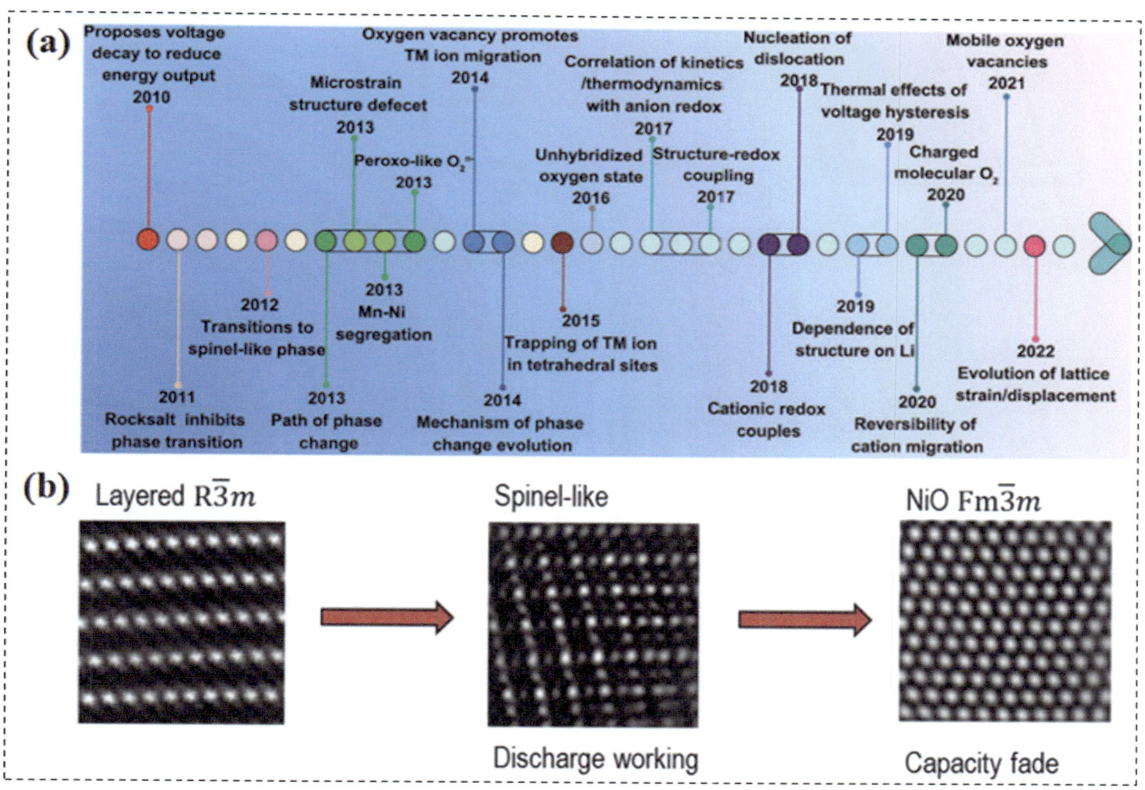

Fig. 9.4 **a** Important steps in developing a low-voltage attenuation response mechanism [11]. **b** Transition metal migration and evolution from layered structure to spinel and rock salt phase structure during the cycling process of layered LMLO cathode [12]

9.4.1 Element Doping

Element doping is a simple and effective method to adjust the body lattice at the atomic scale. Depending on the doping elements, it can be divided into cation doping, anion doping, and composite doping. According to different doping methods, it can also be divided into in-situ doping and non-in-situ doping. In-situ doping usually refers to the addition of doping elements during the synthesis of precursors, while non-in-situ doping usually refers to the doping behavior during the synthesis of cathode materials. Element doping has the advantages of being simple and effective, usually without the need for additional operation steps and equipment. Different element doping and different doping methods ultimately have different effects on LMLO cathode materials. Na^+, K^+, Mg^{2+} and other alkali metal ions have radii close to that of Li^+, and a small amount of doping will preferentially enter the Li sites, which can act as pillars to inhibit the collapse of the layered structure. Al^{3+}, Cr^{3+}, B^{3+}, Ti^{4+}, Nb^{5+}, Zr^{4+} and other cation doping will usually enter the transition metal sites of LMLO, the larger ion radius can expand the lattice parameters and improve the diffusion coefficient of Li^+, and their interaction with the ions of the transition metal layer can alleviate the migration and structural transformation of the transition metal ions [13–16]. Anion doping research mainly focuses on F− and S^{2-} doping, their doping into oxygen sites can increase the bond strength between anions and cations to reduce oxygen loss and slow down voltage decay [17, 18]. Liu et al. [19] reported $Li_{1.2}Mn_{0.56}Ni_{0.16}Co_{0.08}O_2$ The positive gradient W doping, as shown in Fig. 9.5a–c, slows down the disintegration of the crystal and enhances the stability of the interface due to the formation of a Li_2WO_4 coating layer on the material surface. This is mainly because the substitution of W for Mn can stabilize the LLMO structure and regulate the electrochemical activity of Mn. The W-doped LLMO (W@LLMO) has better specific capacity and voltage stability, with a capacity retention rate of 83.2% after 200 cycles. As shown in Fig. 9.5d–h, Li et al. [16] doped the surface of the lithium-rich material $Li_{1.2}Mn_{0.53}Co_{0.13}Ni_{0.13}Ti_{0.01}O_2$

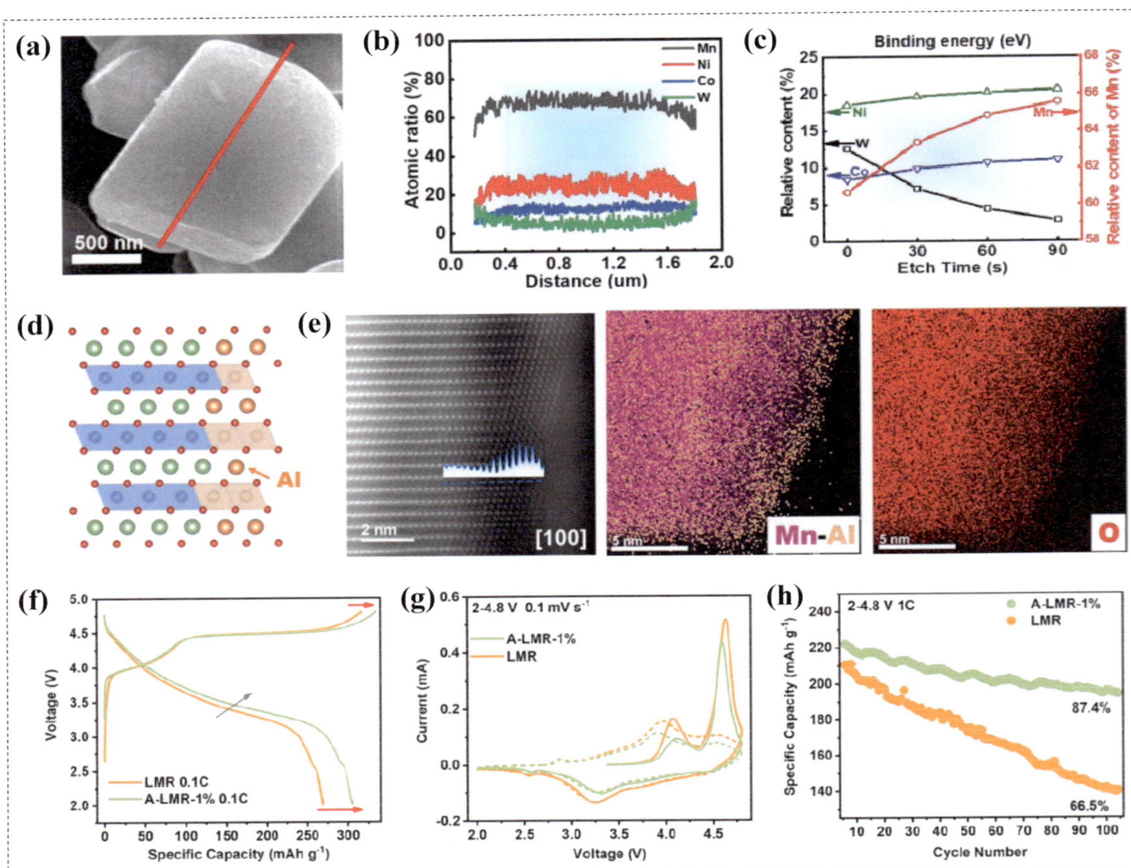

Fig. 9.5 **a** Scanning electron microscope image of W@LLMO, **b** linear scanning EDS signals of Ni, Co, Mn, and W corresponding to (**a**); **c** relationship between the relative content of Ni, Co, Mn, and W based on XPS spectra and etching time [19]. **d** Schematic diagram of the structure of the surface Al-doped LMLO cathode, **e** near-surface HAADF image of Al-doped LMLO cathode particles and corresponding STEM-EDS element mapping; **f–h** comparison of the first cycle charge–discharge curves, CV curves, and cycling performance of undoped and Al-doped LMLO [16]

with Al. Al doping increases the migration barrier of TM ions, and the covalent Al–O bond enhances the reversibility of O redox in the interior and surface of the particles. Therefore, Al surface doping achieves a balance between promoting O redox and stabilizing the LMLO cathode structure, enhancing the electrochemical performance of LMLO.

9.4.2 Morphology Control

The performance of electrode materials largely depends on the size, shape, composition, and structure of the crystals. Therefore, morphology control will be a powerful way to optimize the performance of lithium-ion layered cathode materials. The design of special morphologies is beneficial to improve the kinetic performance of Li^+, maintain the stability of the structure, and reduce the deterioration of the material during cycling. However, controlling the morphology and structure of LMLO cathodes with complex chemical compositions during synthesis is challenging due to the need for long-term calcination at high temperatures. As shown in Fig. 9.6, Fu et al. [20] synthesized lithium-rich layered oxide materials $Li_{1.2}Mn_{0.56}Co_{0.12}Ni_{0.12}O_2$ with four morphological structures: microspheres, micro-rods, nanosheets, and irregular particles, using the solvothermal method or co-precipitation method. They exhibited strong structure-dependent electrochemical performance, and the electrochemical performance of lithium-rich manganese-based materials with different structures was in the order: micro-rods > microspheres > nanosheets > irregular particles. The micro-rods, with their porous structure and 1D shape, showed excellent performance, which may be due to the synergistic effect of the porous structure, 1D shape, nanoscale size, and robust microstructure.

9.4.3 Surface Coating

For lithium-rich manganese-based cathode materials, the reversible oxidation and reduction of oxygen occur inside the material, which does not cause significant structural changes; however, the irreversible release of oxygen occurs on the surface of the material, leading to a rapid decay of LMLO's electrochemical performance. Therefore, surface coating can enhance surface stability to a certain extent and inhibit the corrosion of LMLO electrodes by electrolyte and oxygen release. Surface coating plays a crucial role in controlling charge transfer and Li^+ diffusion, and can effectively enhance rate performance and cycle stability. The coating material usually needs to have high chemical stability, resistance to electrolyte corrosion, and a high lithium ion diffusion coefficient. LMLO surface coating materials can be divided into inactive oxides (such as TiO_2, MgO, Al_2O_3, ZrO_2, and CeO_2, etc.) [21–25], fluorides (such as CaF_2, AlF_3, and LaF_3,

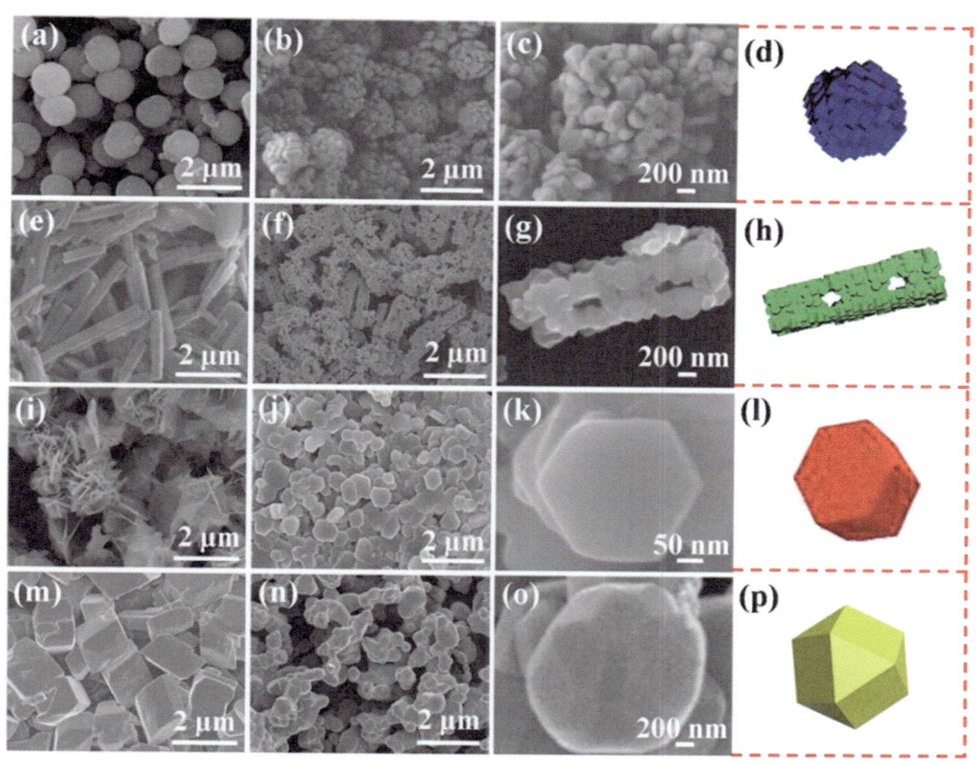

Fig. 9.6 **a** SEM images of spherical, **e** rod-shaped, **i** plate-shaped, and **m** cubic LMLO precursors. SEM images and schematic diagrams of microspheres (**b–d**), micro-rods (**f–h**), nanoplates (**j–1**), and irregular particles (**n–p**) of LMLO cathode [20]

etc.) [26–28], phosphates (such as $FePO_4$, $AlPO_4$, and Li_3PO_4, etc.) [29–31], lithium-ion conductor coatings ($Li_{6.25}La_3Zr_2Al_{0.25}O_{12}$, Li_3VO_4, Li_2RhO_3, etc.) [32–34], and organic coatings, etc. [35, 36]. Different coating materials have different effects, among which oxides have the function of neutralizing acidic substances in the electrolyte; fluoride coating materials have better stability; phosphates have better ionic and electronic conductivity; lithium-ion conductor coatings can improve the diffusion rate of lithium ions and the stability of the material. As shown in Fig. 9.7a–c, Wang et al. [37] introduced molten SeO_2 into polycrystalline $Li_{1.2}Mn_{0.54}Ni_{0.13}Co_{0.13}O_2$ (LMNCO) grain boundaries and particle surfaces. A uniform dual nanolayer structure composed of Li_2SeO_4 and $Li_2Ni_xCo_yO_4$ is presented on the particle surface for the first time. The nanocoating on the outer surface can reduce the contact and side reactions between the cathode material and the electrolyte. Meanwhile, the dual nanolayer structure at the grain boundaries also provides a fast channel for Li^+ transport. Therefore, the capacity retention rate of SeO_2 modified LMNCO reaches 80.0% after 200 cycles at 1 C. Cao et al. [35] reported a binder coating (Fig. 9.7d) where polypropionic acid (PPA) replaces PVDF. The uniform PPA coating serves to protect the cathode particles, stabilize the crystal structure through H^+/Li^+ exchange, and maintain the integrity of the electrode with strong adhesion. Li et al. [38] tightly bound isocyanate-related functional groups to the surface of the lithium-rich cathode through urea decomposition gas heat treatment (as shown in Fig. 9.7e). During the cycling process, the

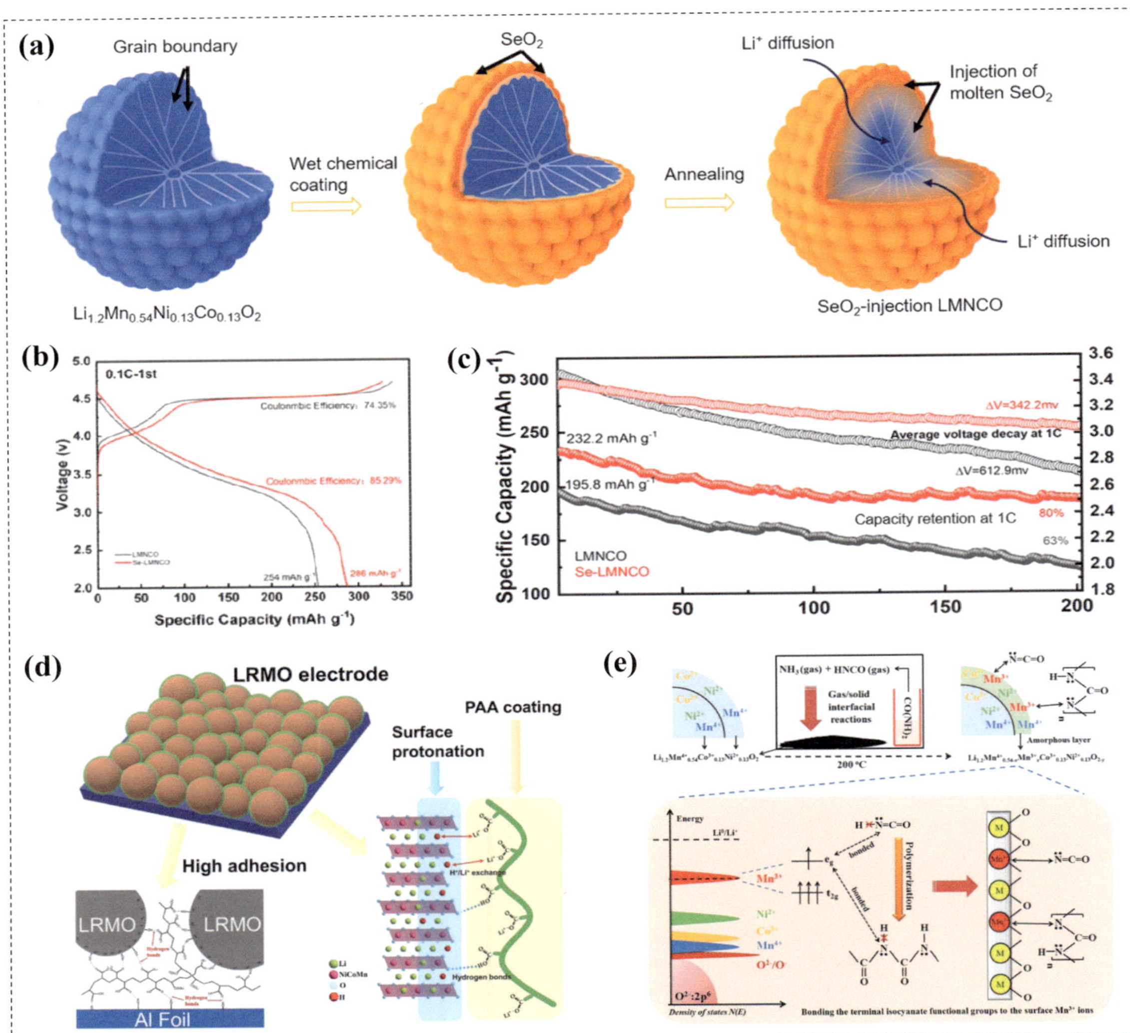

Fig. 9.7 **a** Schematic evolution of SeO_2 surface and grain boundary coating polycrystalline LMNCO cathode; **b–c** Comparison of initial charge–discharge curves and rate performance of original LMNCO and Se-LMNCO at 0.1 C [37]. **d** Schematic of multifunctional protection of LRMO cathode by PAA binder [35]. **e** Schematic of the formation mechanism of surface isocyanate functionalized $Li_{1.2}Mn_{0.54}Co_{0.13}Ni_{0.13}O_2$ oxide prepared by urea decomposition gas heat treatment [38]

surface isocyanate functionalization inhibits the side reaction of PF_5 hydrolysis on the surface of the lithium-rich cathode material to produce $Li_xPF_yO_z$ and HF. Compared with the untreated lithium-rich cathode, the sample with a spinel layer and isocyanate functionalized surface has higher cycle stability.

9.4.4 Electrolyte Additives

When LMLO is charged and discharged at high voltages, the electrolyte tends to decompose, leading to the formation of reactive oxygen species (including the release of molecular oxygen), making the stability of the electrode-solution interface crucial. In LMLO cathode materials, the modification of the material itself has always been the focus of research. However, in addition to the LMLO cathode body surface coating listed above, the in-situ surface treatment of the cathode electrolyte interface (CEI) layer can also be regulated by adding active additives to the electrolyte solution during electrochemical operation. By adding specific additives, the safety performance of the battery can be improved, reducing the risk of combustion and explosion.

Lithium salt electrolyte additives are common additives to improve the performance of cathode materials, such as lithium difluorophosphate ($LiPO_2F_2$), lithium bis(oxalate) borate (LiBOB), lithium difluoro(oxalate)borate (LiODFB), etc. [39–41]. As shown in Fig. 9.8a–d, Choi et al. [42] reported that lithium fluoro(propylene)borate (LiFMDFB) as a dual-function additive protects the lithium-rich cathode by generating a stable cathode electrolyte interface (CEI) and reducing O_2, CO and CO_2 gases. Compared with the baseline electrolyte, the use of LiFMDFB greatly reduced the amount of O_2 and CO gases. The large amount of CO_2 detected below 4.4 V can be explained by the oxidative decomposition of LiFMDFB, including the release of CO_2 during the charging of the lithium-rich cathode. More noticeable CO_2 gas evolution during the discharge of the cathode using the baseline electrolyte is attributed to O_2 promoted EC decomposition. The activation of

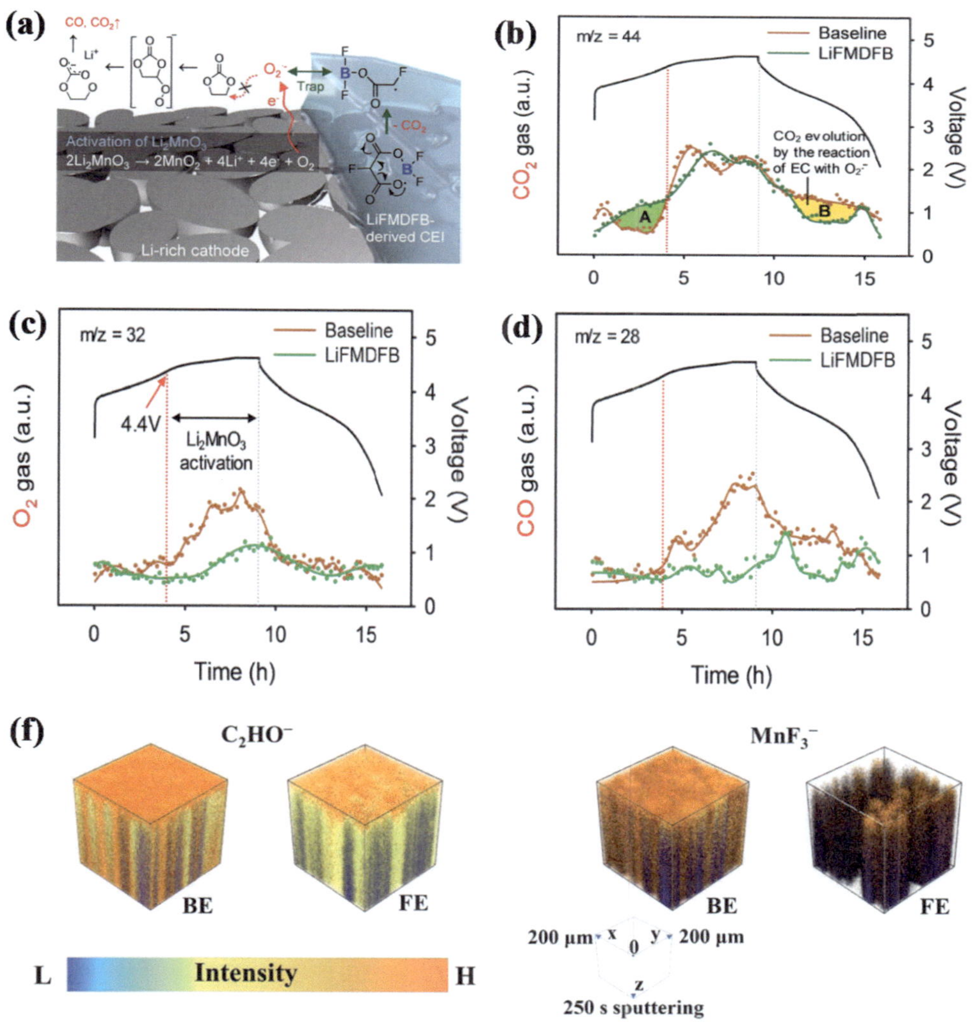

Fig. 9.8 **a** Schematic of the CEI formed on the lithium-rich cathode by LiFMDFB, reducing the evolution of O_2, CO, and CO_2 gases; **c**, **d** In-situ DEMS analysis shows the evolution of **b** O_2 (m/z = 32), **c** CO (m/z = 28), and **d** CO_2 (m/z = 44) gases and corresponding voltage curves during the pre-cycling process at 0.1 C for lithium-rich cathode half-cells with and without LiFMDFB [42]. **f** $Li_{1.13}Mn_{0.517}Ni_{0.256}Co_{0.097}O_2$ cathode after 300 cycles in BE (baseline electrolyte, 1 mol $LiPF_6$ dissolved in equal volume ratio of EC/DEC) and FE (fluorinated electrolyte, 1 mol $LiPF_6$ dissolved in 2:6:2 volume ratio of FEC/FDEC/HFE) as tested by TOF-SIMS, three-dimensional distribution of C_2HO^- and MnF_3^- [49]

Li_2MnO_3 occurs after the construction of the CEI derived from LiFMDFB, so the O_2 produced by the activation of Li_2MnO_2 can stay in the LiFMDFB-derived CEI formed at the cathode below 4.4 V.

In addition, many new types of organic electrolyte solution additives have been successfully tested on LMLO cathodes. These include organic phosphorus compounds, such as triethyl phosphite (TEP) [43], 1,3-divinyltetramethyldisiloxane (DTMS) [44], dimethyl phenylphosphonate (DMPP) [45]; organic borate derivatives, such as trimethylsilyl borate (TMSB) [46], triethyl borate (TEB) [47]; and ionic liquids, etc. [48]. Fluorinated electrolytes have high oxidation onset potentials, high thermal stability, and good anti-combustion capabilities. Yu et al. [49] The surface of O3-type LMLO is regulated using a fully fluorinated electrolyte, and the performance improvement is attributed to the thin, uniform, robust, and compact fluorine-rich cathode electrolyte interface (CEI) (as shown in Fig. 9.8f). It inhibits various types of oxygen-related surface degradation, and more importantly, stabilizes the bulk oxygen reaction. Compared with commercial electrolytes, the fluorine-rich CEI effectively improves the reversibility of LMLO and greatly enhances capacity retention and reduces voltage decay. In addition, Choi et al. [50] reported the use of malonic acid-modified fullerene (MA-C60) with superoxide dismutase activity and water clearance ability as an electrolyte additive to overcome the structural instability of high-capacity electrodes that hinder battery quality. Water clearance inactivates PF_5, leading to long-term stability of the electrode interface structure. Furthermore, the electrolyte with added MA-C60 inactivates reactive oxygen species and builds a chemically robust cathode-electrolyte interface for lithium-rich cathodes.

9.5 Conclusion and Outlook

Overall, this chapter outlines the challenges and progress faced by lithium-rich manganese-based cathode materials. The research focus of LIB cathode materials has always been to achieve higher capacity, therefore LMLO cathode materials have received much attention since their inception. However, the commercialization of LLOs is still hindered by material defects and the resulting performance degradation. (1) The unique redox process of LMLO can cause serious lattice oxygen release and irreversible migration of transition metal (TM) ions to the lithium layer during the cycle. (2) The release of oxygen during the anion oxidation-reduction reaction process can form oxygen vacancies on the surface and subsurface, leading to a decrease in cation valence, the generation of micro-strains inside the particles, and a transition from layered to spinel phase from the surface to the bulk layer in the following cycles. (3) Material defects and side reactions formed during the cycle ultimately lead to low ICE, rapid voltage decay, and capacity decay. (4) The inherent disadvantage of the material is poor diffusion kinetics, so it is difficult to achieve high rate performance. Systematic solutions to these problems can be achieved by optimizing synthesis conditions and methods, adopting doping, surface coating, morphology control and other cathode body modifications, as well as combining electrolyte optimization strategies. However, some other practical issues should also be considered in future research. (1) Further research is needed on the failure mechanism of LMLO cathodes. (2) Current LMLOs are mainly studied through half-cells, so it is difficult to evaluate the real electrochemical performance in actual applications. Therefore, more effort should be put into full cell research. (3) The safety issues of lithium batteries, thermal runaway, gas evolution, and other possible side reactions are all crucial factors that need further exploration. (4) The electrolyte used for LMLO cathodes needs to have a wide electrochemical window, and it cannot be oxidized even at very high voltages (>5.0 V), so there is an urgent need for further optimization of the electrolyte to meet commercial needs. (5) In addition, due to the shortage and high price of lithium, nickel, and cobalt resources, the development of cobalt-free, low-nickel lithium-rich cathode materials and battery recycling technology is imminent.

References

1. Zheng H, Han X, Guo W et al (2020) Recent developments and challenges of Li-rich Mn-based cathode materials for high-energy lithium-ion batteries. Mater Today Energy 18:100518
2. Liu T, Liu J, Li L et al (2022) Origin of structural degradation in Li-rich layered oxide cathode. Nature 606(7913):305–312
3. Cai Z, Wang S, Zhu H et al (2023) Improvement of stability and capacity of Co-free, Li-rich layered oxide $Li_{1.2}Ni_{0.2}Mn_{0.6}O_2$ cathode material through defect control. J Colloid Interface Sci 630:281–289
4. Wei T, Qin X, Lei C et al (2022) Considerable capacity improvement of Mn-based Li-rich cathode $Li_{1.2}Ni_{0.13}Co_{0.13}Mn_{0.54}O_2$ via $KMnO_4$ surface modification. J Alloys Compd 895:162647
5. Li Y, Bai Y, Wu C et al (2016) Three-dimensional fusiform hierarchical micro/nano $Li_{1.2}Ni_{0.2}Mn_{0.6}O_2$ with a preferred orientation (110) plane as a high energy cathode material for lithium-ion batteries. J Mater Chem A 4(16):5942–5951
6. Cai Y, Ku L, Wang L et al (2019) Engineering oxygen vacancies in hierarchically Li-rich layered oxide porous microspheres for high-rate lithium ion battery cathode. Sci China Mater 62(10):1374–1384
7. Zhang H, Liu H, Piper LFJ et al (2022) Oxygen loss in layered oxide cathodes for Li-Ion batteries: mechanisms, effects, and mitigation. Chem Rev 122(6):5641–5681
8. Sharifi-Asl S, Lu J, Amine K et al (2019) Oxygen release degradation in Li-Ion battery cathode materials: mechanisms and mitigating approaches. Adv Energy Mater 9(22):1900551
9. Hy S, Felix F, Rick J et al (2014) Direct in situ observation of Li_2O evolution on Li-rich high-capacity cathode material,

$Li[NixLi_{(1-2x)/3}Mn_{(2-x)/3}]O_2$ ($0 \leq x \leq 0.5$). J Am Chem Soc 136(3):999–1007
10. Wang F-F, Ji Y-R, Chen Y-H et al (2023) Rational construction of graphitic carbon nitride composited Li-rich Mn-based oxide cathode materials toward high-performance Li-ion battery. J Colloid Interface Sci 652:577–589
11. Zeng L, Liang H, Qiu B et al (2023) Voltage decay of Li-Rich layered oxides: mechanism, modification strategies, and perspectives. Adv Func Mater 33(25):2213260
12. Oh P, Ko M, Myeong S et al (2014) A Novel surface treatment method and new insight into discharge voltage deterioration for high-performance $0.4Li_2MnO_3$–$0.6LiNi_{1/3}C_{1/3}Mn_{1/3}O_2$ cathode materials. Adv Energy Mater 4(16):1400631
13. Song J, Ning F, Zuo Y et al (2022) Entropy stabilization strategy for enhancing the local structural adaptability of Li-Rich cathode materials. Adv Mater 35(7):2208726
14. Wang Q, Yao M, Zhu A et al (2023) Semi-metallic superionic layers suppressing voltage fading of Li-Rich layered oxide towards superior-stable Li-Ion batteries. Angew Chem Int Ed 62(37):e202309049
15. Cheng W, Ding J, Liu Z et al (2023) Zn/Ti dual concentration-gradients surface doping to improve the stability and kinetics for Li-rich layered oxides cathode. Chem Eng J 451:138678
16. Li S, Yang L, Liu Z et al (2023) Surface Al-doping for compromise between facilitating oxygen redox and enhancing structural stability of Li-rich layered oxide. Energy Storage Mater 55:356–363
17. Zhang K, Qi J, Song J et al (2022) Sulfuration of Li-Rich Mn-based cathode materials for multianionic redox and stabilized coordination environment. Adv Mater 34(11):2109564
18. Ji X, Xia Q, Xu Y et al (2021) A review on progress of lithium-rich manganese-based cathodes for lithium ion batteries. J Power Sour 487:229362
19. Meng J, Xu L, Ma Q et al (2022) Modulating crystal and interfacial properties by W-gradient doping for highly stable and long life Li-Rich layered cathodes. Adv Func Mater 32(19):2113013
20. Fu F, Yao Y, Wang H et al (2017) Structure dependent electrochemical performance of Li-rich layered oxides in lithium-ion batteries. Nano Energy 35:370–378
21. Dannehl N, Steinmüller SO, Szabó DV et al (2018) High-resolution surface analysis on aluminum oxide-coated $Li_{1.2}Mn_{0.55}Ni_{0.15}Co_{0.1}O_2$ with improved capacity retention. ACS Appl Mater Interfaces 10(49):43131–43143
22. Dong S, He X, Xu Q et al (2023) Nano ZrO_2 modification for enhancement of the electrochemical performance of Li-Rich manganese-based cathodic materials. Langmuir 39(22):7723–7730
23. Ran X, Tao J, Chen Z et al (2020) Surface heterostructure induced by TiO_2 modification in Li-rich cathode materials for enhanced electrochemical performances. Electrochimica Acta 353
24. Shi SJ, Tu JP, Tang YY et al (2013) Enhanced cycling stability of $Li[Li_{0.2}Mn_{0.54}Ni_{0.13}Co_{0.13}]O_2$ by surface modification of MgO with melting impregnation method. Electrochimica Acta 88:671–679
25. Wang B, Lu L, Hu Y et al (2023) A CeO_2-modified Li-rich layered oxide cathode with tunable interfacial oxygen for durable Li-ion batteries. New J Chem 47(5):2173–2176
26. Zheng J, Gu M, Xiao J et al (2014) Functioning mechanism of AlF_3 coating on the Li- and Mn-Rich cathode materials. Chem Mater 26(22):6320–6327
27. Zhu W, Tai Z, Shu C et al (2020) The superior electrochemical performance of a Li-rich layered cathode material with Li-rich spinel $Li_4Mn_5O_{12}$ and MgF_2 double surface modifications. J Mater Chem A 8(16):7991–8001
28. Sun YK, Lee MJ, Yoon CS et al (2012) The role of AlF_3 coatings in improving electrochemical cycling of Li-enriched nickel-manganese oxide electrodes for Li-Ion batteries. Adv Mater 24(9):1192–1196
29. Wu F, Zhang X, Zhao T et al (2015) Multifunctional $AlPO_4$ coating for improving electrochemical properties of low-cost $Li[Li_{0.2}Fe_{0.1}Ni_{0.15}Mn_{0.55}]O_2$ cathode materials for lithium-ion batteries. ACS Appl Mater Interfaces 7(6):3773–3781
30. Wang Z, Liu E, He C et al (2013) Effect of amorphous $FePO_4$ coating on structure and electrochemical performance of $Li_{1.2}Ni_{0.13}Co_{0.13}Mn_{0.54}O_2$ as cathode material for Li-ion batteries. J Power Sour 236:25–32
31. Wang Y, Yu W, Zhao L et al (2023) $AlPO_4$–Li_3PO_4 dual shell for enhancing interfacial stability of Co-free Li-rich Mn-based cathode. Electrochim Acta 462:142664
32. Luo Q, Xie Y, Wu Z et al (2021) Facile Molten Vanadate-assisted surface treatment strategy for Li_2MnO_3 activation of Li-Rich cathode materials. ACS Appl Energy Mater 4(5):4867–4878
33. Wei Y, Cheng J, Li D et al (2023) A structure self-healing Li-rich cathode achieved by lithium supplement of Li-rich LLZO coating. Adv Func Mater 33(22):2214775
34. Zhang K, Jiang Z, Ning F et al (2021) Metal-Ligand π interactions in lithium-rich Li_2RhO_3 cathode material activate bimodal anionic redox. Adv Energy Mater 11(30):2100892
35. Yang J, Li P, Zhong F et al (2020) Suppressing voltage fading of Li-rich oxide cathode via building a well-protected and partially-protonated surface by Polyacrylic acid binder for cycle-stable li-ion batteries. Adv Energy Mater 10(15):1904264
36. Pham HQ, Kim G, Jung HM et al (2017) Fluorinated polyimide as a novel high-voltage binder for high-capacity cathode of lithium-ion batteries. Adv Func Mater 28(2):1704690
37. Wang T, Zeng W, Zhu J et al (2023) SeO_2-infused grain boundaries effectively improve rate and stability performance of Li-rich manganese-based layered cathode materials. Nano Energy 113:108577
38. Fu C, Meng L, Wang J et al (2021) Bonding the terminal isocyanate-related functional group to the surface manganese ions to enhance Li-rich cathode's cycling stability. ACS Appl Mater Interfaces 13(15):17565–17576
39. Jiang B, Li J, Luo B et al (2021) $LiPO_2F_2$ electrolyte additive for high-performance Li-rich cathode material. J Energy Chem 60:564–571
40. Li Y, Li W, Shimizu R et al (2022) Elucidating the effect of borate additive in high-voltage electrolyte for Li-rich layered oxide materials. Adv Energy Mater 12(11):2103033
41. Li S, Li J, Wang P et al (2023) Interface engineering regulation by improving self-decomposition of lithium salt-type additive using ultrasound. Adv Funct Mater 2307180
42. Han J-G, Lee JB, Cha A et al (2018) Unsymmetrical fluorinated malonatoborate as an amphoteric additive for high-energy-density lithium-ion batteries. Energy Environ Sci 11(6):1552–1562
43. Lee DJ, Im D, Ryu Y-G et al (2013) Phosphorus derivatives as electrolyte additives for lithium-ion battery: the removal of O_2 generated from lithium-rich layered oxide cathode. J Power Sour 243:831–835
44. Huang T, Zheng X, Yan C et al (2023) Efficient improvement in electrochemical properties of high-voltage Li-rich Mn-based layered oxide cathode by addition of 1,3-divinyltetramethyldisiloxane to electrolyte. J Power Sour 580:233437
45. Zhu Y, Luo X, Zhi H et al (2017) Structural exfoliation of layered cathode under high voltage and its suppression by interface film derived from electrolyte additive. ACS Appl Mater Interfaces 9(13):12021–12034

46. Chen J, Chen H, Mei Y et al (2022) Polarly modulated solvent strategy for high-voltage cathode materials. Chem Eng J 450:138318
47. Li J, Zhang L, Yu L et al (2016) Understanding interfacial properties between Li-rich layered oxide and electrolyte containing triethyl borate. J Phys Chem C 120(47):26899–26907
48. Huang J, Zhang H, Yuan X et al (2023) Regulating robust interphase using a functional ionic liquid additive with bi-electrode affinity to stabilize the high-voltage lithium-Lithium-rich metals batteries. Chem Eng J 464:142578
49. Zhang X, Zhao J, Lee GH et al (2022) Bulk oxygen stabilization via electrode-electrolyte interphase tailored surface activities of Li-rich cathodes. Adv Energy Mater 13(2):2103033
50. Han JG, Hwang C, Kim SH et al (2020) An antiaging electrolyte additive for high-energy-density lithium-ion batteries. Adv Energy Mater 10(20):2000563

10 Spinel Lithium Nickel Manganese Oxide Cathode

High-voltage spinel $LiNi_{0.5}Mn_{1.5}O_4$ (LNMO) has advantages such as high working voltage (~4.7 V vs. Li/Li^+), high theoretical energy density (~650 Wh kg^{-1}), fast lithium-ion diffusion kinetics, and lower cost. It is the most promising cathode material for high-energy-density lithium-ion batteries used in plug-in hybrid electric vehicles and pure electric vehicles, with broad application prospects and huge commercial value [1–3]. However, the high working voltage of LNMO challenges the electrochemical stability of other components in the battery and triggers various interface side reactions, leading to irreversible capacity loss, poor cycle stability, and safety issues, especially under high temperature conditions [4–7]. Therefore, understanding its inherent surface characteristics and the interface interaction mechanism between various components in the electrochemical system is a key requirement for developing substantially enhanced LNMO-based batteries. In recent years, efforts have been made to improve the cycle stability of high-voltage LNMO cathodes, and some good results have been achieved, which helps us deepen our understanding of the degradation mechanism of LNMO.

10.1 Spinel $LiNi_{0.5}Mn_{1.5}O_4$ Crystal Structure

The crystal structure of spinel LNMO is divided into a face-centered cubic structure with the *Fd*-3 m space group and a simple cubic structure with the $P4_332$ space group (Fig. 10.1a, b) [8–10]. Generally, we refer to LNMO with the Fd-3m space group as a disordered structure and LNMO with the $P4_332$ space group as an ordered structure based on the different positions occupied by Mn/Ni in the lattice. Specifically, Ni and Mn randomly occupy the 16d octahedral sites in a 1:3 molar ratio in the disordered structure; in the ordered structure, Ni orderly replaces the lattice position of Mn: Ni occupies the 4b position while Mn occupies the 12d position. The typical XRD patterns of ordered and disordered LNMO are shown in Fig. 10.1c. The diffraction peaks of the rock salt phase $Ni_{1-x}Li_xO$ impurities can be seen in the face-centered cubic structure with the Fd-3m space group, while many tiny impurity peaks are present in the simple cubic structure with the $P4_332$ space group [11].

Due to the similar scattering factors of Ni and Mn, the ordered and disordered XRD spectra are almost identical, making it impossible to accurately distinguish between the Fd-3m and $P4_332$ structures through XRD spectra.

10.2 Preparation Method of Spinel $LiNi_{0.5}Mn_{1.5}O_4$ Cathode Material

Regardless of the method used to prepare LNMO precursors, such as the solid phase method, sol-gel method, molten salt method, etc., the annealing temperature in the final step of synthesis determines the chemical properties of LNMO crystals. Studies have shown that when the annealing temperature is below 700 °C, LNMO crystals tend to form an ordered phase; otherwise, their crystal structure will become disordered. Generally, due to the presence of some Mn^{3+} in disordered spinel LNMO, disordered spinel LNMO has higher electronic conductivity than the ordered structure (Fig. 10.1d). At higher sintering temperatures, a small amount of oxygen will escape from the LNMO material lattice, leading to a deficiency of oxygen atoms and the formation of $LiNi_{0.5}Mn_{1.5}O_{4-x}$, thereby reducing some Mn^{4+} to Mn^{3+} to maintain electrical neutrality. Therefore, LNMO cathode materials with Fd-3m space group are superior to those with $P4_332$ space group. However, due to the disproportionation of Mn^{3+}–Mn^{4+} and Mn^{2+} during the battery cycling process; Mn^{2+} is easily soluble in the electrolyte, and LNMO crystals can generate electrochemically inert impurity phases at high temperatures, these lead to poorer stability and cycling performance of LNMO cathode materials with Fd-3m space group.

X. Li et al., *Electrode Development for Lithium-ion Batteries*, https://doi.org/10.1007/978-981-96-3021-9_10

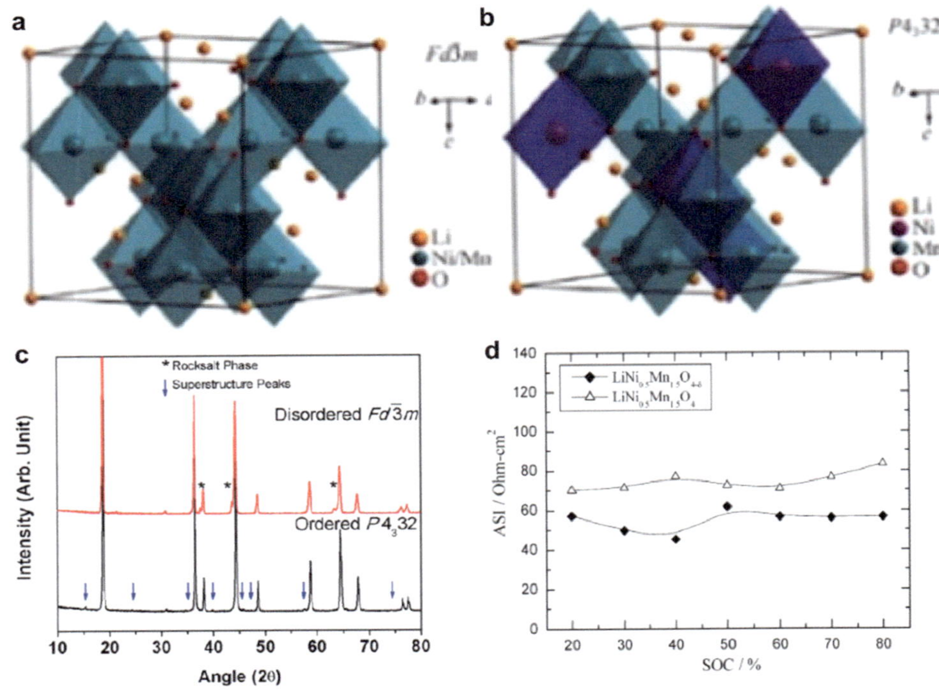

Fig. 10.1 LNMO cathode **a** Fd-3m (disordered) and **b** $P4_332$ (ordered) crystal structure schematic [10]; **c** XRD spectra of disordered and ordered structures of LNMO cathode [11]; **d** Area specific impedance of disordered and ordered structures of LNMO [8]

10.2.1 Solid Phase Method

The solid phase method includes ball milling, mechanical solid phase method, etc., its preparation process is simple, low cost, and time-saving [12–14]. However, LNMO synthesized by the solid-phase method has many oxygen defects and magazines, such as NiO and $Li_xNi_{1-x}O$, etc. In addition, the solid-phase method also has the disadvantages of active material agglomeration, uneven particle size distribution, and uncontrollable particle growth [15]. Therefore, Yin et al. [16] used a high-temperature solid-phase method and complexation reaction to prepare LNMO cathode material with high surface orientation (Fig. 10.2a). Firstly, a bimetallic organic framework containing Ni^{2+} and Mn^{2+} ions was synthesized by self-assembly method, and then calcined with lithium acetate. The sample prepared by this method has low magazine content, uniform grain size, superior orientation of (111) plane and (110) plane, and extremely high cycle stability and rate performance. Liu et al. [14] further studied the lithium manganate synthesized from $MnCO_3$, Mn_3O_4, Mn_2O_3 or MnO_2. The presence of a small amount of Mn^{3+} in the spinel crystal structure significantly affects its conductivity and electrochemical performance. Among them, LNMO-1 ($MnCO_3$ as the source of manganese) has the best electrochemical performance, the highest specific capacity, and the best cycle performance (Fig. 10.2b). Luo et al. [13] synthesized LNMO hollow microspheres by a simple solid-phase method (Fig. 10.2d). The electrochemical test results show that at a high rate of 5 C, the cathode material still maintains 93.4% of the initial capacity at 0.1 C, which is attributed to the large porosity and small secondary particles of the LNMO cathode material. Isozumi et al. [10] prepared LNMO cathode material by ball milling method, followed by a series of heat treatments. The prepared LNMO has a high discharge capacity.

10.2.2 Sol-Gel Method

The samples prepared by the sol-gel method usually have small particle sizes, good crystallinity, uniform dispersion, and excellent electrochemical performance, so this preparation method is widely used [17]. However, it is much more complicated and time-consuming compared to other synthesis methods using organic reagents, leading to increased costs. Zhu et al. [18] synthesized original LNMO and LNMO@ $Li_{0.33}La_{0.56}TiO_3$ (LLTO) composite material, due to the inherent characteristic of LLTO having a large ionic conductivity, reduces the interface charge transfer resistance, leading to an increase in the migration rate of lithium ions in the composite material (Fig. 10.3a, b). Piao et al. [19] also obtained the original LNMO by the sol-gel method, which had an initial capacity of 131.8 mAh g^{-1}, and after 100 cycles, the capacity retention rate was 92.1% (Fig. 10.3c). Bi et al. [20] used lithium acetate, nickel acetate, and manganese acetate as raw materials, mixed in ethanol, added oxalic acid to synthesize the LNMO precursor, placed the precursor in an alumina crucible, and obtained

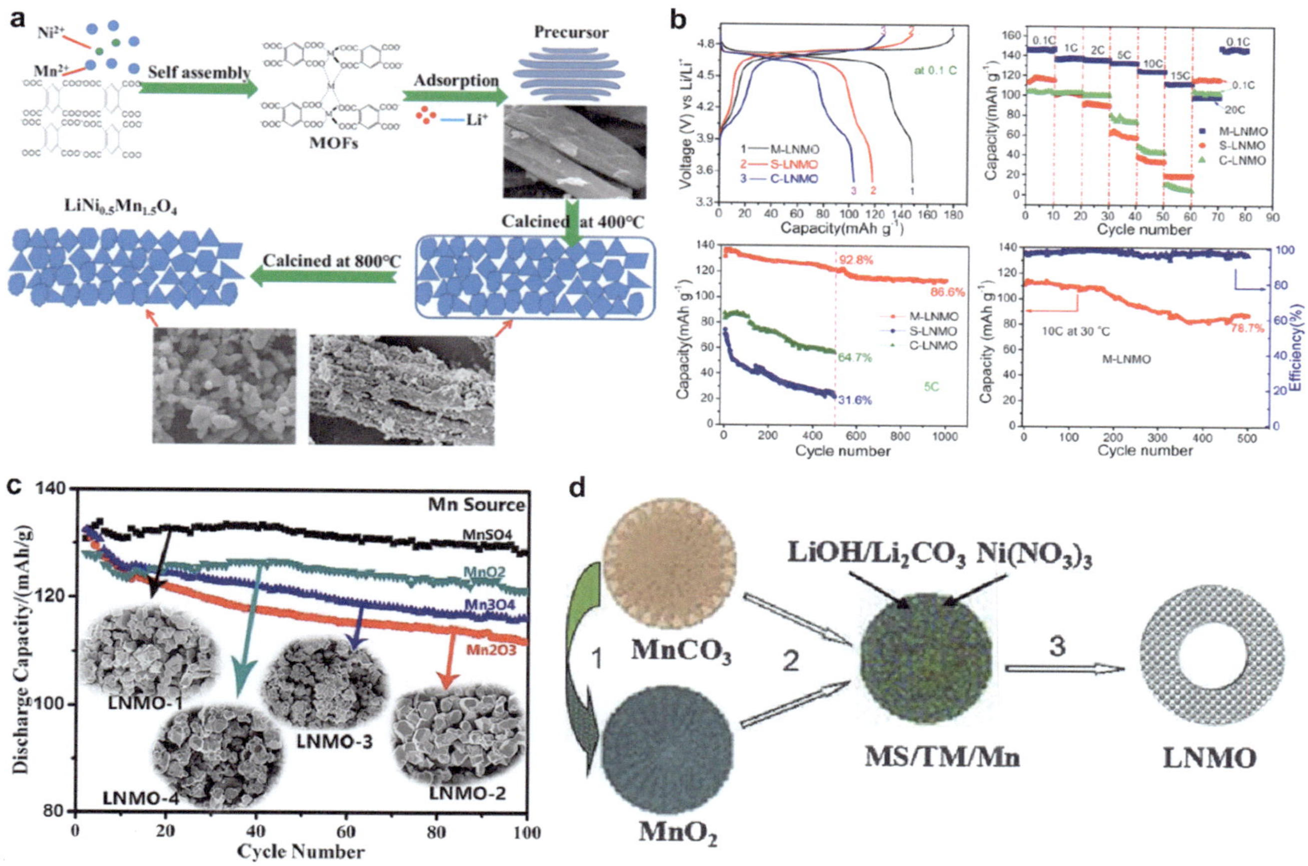

Fig. 10.2 **a** Schematic diagram of the solid-phase synthesis of LNMO from a bimetallic organic framework; **b** electrochemical performance of modified LNMO [16]; **c** electrochemical performance of LNMO synthesized from different manganese sources [14]; **d** schematic diagram of the synthesis process of hollow $LiNi_{0.5}Mn_{1.5}O_4$ microspheres [13]

LNMO by calcination in air. Sha et al. [21] first introduced urea as a chelating agent to prepare LNMO. Under the conditions of 5 C current density and high temperature of 55 °C, the material can maintain good cycle stability, with a capacity retention rate of 91.4% after 150 cycles, and a capacity retention rate of 87.5% after 200 cycles at a large current discharge rate of 20 C. Xu et al. [22] synthesized the precursor gel, then annealed the gel in air at 550 °C, and then further calcined it in air to obtain the original LNMO cathode material with good performance (Fig. 10.3d).

10.2.3 Molten Salt Method

Both the solid-phase method and wet chemistry method must undergo high-temperature annealing, so they are prone to produce magazines caused by oxygen defects, such as NiO and $Li_xNi_{1-x}O$ [23]. The molten salt method is a simple preparation method based on low-melting-point salts, which can be used to prepare pure phase LNMO cathode materials. Kim et al. [24] prepared $LNMO_{4-\delta}$ using the LiCl–KCl flux growth method, and reacted $LNMO_{4-\delta}$ crystals with different concentrations of LiF under KCl flux (800 °C, 20 h) to obtain LNMOF samples with different fluorine substitution amounts (Fig. 10.4a–e). Mei et al. [25] used the molten salt and programmed temperature method to synthesize hollow microspheres of LNMO, which still had a discharge capacity of 140 mAh g^{-1} after 100 charge–discharge cycles. Deng et al. [26] first prepared MnO_2 Microspheres, and then the double-shell LNMO hollow microspheres of the rock salt phase impurity synthesized. The double-shell structure can greatly reduce the volume effect of the electrode during the cycle. In summary, although the molten salt method has many advantages, it is not widely used in LNMO synthesis due to high calcination temperature and high energy consumption.

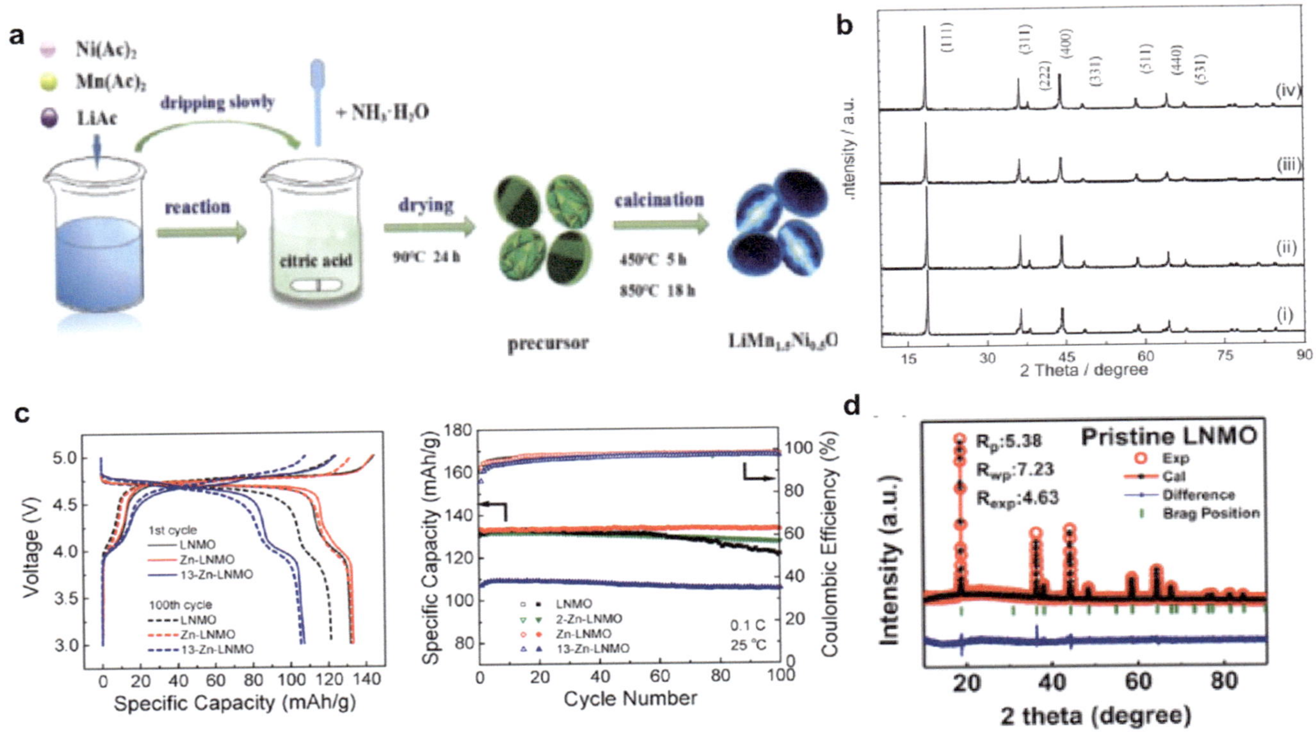

Fig. 10.3 **a** Schematic diagram of LNMO synthesized by sol-gel method; **b** XRD spectra of LNMO and its modified [18]; **c** first and 100th charge–discharge curves of LNMO at 0.1 C and cycle performance at 0.1 C [19]; **d** fine structure of the original LNMO [22]

10.3 Main Problems with Spinel $LiNi_{0.5}Mn_{1.5}O_4$

LNMO, as a high-voltage positive electrode material, has a voltage platform of 4.7 V. This is undoubtedly one of the most prominent highlights of this material, because a high voltage platform means a higher energy density. However, when the charging voltage exceeds 4.3 V (vs. Li^+), the currently commercially available carbonate liquid electrolyte will continuously decompose, leading to a large amount of solid electrolyte interphase (SEI) generation, exacerbating the dissolution of manganese, and destroying the electrode structure [27, 28]. In other words, the side reactions between the electrolyte and the electrode material are the main reasons for the capacity decline of lithium-ion batteries. The development of high-performance solid electrolytes may be the most effective way to optimize the recyclability of LNMO cathodes, but due to their slow ion diffusion, severe cycle polarization, high interface impedance, high raw material costs, etc., they are currently still unable to meet practical application needs [29, 30]. Especially in practical applications, the cycle stability is not ideal, so the series of problems brought about by the high voltage performance of LNMO still need to be explored and solved.

10.4 Modification Strategies for Spinel $LiNi_{0.5}Mn_{1.5}O_4$

The high-voltage spinel LNMO cathode has broad application prospects due to its high discharge platform (~4.7 V), high energy density (650 Wh kg^{-1}), fast Li^+ deintercalation kinetics, and low cost. However, its poor cycle stability, rapid capacity decay, and easy reaction with the electrolyte hinder its commercial application. In recent years, a large amount of research has been conducted on LNMO to improve its cycle stability. The research and development of LNMO for high-energy batteries generally focus on the following four aspects: (1) Morphology size regulation; (2) Electrolyte modification; (3) Ion doping modification; (4) Surface coating, etc. The following will discuss these four main modification strategies in detail.

10.4.1 Particle Morphology and Structure Regulation

Due to the long diffusion path inside the bulk LNMO, the deintercalation of Li^+ in the lattice has experienced a great impedance, leading to the formation of Li-rich and Li-poor in single crystal particles, causing serious grain stress and

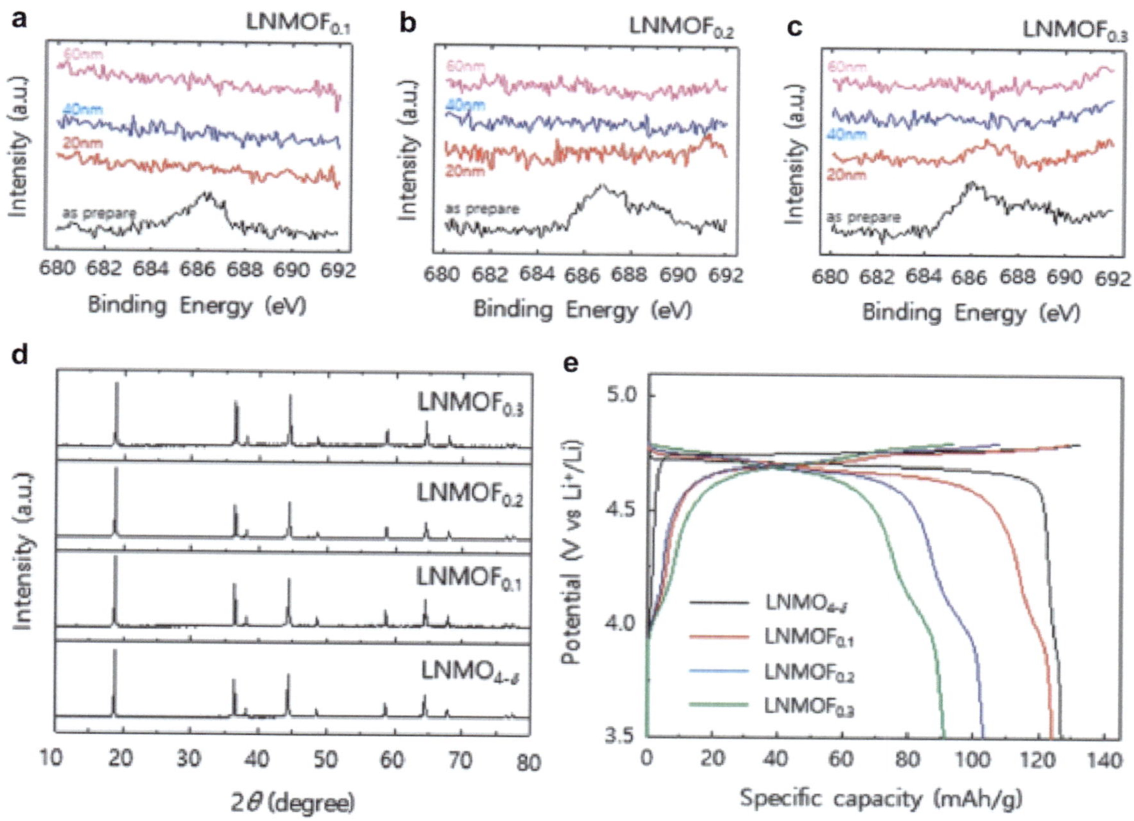

Fig. 10.4 **a–c** $LNMOF_{0.1}$, $LNMOF_{0.2}$ and $LNMOF_{0.3}$ with different Ar sputtering etching thicknesses; **d** the XRD spectra of the prepared samples match well with the reference data of ICDD PDF 70–8650 ($LiNi_{0.5}Mn_{1.5}O_4$), with a small offset with the addition of fluorine; **e** constant current charge and discharge curves at cut-off voltage 3.5 ~ 4.8 (0.2 C, 23 °C) [24]

grain boundary dislocations, severely damaging the stability of the structure. Therefore, the preparation of structurally stable LNMO materials needs to avoid huge structures. In contrast, nano-sized LNMO materials are attractive because they not only provide a large surface area for electrochemical reactions but also have short Li^+ diffusion paths. Studies have shown that different microscopic structures and different surface states of LNMO can be obtained through different synthesis methods, thereby improving its electrochemical performance [29]. At present, scholars have developed nanostructured LNMO materials, such as nanoparticles (0D), nano-nanorods (1D), nanorods (1D), and nanosheets (2D), achieving excellent electrochemical performance. Chen et al. [30] synthesized a Mn_3O_4 precursor with a particle size of 7 nm by chemical solution method, and sintered it with lithium salt to obtain LNMO particles with a particle size of about 350 nm. This material has a capacity retention rate of 98% after 500 cycles at 10 C (Fig. 10.5a–c). Electrospinning technology is widely used in battery research to manufacture one-dimensional structures. For example, Xu et al. [31] prepared nano-fiber LNMO with a diameter of only 50–100 nm by electrospinning. The LNMO with this morphology has an initial discharge capacity of 130 mAh g^{-1} at a current density of 27 mA g^{-1} (Fig. 10.5d, e). Zhao et al. [32] prepared LNMO nanorods using MnO_2 nanorods as the manganese source. During cycling at 1 C, the first week discharge capacity of this material reached 125.2 mAh g^{-1}, and after 200 cycles, the capacity was still as high as 119.8 mAh g^{-1}. In addition, two-dimensional nanostructured LNMO has also been proven to have good electrochemical performance. Chen et al. [33] prepared structurally ordered, rate-performance excellent $LiNi_{0.5}Mn_{1.5}O_4$ nanosheets by co-precipitation method. This material provides discharge capacities of 141, 90, and 60 mAh g^{-1} at 1 C, 20 C, and 30 C rates, respectively. Notably, Liu et al. [34] produced mixed microsphere-nanorod LNMO by hydrothermal calcination method. The mixed morphology LNMO structure has smaller deformation, and after 1000 cycles at 10 C, the reversible capacity is as high as 107.8 mAh g^{-1}. Although nanostructures can greatly enhance the rate capability of LNMO, they also have some drawbacks, such as lower compaction density leading to lower volumetric energy density; high specific surface area can produce serious surface side reactions, resulting in poorer cycle stability. Compared with nanostructures, nano/micro hierarchical design is a more promising structure

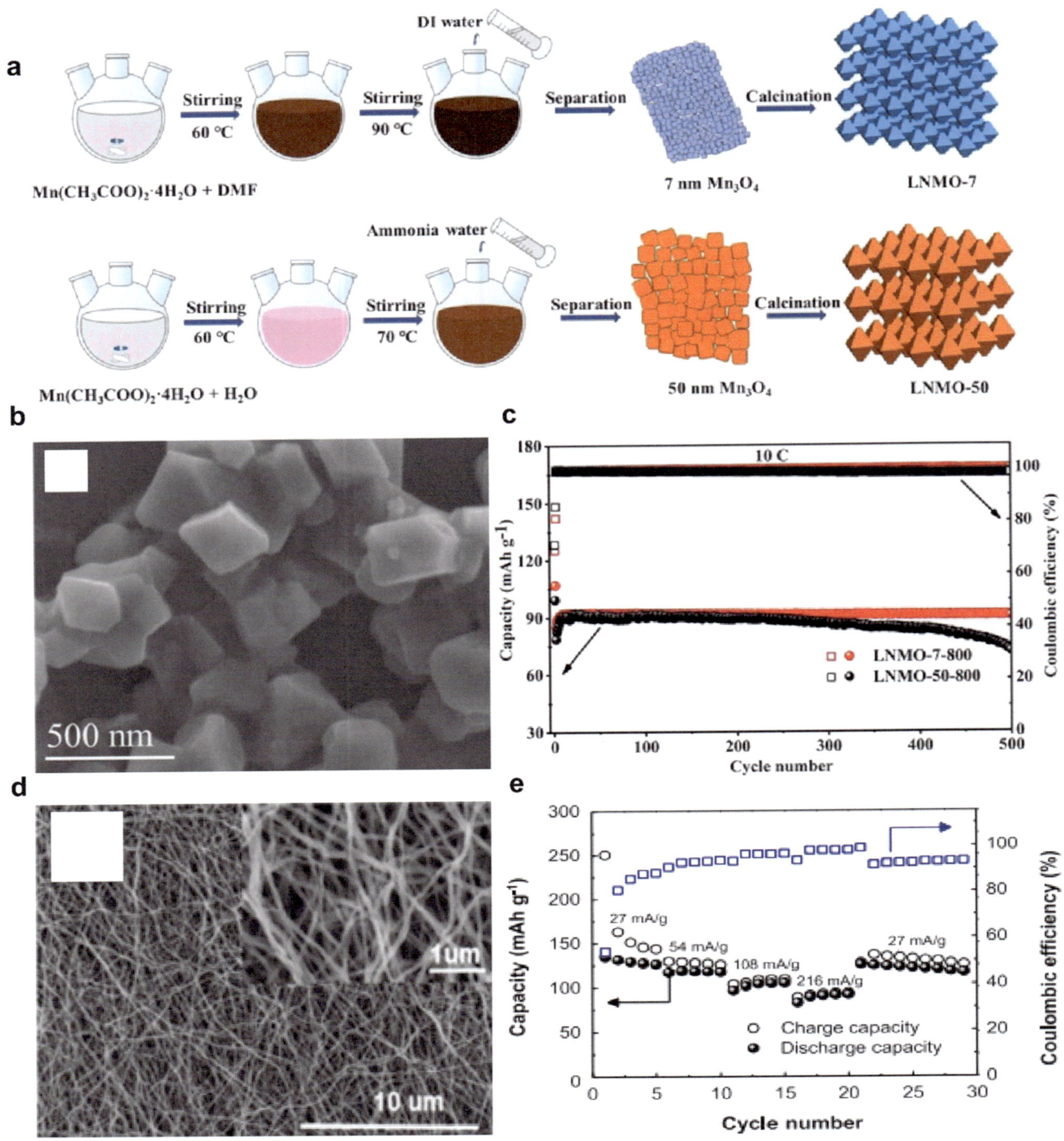

Fig. 10.5 **a** Schematic diagram of the formation process of Mn_3O_4 nanocrystals of different sizes and $LiNi_{0.5}Mn_{1.5}O_4$; **b** SEM image of LNMO nanoparticles; **c** cycle stability of LNMO [30]. **d** SEM image of LNMO nanorods; **e** relationship between the cycle capacity and current density of spinel $LiNi_{0.5}Mn_{1.5}O_4$ nanorods [31]

for optimizing LNMO performance, because it has short ion diffusion distance, large surface area, ensuring sufficient contact between the electrode and the electrolyte, and providing interconnected ion diffusion pathways within the pressure that the active material can withstand. Sun et al. [35] prepared LNMO with a nano-micro hierarchical structure resembling a bayberry shape through a solvent thermal method. The LNMO precursor has a uniform microscale spherical structure assembled from ultra-thin nanoplates. This is because during the ethylene glycol-assisted hydrothermal process, ethylene glycol will produce a strong steric hindrance effect, which can effectively prevent the initial crystal nucleus from agglomerating, thus ensuring a uniform particle size distribution of the product. Therefore, the Brunauer-Emmett-Teller surface area of the bayberry-like LNMO reached 57 $m^2\ g^{-1}$. This layered structure provides a rich electrode/electrolyte contact area to promote electrochemical reactions and ensure that the electrolyte can fully penetrate the LNMO cathode. Under the discharge conditions of 30 C ultra-high rate, the capacity retention rate is still as high as 84% after 1200 cycles. This excellent cycling performance can be attributed to the fact that the nano/micro hierarchical structure can reduce the volume change caused by the deintercalation of Li^+. In summary, LNMO particles synthesized by different methods have different morphological characteristics, which undoubtedly is an important factor affecting their electrochemical performance.

10.4.2 Electrolyte Modification

Traditional liquid electrolytes have become a constraint on the development of liquid electrolytes due to their poor thermal stability, low ion selectivity, and poor safety [36]. At present, the mainstream electrolyte composition is $LiPF_6$, and ethyl carbonate (EC), diethyl carbonate (DEC), dimethyl carbonate (DMC) or their mixtures are used as electrolyte solvents. At high potentials, $LiPF_6$ easily reacts with trace water in the electrolyte to produce HF and PF_5. HF and PF_5 from the electrolyte. It can corrode the positive electrode material, causing structural damage to the positive electrode material, and then the transition metal dissolves into the electrolyte. Therefore, the performance of the liquid electrolyte can be improved by adding electrolyte additives. For example, Andreeas et al. [37] studied the effects of five additives: lithium bis(oxalatophosphate), lithium difluoro(oxalatophosphate), 1-vinyl-1,2,4-triazole, 1-vinylimidazole, and dimethyl-2,5-dioxahexane dicarboxylate on the performance of LNMO. The results showed that all additives could significantly improve the cycle performance of the battery. Rath et al. [38] used lithium bis(trifluoromethanesulfonyl)imide (LiTFSI) to replace $LiPF_6$, revealing that the concentration of LiTFSI and the ratio of ionic liquids severely affected the rate of irreversible side reactions at high potentials. Simply adjusting the concentration of LiTFSI in the electrolyte can effectively suppress the corrosion problem of TFSI-based ionic liquid/carbonate solvent mixed electrolyte on aluminum. The LNMO battery using this electrolyte can maintain good rate performance even under a 3–5 V electrochemical window (Fig. 10.6a). Huang et al. [39] added tributyl borate (TBB) to the electrolyte, which can enhance the stability of the LNMO-electrolyte interface, effectively inhibit the oxidative decomposition of carbonate solvents and the dissolution of TMs. However, relying on additives cannot fundamentally solve the defects of liquid electrolytes. The main solution for the oxidative decomposition and side reactions of liquid electrolytes under high pressure is to modify and develop solid electrolytes. Lithium-ion solid-state batteries are the key to the future development of lithium batteries. Solid electrolytes are mainly divided into polymer solid electrolytes, garnet solid electrolytes,

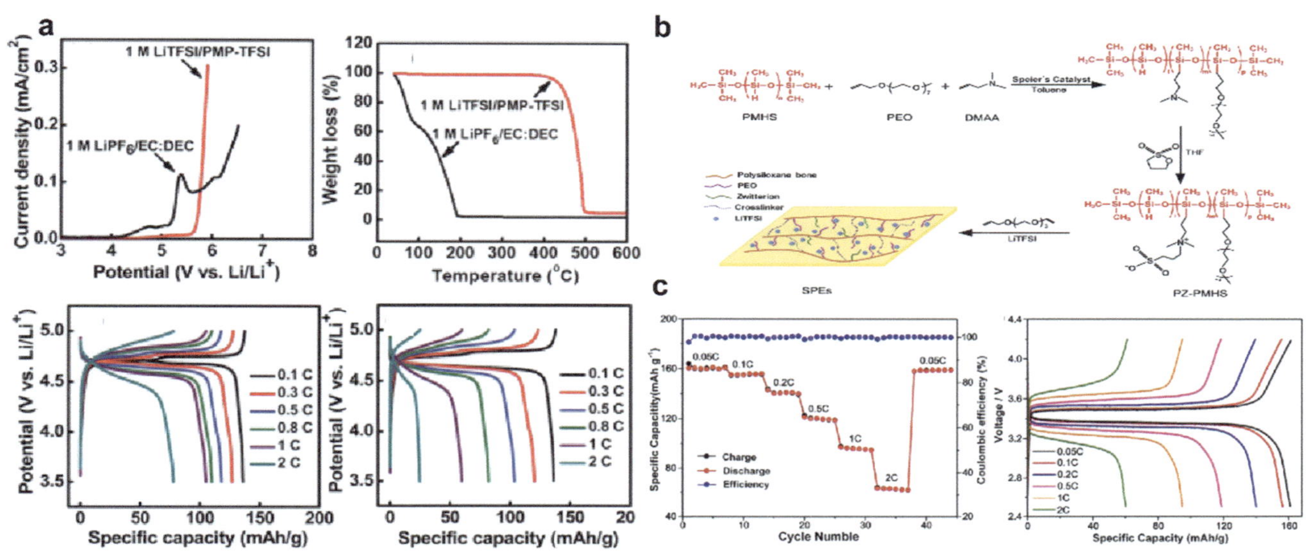

Fig. 10.6 **a** LSV scans of Pt electrode recorded in 1 M $LiPF_6$/EC:DEC and 1 M LiTFSI/PMP-TFSI electrolytes, with a potential scan rate of 1 mV s^{-1} (top); charge–discharge curves of LNMO at different rates (bottom) [38]; **b** schematic diagram of the synthesis process of solid polymer electrolyte; **c** magnification performance and charge–discharge curve [40]

sulfide crystalline solid electrolytes, and glass solid electrolytes. Liu et al. [40] used zwitterionic betaine sulfonate (ZI) and polyethylene oxide (PEO) to modify polysiloxane (Fig. 10.6b). The prepared solid electrolyte has a wide electrochemical window. The LNMO half-cell with solid electrolyte has an initial capacity of 131.5 mAh g^{-1} at a 0.1 C rate and excellent cycle performance (Fig. 10.6c). Sun et al. [41] doped nano-SiO_2 and Al_2O_3 into poly(methyl methacrylate-acrylonitrile-ethyl acrylate), synthesizing a new type of solid electrolyte. The half-cell with this new solid electrolyte showed excellent rate performance and capacity retention. Interestingly, they found a synergistic effect between nano-SiO_2 and nano-Al_2O_3.

10.4.3 Ion Doping Modification

The doping modification of LNMO cathode material can effectively improve its cycle stability. This is because the doping ions not only change the chemical composition of the LNMO material, but also change the crystal structure, lattice parameters, and orderliness of LNMO. Therefore, the method of ion doping modification is not only an effective way to improve the electrochemical performance of LNMO, but also conducive to large-scale production. There are currently three main doping methods: (1) metal cation doping; (2) non-metal ion doping; (3) co-doping of multiple elements. Different doping ions occupy different positions, producing different modification effects. At present, there are various synthesis methods for LNMO materials doped with different metal elements, mainly including Na^+, Mg^+, Si^{2+}, Al^{3+}, Ga^{3+}, Ge^{3+}, Ti^{4+}, V^{5+}, Fe^{3+}, Co^{3+}, Cu^{2+}, Zn^{2+} and so on. For example, Guo et al. [42] selectively doped Mg into the Fd-3m structure of LNMO at the 8a and 16c sites through a simple solid-phase reaction strategy (Fig. 10.7a). Since the Mg ion has a radius similar to the ions at the tetrahedral and octahedral sites, severe structural deformation is avoided. The LNMO doped with Mg ions exhibits excellent electrochemical performance, with about 86% and 87% capacity retention after 1500 cycles at

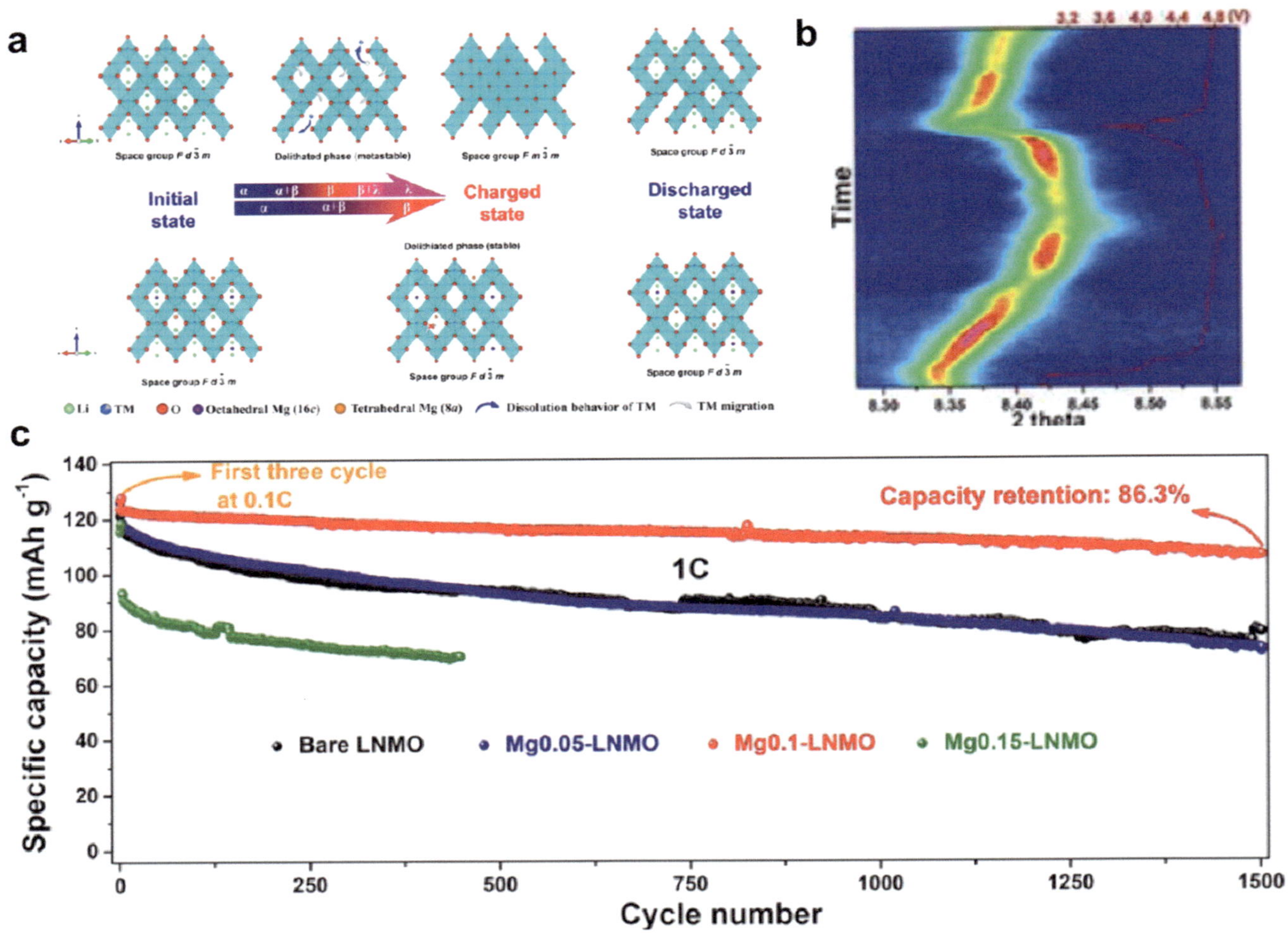

Fig. 10.7 **a** Schematic diagram of the structural evolution of undoped LNMO and Mg-doped LNMO during charge–discharge cycles; **b** in-situ XRD schematic diagram; **c** long cycle data of undoped and different amounts of Mg-doped LNMO [42]

1 C and 2200 cycles at 10 C, respectively (Fig. 10.7b, c). Wang et al. [43] first used Na^+ doping to apply alkali metal doping to LNMO modification. The research results show that Na^+ can reduce particle size and increase the disorder of Ni/Mn arrangement. In addition, Na^+ doped samples provide additional electron transition paths to reduce the impedance of LNMO. Therefore, Na^+ ion doping can effectively reduce electrode polarization and improve the magnification performance of LNMO cathode material. During high-temperature annealing, oxygen in LNMO will escape to form oxygen vacancies, and Mn^{4+} will also be reduced to Mn^{3+}. Non-metal doping is also an effective means to improve the electrochemical performance of LNMO. Unlike metal dopants that only have cations, non-metal dopants have both anions and cations. The reported non-metal dopants include F^-, Cl^-, S^{2-}, B^{3+}, P^{3+}, S^{4+}, etc. For example, LNMO doped with B^{3+} contains a B–O bond with a key energy as high as 809 kJ mol^{-1}, which enhances the structural stability of LNMO and thus improves its cycling performance. Wei et al. [44] synthesized B^{3+} doped $LiNi_{0.495}B_{0.01}Mn_{1.495}O_4$, which showed good cycling stability, with a capacity retention rate of 83.3% after 500 cycles at a current density of 3 C. In addition, their data showed that LNMO doped with B^{3+} has excellent rate performance, which is due to B^{3+} accelerating the diffusion rate of Li^+. Luo et al. [45] synthesized LNMO spinel by gradient doping. The results showed that LNMO crystals doped with F– have larger grain sizes and larger lattice parameters, and a fluorine-rich surface is formed. This surface inhibits the side reactions between the electrolyte and the positive electrode surface, forming a thin and uniform CEI on the positive electrode surface. Therefore, LNMO samples with F^- gradient doping have long-term cycling stability and excellent rate performance. In addition, the synergistic effect of different elements can solve multiple problems of LNMO materials at the same time. Therefore, co-doping is also an important method to improve the electrochemical performance of spinel LNMO. Li et al. [46] synthesized Cu^{2+} and Al^{3+} co-doped LNMO, which showed good cycling stability and rate performance. This is due to the synergistic effect of Cu^{2+} and Al^{3+}. The introduction of these two elements not only increases the disorder of cation arrangement, but also produces different effects. Al^{3+} enriches the surface to form a protective layer, inhibiting the dissolution of transition metals; at the same time, Cu^{2+} in the bulk provides larger diffusion channels, accelerating the diffusion of Li^+. Yao et al. [47] co-doped trace amounts of Cr–Fe–Cu elements in LNMO through a simple solution blow spinning technique to achieve excellent cycling stability and rate performance of thick-loaded LNMO electrodes. Among them, Cu^{2+} with a larger diameter acts as a structural pillar to stabilize the LNMO structure and inhibit the dissolution of transition metals; Fe^{3+} and Cr^{3+} improve the thermal stability of the material and inhibit the release of oxygen; at the same time, multi-element doping effectively increases the cation disorder in the lattice, thereby promoting the diffusion of Li^+. Therefore, $LiNi_{0.45}Cr_{0.0167}Fe_{0.0167}Cu_{0.0167}$ doped with multiple elements of Cr–Fe-Cu shows excellent performance. $Mn_{1.5}O_4$ (CFC0.5-LNMO) maintains a capacity retention rate of 78.2% after 300 cycles at 0.5 C. In summary, LNMO doped with one or more elements can be achieved in different ways. By utilizing the valence states and ionic radii of different dopants, substitution at different positions in the LNMO lattice can be achieved, thereby achieving ideal electrochemical performance.

10.4.4 Surface Coating Modification

The most significant drawback of LNMO is the instability of the electrode surface and the side reactions produced when in contact with the electrolyte under high pressure, which leads to the dissolution of Mn^{3+}, destroying the original structure of LNMO. Therefore, surface coating modification of LNMO is undoubtedly an effective way to stabilize the LNMO surface, reduce side reactions with the electrolyte, and inhibit the dissolution of transition metal ions. To improve its cycle stability without reducing the discharge capacity and rate performance of LNMO, the coating layer should have the following characteristics: (1) a high Li^+ diffusion coefficient; (2) the coating material should match well with the lattice of the body material; (3) the coating layer will not increase the overall resistance of the material. Currently, commonly used coating materials mainly include the following: metal oxides, metal fluorides, phosphates, etc. For example, Courbaron et al. [48] prepared a simple, fast, and cheap Al_2O_3 uniform coating using supercritical fluid chemical deposition to modify the surface of spinel LNMO. Experimental results show that the Al_2O_3 modified LNMO still exhibits good performance even under high load and high rate conditions, with a reversible capacity of 120 mAh g^{-1} at a current density of 4 C (Fig. 10.8a). Metal fluorides have stronger HF corrosion resistance than metal oxides. Li et al. [49] prepared AlF_3 coated LNMO, which showed a higher discharge specific capacity and capacity retention rate. Experimental data show that the capacity is 114.8 mAh g^{-1} after 40 cycles at a current density of 300 mAh g^{-1}, while the capacity of the unmodified LNMO under the same conditions is only 84.3 mAh g^{-1}. The above improvement in rate performance is mainly attributed to the AlF_3 layer effectively blocking the erosion of HF on the LNMO surface. Recently, Haber et al. [50] also proved that AlF_3 modified LNMO has a faster ion

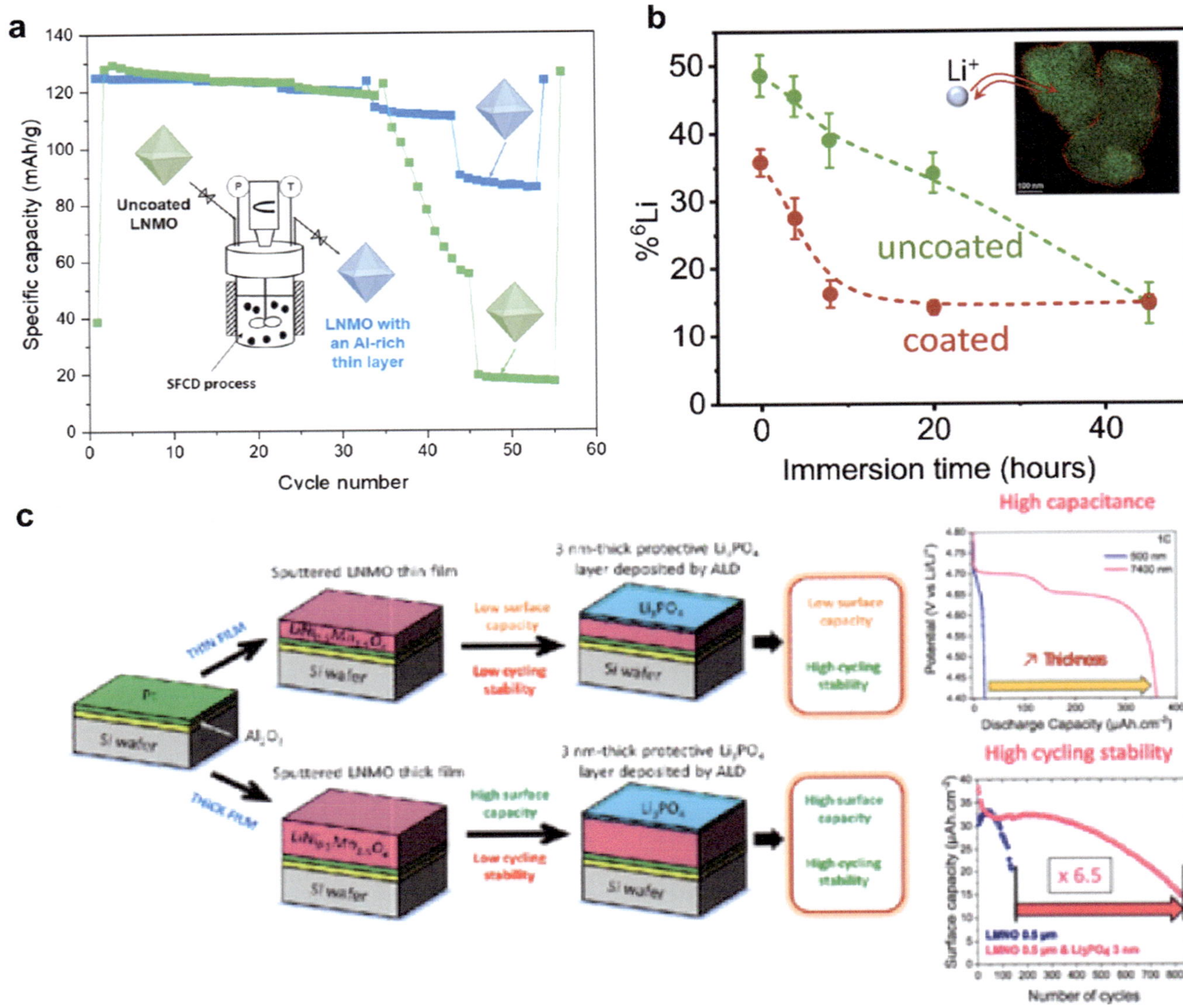

Fig. 10.8 **a** Supercritical fluid chemical deposition (SFCD) route and rate performance of original and modified LNMO electrodes [48]; **b** nuclear magnetic resonance shows faster cathode ion exchange after coating [50]; **c** strategy schematic for improving surface capacity and cycle stability, and long cycle performance of modified and unmodified LNMOD [51]

exchange rate than the original material, thereby improving the interfacial ion transport dynamics (Fig. 10.8b). In addition, Hallot et al. [51] used atomic layer deposition technology to deposit 7400 A lithium-ion conductive layer Li_3PO_4 of nm thickness deposited on the surface of LNMO. After modification by Li_3PO_4, the LNMO provides a higher reversible capacity and significantly improved cycle life, which is 6.5 times that of the original LNMO electrode (Fig. 10.8c). In summary, the significance of the surface coating is to maintain the stability between interfaces, suppress the side reactions of LNMO and electrolyte under high pressure, reduce the dissolution of Mn, and protect the material from HF corrosion. Different types of coating materials have different protective effects. For example, polymer coatings can alleviate the volume change of LNMO during the cycle; Li^+ and electron conductive coatings can improve the rate performance of the material, etc.

10.5 Conclusion

As a 5 V high-voltage positive electrode material, LNMO has advantages such as high energy density, excellent high current discharge performance, good safety performance, and low cost compared to existing commercial 4 V positive electrode materials $LiCoO_2$ (LCO) and $LiFePO_4$ (LFP). In addition, from the perspective of resource cost, since the material does not contain Co elements, LNMO has advantages that LCO and ternary materials cannot match. However, the commercial application of LNMO in LIBs is not only hindered by its own subtle structural evolution, but also faces major challenges in the surface/interface when using commercially available components in LIBs (such as liquid organic electrolytes, polymer separators, conductive additives, adhesives, etc.) under high working voltage. Therefore, the surface and interface chemical properties of high-voltage LNMO lithium-ion batteries are key factors determining their structure and electrochemical stability. In recent years, problems such as electrolyte decomposition, electrode erosion, surface structure rearrangement, dissolution of transition metal ions and deposition on the negative electrode surface, formation of unstable solid electrolyte interface films, etc., have been discovered and have attracted widespread attention from researchers. As a result, many researchers have proposed strategies to overcome these problems. Traditional strategies are to coat and dope on the LNMO positive electrode, and explore high-voltage electrolytes, etc. Despite this, there is still a long way to go for the commercial application of LNMO in high-power lithium batteries. In the future, all-solid-state batteries are the inevitable development trend of high energy/power density and safe lithium-ion batteries [52, 53]. Solid electrolytes in all-solid-state batteries can solve inherent problems of traditional batteries: electrolyte decomposition, dissolution of transition metals, growth of lithium dendrites, and the flammability of liquid organic electrolytes. In order to realize the commercial application of LNMO in high-power lithium-ion batteries, future work will focus on exploring solid polymer electrolytes with high voltage and low operating temperature, improving ion conductivity, reducing interface impedance, and optimizing the interface compatibility of the electrode/solid polymer electrolyte. At the same time, in order to observe complex interface reactions and interface resistance in real time without damaging, oxidizing, and contaminating the material surface, it is necessary to develop advanced in-situ characterization techniques.

References

1. Tarascion J-M, Armand M (2001) Issues and challenges facing rechargeable lithium batteries. Nature 414:359–367
2. Yang ZG, Zhang JL, Kintner-Meyer MC et al (2011) Electrochemical energy storage for green grid. Chem Rev 111:3577–3613
3. Diouf B, Pode R (2015) Potential of lithium-ion batteries in renewable energy. Renew Energy 76:375–380
4. Qiao Y, Zhou Z, Chen Z et al (2018) Visualizing ion diffusion in battery systems by fluorescence microscopy: a case study on the dissolution of $LiMn_2O_4$. Nano Energy 45:68–74
5. Ramakrishnan S, Park B, Wu J et al (2020) Extended interfacial stability through simple acid rinsing in a Li-rich oxide cathode material. J Am Chem Soc 142:8522–8531
6. Song J, Shin DW, Lu Y et al (2012) Role of oxygen vacancies on the performance of $Li[Ni_{0.5-x}Mn_{1.5+x}]O_4$ (x = 0, 0.05, and 0.08) spinel cathodes for lithium-ion batteries. Chem Mater (24):3101–3109
7. Maleki Kheimeh Sari H, Li X (2019) Controllable cathode–electrolyte interface of $Li[Ni_{0.8}Co_{0.1}Mn_{0.1}]O_2$ for lithium ion batteries: a review. Adv Energy Mater (9):1901597
8. Kim J-H, Myung S-T, Yoon CS et al (2004) Comparative study of $LiNi_{0.5}Mn_{1.5}O_4$ and $LiNi_{0.5}Mn_{1.5}O_4$ cathodes having two ccrystallographic structures: Fd-3m and $P4_332$. Chem Mater (16):906–914
9. Liu D, Hamel-Paquet J, Trottier J et al (2012) Synthesis of pure phase disordered $LiMn_{1.45}Cr_{0.1}Ni_{0.45}O_4$ by a post-annealing method. J Power Sour (217):400–406
10. Chen J, Huang Z, Zeng W et al (2021) Synthesis, modification, and lithium-storage properties ofspinel $LiNi_{0.5}Mn_{1.5}O_4$. ChemElectroChem (8):608–624
11. Manthiram A, Chemelewski K, Lee E-S (2014) A perspective on the high-voltage $LiMn_{1.5}Ni_{0.5}O_4$ spinel cathode for lithium-ion batteries. Energy Environ Sci (7):1339
12. Ding XK, Zhang LL, Yang XL et al (2017) Anthracite-derived dual-phase carbon-coated $Li_3V_2(PO_4)_3$ as high-performance cathode material for lithium ion batteries. ACS Appl Mater Interfaces 9:42788–42796
13. Luo H, Nie P, Shen L et al (2015) Synthesis of $LiNi_{0.5}Mn_{1.5}O_4$ hollow microspheres and their lithium-storage properties. ChemElectroChem (2):127–133
14. Liu Y, Lu Z, Deng C et al (2017) Preparation and electrochemical properties of high-voltage spinel $LiNi_{0.5}Mn_{1.5}O_4$ synthesized by using different manganese sources. ChemElectroChem (4):1205–1213
15. Liu G, Wen L, Liu Y (2019) Spinel $LiNi_{0.5}Mn_{1.5}O_4$ and its derivatives as cathodes for high-voltage Li-ion batteries. J Solid State Electrochem (14):2191–2202
16. (2018) Synthesis and electrochemical properties of $LiNi_{0.5}Mn_{1.5}O_4$ for Li-ion batteries by the metal−organic framework method. ACS Appl Mater Interfaces (10):13625−1363
17. Santhanam R, Rambabu B (2010) Research progress in high voltage spinel $LiNi_{0.5}Mn_{1.5}O_4$ material. J Power Sour (195):5442–5451
18. Zhu Y, Yi T, Li X et al (2019) Improved rate performance of $LiNi_{0.5}Mn_{1.5}O_4$ as cathode of lithium-ion battery by $Li_{0.33}La_{0.56}TiO_3$ coating. Mater Lett (239):56–58
19. Piao J, Gu L, Wei Z et al (2019) Phase Control on surface for the stabilization of high energy cathode materials of lithium ion batteries. J Am Chem Soc 141:4900–4907
20. Bi K, Zhao S, Huang C et al (2018) Improving low-temperature performance of spinel $LiNi_{0.5}Mn_{1.5}O_4$ electrode and $LiNi_{0.5}Mn_{1.5}O_4/Li_4Ti_5O_{12}$ full-cell by coating solid-state electrolyte Li–Al–Ti–P–O. J Power Sour (389):240–248
21. Sha O, Wang S, Qiao Z et al (2012) Synthesis of spinel $LiNi_{0.5}Mn_{1.5}O_4$ cathode material with excellent cycle stability using urea-based sol-gel method. Mater Lett (89):251–253
22. Xu T, Li Y, Wang D et al (2018) Enhanced electrochemical performance of $LiNi_{0.5}Mn_{1.5}O_4$ cathode material by YPO_4 surface modification. ACS Sustain Chem Eng (6):5818–5825
23. Feng ZH, Guo C et al (2013) An organic coprecipitation route to synthesize high voltage $LiNi_{0.5}Mn_{1.5}O_4$. ACS Appl Mater Interfaces (5):10227–10232
24. Kim D-W, Shiiba H, Zettsu N et al (2017) Full picture discovery for mixed-fluorine anion effects on high-voltage spinel lithium nickel manganese oxide cathodes. NPG Asia Mater (9):e398–e398
25. Mei T, Pi W, Zhang L et al (2016) Synthesis of shell-in-shell $LiNi_{0.5}Mn_{1.5}O_4$ hollow microspheres and their enhanced performance for lithium ion batteries. Mater Lett (173):141–144

26. Deng S, Mao D, Wang H et al (2016) Preparation and electrochemical properties of double-shell $LiNi_{0.5}Mn_{1.5}O_4$ hollow microspheres as cathode materials for Li-ion batteries. RSC Adv (6):45369–45375
27. Huang J-Y, Cheng C-Y, Lai Y-M et al (2023) Engineering cathode–electrolyte interface of high-voltage spinel $LiNi_{0.5}Mn_{1.5}O_4$ via halide solid-state electrolyte coating. ACS Appl Mater Interfaces 2023(x):xxx–xxx
28. Wang B, Son S-B, Badami P et al (2023) Understanding and $_{0.5}Mn_{1.5}O_4$. Batteries (9):435
29. Yi T, Mei J, Zhu Y et al (2016) Key strategies for enhancing the cycling stability and rate capacity of $LiNi_{0.5}Mn_{1.5}O_4$ as high-voltage cathode materials for high power lithium-ion batteries. J Power Sour (316):85–105
30. Chen R, Mei J, Xu J et al (2022) Preparation of $LiNi_{0.5}Mn_{1.5}O_4$ cathode materials by using different-sized Mn_3O_4 nanocrystals as precursors. J Solid State Electrochem (26):1359–1368
31. Xu R, Zhang X, Chamoun R et al (2015) Enhanced rate performance of $LiNi_{0.5}Mn_{1.5}O_4$ fibers synthesized by electrospinning. Nano Energy (15):616–624
32. Zhao H, Li F, Shu X et al (2018) Environment-friendly synthesis of high-voltage $LiNi_{0.5}Mn_{1.5}O_4$ nanorods with excellent electrochemical properties. Ceram Int (44):20575–20580
33. Chen Z, Wang X, Tian X et al (2019) Synthesis of ordered $LiNi_{0.5}Mn_{1.5}O_4$ nanoplates with exposed {100} and {110} crystal planes and its electrochemical performance for lithium ions batteries. Solid State Ionics (333):50–56
34. Liu H, Liang G, Gao C et al (2019) Insight into the improved cycling stability of sphere-nanorod-like micro-nanostructured high voltage spinel cathode for lithium-ion batteries. Nano Energy 66:104100
35. Sun W, Li Y, Liu Y et al (2018) Hierarchical waxberry-like $LiNi_{0.5}Mn_{1.5}O_4$ as an advanced cathode material for lithium-ion batteries with a superior rate capability and long-term cyclability. J Mater Chem A (6):14155–14161
36. Wang Q, Yao Z, Zhao C et al (2020) Interface chemistry of an amide electrolyte for highly reversible lithium metal batteries. Nat Commun 11:4188
37. Hofmann A, Howeling A, Bohn N et al (2019) Additives for cycle life improvement of high-voltage LNMO-based Li-ion cells. ChemElectroChem 6:5255–5263
38. Rath PC, Wu C-J, Patra J et al (2019) Hybrid electrolyte enables safe and practical 5 V $LiNi_{0.5}Mn_{1.5}O_4$ batteries. J Mater Chem A (7):16516–16525
39. Huang T, Zheng X, Pian Y et al (2019) Effect of tributyl borate on electrochemical performance at an elevated temperature of high-voltage $LiNi_{0.5}Mn_{1.5}O_4$ cathode. (11):26872–26879
40. Liu M, Jin B, Zhang Q et al (2018) High-performance solid polymer electrolytes for lithium ion batteries based on sulfobetaine zwitterion and poly (ethylene oxide) modified polysiloxane. J Alloy Comp 742:619–628
41. Sun P, Liao Y, Luo X et al (2015) The improved effect of co-doping with nano-SiO_2 and nano-Al_2O_3 on the performance of poly (methyl methacrylate-acrylonitrile-ethyl acrylate) based gel polymer electrolyte for lithium ion batteries. RSC Adv 5:64368–64377
42. Liang G, Wu Z, Didier C et al (2020) A long cycle-life high-voltage spinel lithium-ion battery electrode achieved by site-selective doping. Angew Chem Int Ed 59:10594–10602
43. Wang J, Lin W, Wu B et al (2014) Syntheses and electrochemical properties of the Na-doped $LiNi_{0.5}Mn_{1.5}O_4$ cathode materials for lithium-ion batteries. Electrochim Acta (145):245–253
44. Wei A, Mu J, He R et al (2021) Enhancing electrochemical performance and structural stability of $LiNi_{0.5}Mn_{1.5}O_4$ cathode material for rechargeable lithium-ion batteries by boron doping. Ceram Int (47):226–237
45. Luo Y, Li H, Lu T et al (2017) Fluorine gradient-doped $LiNi_{0.5}Mn_{1.5}O_4$ spinel with improved high voltage stability for Li-ion batteries. Electrochim Acta (238):237–245
46. Li S, Wei Y, Wang P et al (2020) Synergism of Cu and Al co-doping on improvements of structural integrity and electrochemical performance for $LiNi_{0.5}Mn_{1.5}O_4$. J Alloys Compd (820):153140
47. Tian T, Lu L, Yin Y et al (2022) Trace doping of multiple elements enables stable cycling of high areal capacity $LiNi_{0.5}Mn_{1.5}O_4$ cathode. Small (18):2106898
48. Courbaron G, Peitit E, Serano-Sevillano J et al (2023) Improved electrochemical performance for high-voltage spinel $LiNi_{0.5}Mn_{1.5}O_4$ modified by supercritical fluid chemical deposition. ACS Appl Mater Interfaces (15):2812–2824
49. Li J, Zhang Y, Li J et al (2011) AlF_3 coating of $LiNi_{0.5}Mn_{1.5}O_4$ for high-performance Li-ion batteries (17):671–675
50. Haber S, Solomatin N, Shapira A et al (2023) Lithium exchange across a lithium-less coating for high energy cathodes. J Power Sour 560:232693
51. Hallot M, Caja-Munoz B, Leviel C et al (2021) Atomic layer deposition of a nanometer-thick Li_3PO_4 protective layer on $LiNi_{0.5}Mn_{1.5}O_4$ films: dream or reality for long-term cycling?. ACS Appl Mater Interfaces (13):15761–15773
52. Zhang J, Yue L, Hu P et al (2014) Taichi-inspired rigid-flexible coupling cellulose-supported solid polymer electrolyte for high-performance lithium batteries. Sci Rep 4:6272
53. Zhang J, Zhao J, Yue L et al (2015) Safety-reinforced poly (propylene carbonate) based all-solid-state polymer electrolyte for solid polymer lithium batteries. Adv Energy Mater 5:1501082

MIX
Papier aus verantwortungsvollen Quellen
Paper from responsible sources
FSC® C105338

If you have any concerns about our products,
you can contact us on
ProductSafety@springernature.com

In case Publisher is established outside the EU,
the EU authorized representative is:
Springer Nature Customer Service Center GmbH
Europaplatz 3, 69115 Heidelberg, Germany

Printed by Libri Plureos GmbH
in Hamburg, Germany